SURFACE ACTIVITY OF PETROLEUM DERIVED LUBRICANTS

SURFACE ACTIVITY OF PETROLEUM DERIVED LUBRICANTS

Lilianna Z. Pillon

CRC Press
Taylor & Francis Group
Boca Raton London New York

CRC Press is an imprint of the
Taylor & Francis Group, an **informa** business

CRC Press
Taylor & Francis Group
6000 Broken Sound Parkway NW, Suite 300
Boca Raton, FL 33487-2742

First issued in paperback 2018

© 2011 by Taylor and Francis Group, LLC
CRC Press is an imprint of Taylor & Francis Group, an Informa business

No claim to original U.S. Government works

ISBN-13: 978-1-4398-0340-0 (hbk)
ISBN-13: 978-1-138-37409-6 (pbk)

Library of Congress Cataloging-in-Publication Data

Pillon, Lilianna Z.
    Surface activity of petroleum derived lubricants / Lilianna Z. Pillon.
        p. cm.
    Includes bibliographical references and index.
    ISBN 978-1-4398-0340-0 (alk. paper)
    1. Synthetic lubricants. 2. Petroleum--Refining. 3. Surface chemistry. I. Title.

TP698.P55 2010
665.5'385--dc22   2009024348

Visit the Taylor & Francis Web site at
http://www.taylorandfrancis.com

and the CRC Press Web site at
http://www.crcpress.com

To David, Sylvia, Monica, and Samantha for their
continuing encouragement and support

Contents

Preface

The term "lubrication" is described as the control of friction and wear between moving surfaces in contact by the introduction of a lubricant. A typical lubricant contains petroleum-derived base stocks and additives. The literature reports that lube oils are by-products of the crude oil-refining process; however, only some crude oils are suitable to produce lube oil base stocks and the finished lubricants represent a high added value group of petroleum products and require high technical expertise to formulate them. The lubricant industry formulates products to meet the performance requirements of different applications, which can range from severe low-temperature fluidity specifications to high-temperature oxidation resistances. Lubricants also need to meet industry surface activity specifications, which describe their properties at the oil/air interface, such as the resistance to foaming and air entrainment; at the oil/water interface, such as emulsification or demulsification; and at the oil/metal interface, such as rust and corrosion inhibition and wear prevention.

While there are many books on crude oil refining, I wrote Chapter 1 (Conventional refining of crude oils) and Chapter 2 (Nonconventional processing of base stocks) to provide readers with an easy access to information. The selection of base stocks depends on whether the level of performance required can be achieved by the use of conventionally refined petroleum-derived base stocks, called mineral base stocks. The majority of lubricant products use mineral base stocks produced by vacuum distillation, solvent extraction, and solvent dewaxing of lube oil distillates. Some lubricant applications require higher-quality base stocks and the use of nonconventional processing based on severe hydrotreating, hydrocracking, or hydroisomerization of different petroleum-derived feedstocks. Many different synthetic fluids are also available; however, their cost versus performance is a key issue. When using synthetic oil instead of petroleum-derived base stocks, the surface activity requirements of some lubricants and their performances need to be carefully compared.

While hundreds of additives are available to improve the properties of the petroleum-derived lube oil base stocks, their effectiveness is limited as shown in Chapter 3 (Low temperature fluidity of base stocks), Chapter 4 (Oxidation stability of base stocks), and Chapter 5 (Interfacial properties of base stocks). To formulate the lubricants, the proper base stocks and the proper additives need to be selected as discussed in Chapter 6 (Lubricant formulation). The finished lubricant products need to meet specific performance targets as discussed in Chapter 7 (Effects of additives on surface activity of turbine oils), Chapter 8 (Effects of base stocks on surface activity of hydraulic oils), and Chapter 9 (Surface activity of engine oils). When formulating the lubricants, the additive interactions (Chapter 10), the testing (Chapter 11), and the storage (Chapter 12) is also important. This book will be of interest to any petroleum chemist or engineer involved in the refining of crude oils, in the processing of lube oil base stocks, or in the formulation of finished lubricants, their marketing, testing, and new additive development.

Author

Lilianna Z. Pillon received her MSc in chemistry from the University of Lodz, Poland, and her PhD from the University of Windsor, Canada. She was awarded a National Research Council (NRC) of Canada Postdoctoral Fellowship and studied polymer blends at the NRC Industrial Materials Research Institute in Boucherville, Quebec. She joined Polysar Ltd., Latex R&D Division, where she developed an interest in polymer emulsions and chemical interactions. She also worked for Imperial Oil, Research Department in the lube oil base stock processing and quality area followed by work in the industrial oil group and the engine oil group. She was awarded patents related to the interfacial properties of lube oil base stocks and the use of surface active additives. She is the author of the book *Interfacial Properties of Petroleum Products,* published by CRC Press in 2007.

1 Conventional Refining of Crude Oils

1.1 PROPERTIES AND COMPOSITIONS OF CRUDE OILS

Petroleum and the equivalent term crude oil is a mixture of gaseous, liquid, and solid molecules that occurs in rock deposits found in different parts of the world. The initial pressure is sufficient to move oil to the production wells and is called the primary oil recovery. Molecules found in crude oils can exist in various states of matter—gas, liquid, or solid—depending on their chemistry, molecular weight, temperature, solvency, and pressure. With a decrease in the temperature, the clusters of wax crystals can grow until no flow is observed, which is reported as the pour point. Crude oils contain hydrocarbons and heteroatom molecules that in turn contain sulfur (S), nitrogen (N), and oxygen (O), and even some metals such as nickel (Ni), vanadium (V), and iron (Fe). Some crude oils are acidic, as indicated by a high total acid number (TAN), and have a significant carbon residue (CCR) and ash content when exposed to high temperatures.

The properties and composition of typical crude oils obtained using primary recovery are shown in Table 1.1. When the initial pressure decreases, it is necessary to increase the pressure by injecting water, which is called the secondary recovery or water flooding. Crude oil recovery from porous sedimentary rocks depends on the efficiency with which oil is displaced by some other fluids. The effect of brine composition on oil recovery by water flooding was studied. The injection of brine containing 2% of KCl and 2% of NaCl increased the oil recovery by 18.8% over the injection of distilled water (Bagci, Kok, and Turksoy, 2001). The total recovery by the primary and secondary recoveries is usually less than 40% of the original oil and the remaining oil is recovered during the tertiary oil recovery also known as enhanced oil recovery (EOR). The literature reports on several proven techniques for the EOR, such as surfactant–polymer flooding, foam flooding, CO_2 flooding, caustic solution flooding, microbial method, steam injection, and thermal combustion. It also reports that the most promising EOR technique is the use of surfactants (Ling, Lee, and Shah, 1986).

The typical properties and composition of crude oils obtained using EOR are shown in Table 1.2. Crude oil is a mixture of organic compounds ranging in size from simple gaseous molecules to high molecular weight wax and asphaltenes. Wax type molecules reduce the flowability or pumpability of crude oils. For some crude oils, a significant increase in acids will cause a high TAN. Some crude oils contain a large amount of metals and inorganic salts and an increase in CCR will be observed. The literature reports on the variations in wax, sulfur, metals, salts, and asphaltenes of Chinese crude oils (Liu, Xu, and Gao, 2004).

TABLE 1.1

Properties and Composition of Crude Oils Obtained Using Primary Recovery

Primary Recovery	Crude Oil #1	Crude Oil #2	Crude Oil #3
Gravity (API)	10.4	15	10.4
Viscosity at 38°C (cP)	30,200	786	20,500
Pour point (°C)	18	−4	13
Carbon (wt%)	85.4	86.9	85.7
Hydrogen (wt%)	11.2	11.7	11.1
Sulfur (wt%)	2.5	1.2	2.0
Nitrogen (wt%)	1.2	0.8	1.0
Nickel (ppm)	51	49	69
Vanadium (ppm)	127	22	119
Iron (ppm)	8	30	17
Water (wt%)	<0.1	<0.2	<0.1
TAN (mg KOH/g)	3.8	3.2	3.4
CCR (wt%)	10.1	9.4	15.2
Ash (wt%)	<0.01	<0.01	<0.01

Source: Speight, J.G., *The Chemistry and Technology of Petroleum*, 4th ed., CRC Press, Boca Raton, FL, 2006. With permission.

The variations in wax, sulfur, metals, salts, and asphaltenes of some Chinese crude oils are shown in Table 1.3. The first step in refining of crude oils is desalting, which is required to remove metals and water soluble inorganic salts. During the desalting process, the crude oil is washed with 3–10 vol% of water, and a combination of a demulsifier and an electrostatic desalting system is used to separate the salty water. The presence of asphaltenes and clay particles, at a specific ratio, was reported to increase the stability of emulsions, and, in some cases, wetting agents are added to improve the water wetting properties of solids (Gary and Handwerk, 2001). According to the literature, heavy naphthenic crude oils were reported to form stable emulsions. While many different chemistry demulsifiers were developed, desalting of heavy crude oils is difficult due to a high viscosity effect. The demulsification process was reported to be affected by the viscosity of crude oils, the temperature, and the concentration of the demulsifier (Al-Sabagh et al., 2002).

The effects of desalting on removal rates of different metals from Chinese crude oils are shown in Table 1.4. The removal rates for nickel and iron were relatively low when compared to the high removal rates of Na and Ca, and even Mg and V, thus confirming the importance of their chemistry and water solubility. The literature reports that iron can exist as ferric chloride, oxide, sulfide, and oil-soluble naphthenate, and can also form complexes with porphyrins. Only a small amount of iron can be removed from crude oils during the desalting process. The deferric agents, which are organophosphates and polyamine carboxylates, were reported

TABLE 1.2
Properties and Composition of Crude Oils Obtained Using EOR

EOR—Process	Crude Oil #1—Steam Injection	Crude Oil #2—Steam Injection	Crude Oil #3—Thermal
Gravity (API)	24.5	12.9	11.2
Viscosity at 38°C (cP)	2,210	8,650	12,300
Pour point (°C)	13	10	13
Carbon (wt%)	85.4	86.9	85.8
Hydrogen (wt%)	11.2	11.4	11.4
Sulfur (wt%)	2.1	1.3	2.2
Nitrogen (wt%)	0.9	0.9	0.9
Nickel (ppm)	58	66	30
Vanadium (ppm)	104	37	125
Iron (ppm)	12	29	80
Water (wt%)	<0.1	<0.1	<0.2
TAN (mg KOH/g)	3.4	2.5	3.3
CCR (wt%)	17.1	16.0	17.5
Ash (wt%)	<0.01	<0.01	<0.01

Source: Speight, J.G., *The Chemistry and Technology of Petroleum*, 4th ed., CRC Press, Boca Raton, FL, 2006. With permission.

TABLE 1.3
Variations in Wax, Sulfur, Metals, Salts, and Asphaltenes of Chinese Crude Oils

	Chinese Crude Oils		
	Luning	Rocalya	Shengli
Density at 20°C (g/cm³)	0.9108	0.9263	0.9156
Freezing point (°C)	21	<−30	15
Wax content (wt%)	12.3	10.8	9.7
Sulfur (wt%)	0.2	0.5	0.9
Calcium (Ca) (ppm)	31	5	438
Sodium (Na) (ppm)	19	<1	1
Nickel (Ni) (ppm)	16	28	63
Iron (Fe) (ppm)	10	38	53
Vanadium (V) (ppm)	4	4	2
Magnesium (Mg) (ppm)	1	1	13
Salt content (mg/L)	7	14	15
Asphaltenes (wt%)	2.5	0.9	1.8

Source: Reprinted from Liu, G. et al., *Energy Fuels*, 18, 918, 2004. With permission.

TABLE 1.4

Effects of Desalting on the Removal Rates of Metals from Different Crude Oils

	Desalted Crude Oils (%)		
	Luning	Rocalya	Shengli
	Metal Removal Rates		
Na	95	99	97
Ca	76	79	88
Mg	44	68	65
V	43	38	40
Ni	21	25	24
Fe	20	3	16

Source: Reprinted from Liu, G. et al., *Energy Fuels*, 18, 918, 2004. With permission.

to be effective in decreasing the iron content of desalted crude oils. However, the efficiency of deferric agents was also reported to vary depending on their chemistry and concentration (Liu, Xu, and Gao, 2004). While many different chemistry additives, including demulsifiers, are available, the task of selecting effective additives is often difficult due to a variation in the composition of crude oils and in the use of surfactants, which can affect their interfacial properties (e.g., emulsion stability).

1.2 PROPERTIES AND COMPOSITIONS OF PETROLEUM FRACTIONS

Pipelines transport natural gas from the wellhead to gas-processing plants, and crude oils to storage tanks. The properties, compositions, and stabilities of crude oils from different oil fields, and even from the same oil field, can vary. Upon cooling, the paraffin hydrocarbons can form big crystals known as macrocrystalline wax. The naphthene hydrocarbons form small crystals known as microcrystalline wax. The literature reports that the tank sludge of Bombay High crude oil contained high melting point crystalline waxes. High temperature gas chromatography (HTGC) analysis indicated the presence of alkanes ranging from C_{21} to C_{75}, along with many different isomers. The most predominant alkanes were identified to be C_{40} and C_{67} (Kumar, Gupta, and Agrawal, 2003). Tank bottom waxes, separating and accumulating on the bottoms of tanks used for storing waxy crude oils, were reported to be high melting and microcrystalline in type (Shell, 2001). Crude oil is a mixture of compounds having different boiling points, and, after desalting, distillation is the simplest way to separate them into different petroleum fractions.

The content of petroleum fractions obtained during atmospheric distillation of different crude oils is shown in Table 1.5. Around the year 1940, it was discovered

TABLE 1.5

Effect of Different Crude Oils on Atmospheric Petroleum Fractions

	Crude Oils			
Source	Gasoline and Gas	Kerosene	Gas Oil	Atmospheric Residua
	Content %			
California	36.6	4.4	36.0	23.0
Pennsylvania	47.4	17.0	14.3	1.3
Oklahoma	47.6	10.8	21.6	20.0
Texas	44.9	4.2	23.2	27.9
Iraq	45.3	15.7	15.2	23.8
Iran	45.1	11.5	22.6	20.8
Kuwait	39.2	8.3	20.6	31.9
Bahrain	26.1	13.4	34.1	26.4
Saudi Arabia	34.5	8.7	29.3	27.5

Source: Reproduced from Speight, J.G., *The Chemistry and Technology of Petroleum*, 4th ed., CRC Press, Boca Raton, FL, 2006. With permission.

that n-paraffins formed solid adducts with urea in a methyl alcohol solution. The branched chain paraffins, having a similar molecular weight, were found not to form solid adducts with urea, which allowed the separation of the n-paraffins from *iso*-paraffins. The n-paraffin–urea adducts were used to separate n-paraffins, C_6 to C_{20}, from hydrocarbon mixtures. The separation was based on the fact that there was sufficient room inside the crystal molecules of urea for n-paraffins but branched chain paraffins were too big (Goldstein, 1958). Crude oils vary in composition and are classified as paraffinic, paraffinic–naphthenic, naphthenic, paraffinic–naphthenic–aromatic, and aromatic (Speight, 2006). The atmospheric gas oil, produced from paraffinic crude oil, was reported to have a high cloud point of 19°C and a high pour point of 9°C. Using the urea adduction method, the paraffinic gas oil was analyzed and found to contain 18.4 wt% of n-paraffins and only 1 wt% of *iso*-paraffins (Khidr and Mohamed, 2001).

The effect of different crude oil types on composition of 250°C–300°C petroleum fraction is shown in Table 1.6. The petroleum fraction, from paraffinic type crude oil and having a boiling range of 250°C–300°C, was reported to contain 46%–61% of paraffins, 22%–32% of naphthenes, 12%–25% of aromatics, and 1%–10% of waxes. The petroleum fraction, from naphthenic type crude oil and also having a 250°C–300°C boiling range, was reported to contain 15%–26% of paraffins, 61%–76% of naphthenes, 8%–13% of aromatics, and a trace of wax. The petroleum fraction, from aromatic type crude oil and having a 250°C–300°C boiling range, was reported to contain only 0%–8% of paraffins, 57%–78% of naphthenes, 20%–25% of aromatics, and some wax. The literature reports on solid paraffins, boiling

TABLE 1.6
Effect of Different Types of Crude Oils on
Composition of 250°C–300°C Fraction

Petroleum Fraction	Crude Oil Type		
(250°C–300°C)	Paraffinic	Naphthenic	Aromatic
Paraffins (%)	46–61	15–26	0–8
Naphthenes (%)	22–32	61–76	57–78
Aromatics (%)	12–25	8–13	20–25
Wax (%)	1–10	Trace	<1
Asphalt (%)	0–6	0–6	0–20

Source: Speight, J.G., *The Chemistry and Technology of Petroleum*, 4th ed., CRC Press, Boca Raton, FL, 2006. With permission.

TABLE 1.7
Boiling Range of Petroleum Fractions and Some
Petroleum Products

Petroleum Fractions	Boiling Range (°C)	Petroleum Products	Boiling Range (°C)
Light naphtha	−1–150		
Heavy naphtha	150–205	Gasoline	−1–180
Kerosene	205–260	Jet fuel	175–290
LGO	260–315	Stove oil	205–290
Heavy gas oil	315–425	Diesel	175–370
VGO	425–600		
Vacuum residua	>600		

Source: Speight, J.G., *The Chemistry and Technology of Petroleum*, 4th ed., CRC Press, Boca Raton, FL, 2006. With permission.

above 270°C, separated from the Mukta crude oil by urea adduction and analyzed by gas chromatography (GC), infrared (IR), and nuclear magnetic resonance (NMR). With an increase in the boiling range, their average carbon number was reported to increase from C_{17} to C_{30} and the CH_2/CH_3 ratio increased from 7.9 to 13.8 (Khan et al., 2000). Some refineries separate crude oils in two separate steps: first, the fractionation of the crude oil at atmospheric pressure; second, the fractionation of the atmospheric residue under vacuum.

The boiling range of different atmospheric and vacuum petroleum fractions and some petroleum products is shown in Table 1.7. Crude oil can be separated by atmospheric distillation into a variety of different generic fractions and the main products are naphtha or straight run gasoline, kerosene, stove oil, diesel, and gas oil. The reduced crude can be fractionated in a vacuum distillation unit into light and heavy

TABLE 1.8

Effect of Paraffinic and Aromatic Type Crude Oils on the Composition of VGO

	VGO	
	Paraffinic Crude Oil	**Aromatic Crude Oil**
Boiling range (°C)	370–550	370–550
Total saturates (wt%)	76.9	46.5
n-Paraffins (wt%)	53.4	29.5
Aromatics (wt%)	23.1	53.5

Source: Reprinted from Srivastava, S.P. et al., *Petrol. Sci. Technol.*, 20(3–4), 269, 2002. With permission.

vacuum gas oil (VGO). Distillation of crude oil under reduced pressure produces the vacuum residue, which is a solid and has a bp > 565°C (Speight, 2006). The urea adduction method was reported to be effective in separating the n-paraffins from *iso*-paraffins and cyclo-paraffins present in VGOs produced from different types of crude oils (Srivastava et al., 2002).

The effect of the paraffinic and aromatic type crude oils on the composition of VGO having the same boiling range is shown in Table 1.8. The VGO from paraffinic base crude oil was reported to have a higher saturate content of 76.9 wt%, including 53.4 wt% of n-paraffins and an aromatic content of 23.1 wt%. The VGO from aromatic base crude oil was reports to have a lower saturate content of 46.5 wt%, including a lower n-paraffinic content of 29.5 wt% and a higher aromatic content of 53.5 wt%. The literature also reports that the crystallization of n-paraffins and agglomeration of high viscosity molecules, at low temperatures, leads to gel formation (Srivastava et al., 2002).

The properties and heteroatom content of different middle distillates (MDs) and VGOs are shown in Table 1.9. With an increase in the boiling range of MDs, a decrease in the saturate content with an increase in the aromatic content, from 42.4 to 59.8 wt%, an increase in the sulfur content, from 0.1 to 0.25 wt%, and some increase in the nitrogen content, from 0.05 to 0.08 wt%, was reported. While some cracking might take place during distillation, no presence of olefins in MDs was reported (Severin and David, 2001). With a further increase in the boiling range of VGO, over 400°C, a further increase in the sulfur content to 2.3 wt% and some increase in the nitrogen content to 0.1 wt% was reported. VGO was also reported to have a high polar content of 9 wt% indicating the presence of other heteroatom-containing molecules. The oxygen content of petroleum oils is difficult to measure; however, based on the high polar content of VGO, the oxygen content could be very significant. The surface activity of MDs toward alumina was studied. Different MDs were dissolved in chloroform, adsorbed onto neutral alumina, and eluted with different polarity solvents ranging from nonpolar n-hexane to toluene, 10% tetrahydrofuran (THF) in $CHCl_3$, and highly polar 10% ethanol in THF (Severin and David, 2001).

TABLE 1.9

Properties and Heteroatom Content of Different MDs and VGOs

	Petroleum Fractions		
	MD #1	MD #2	VGO
Boiling range (°C)	144–379	182–385	>400
Density at 15°C (g/mL)	0.8548	0.8659	0.8720
Saturates (vol%)	57.6	40.2	49.9
Aromatics (vol%)	42.4	59.8	41.1
Olefins (vol%)	0	0	—
Polars (wt%)	—	—	9.0
Carbon (wt%)	86.6	86.9	85.1
Hydrogen (wt%)	13.1	12.7	12.5
Sulfur (wt%)	0.1	0.25	2.3
Nitrogen (wt%)	0.05	0.08	0.1
Flash point (°C)	60	86	—

Source: Reprinted from Severin, D. and David, T., *Petrol. Sci. Technol.*, 19(5–6), 469, 2001. With permission.

TABLE 1.10

Surface Activity of Heterocompounds Found in MDs

		Neutral Alumina	
n-Hexane	Toluene	10% THF in CHCl$_3$	10% Ethanol in THF
Paraffins	Benzene	Aniline	Diphenylpyridines
	Indanes	Naphthylamine	
	Naphthalenes	Diphenylaniline	Benzaldehyde
	Biphenyls	Pyridines	2-Ethylhexanol
	Fluorenes	Diphenylpyridine	*p*-Octylanisol
	Phenantrenes	Quinolines	
	Pyrenes		
	Chrysenes	Indoles	
		Carbazoles	
	Dibenzothiophenes	Acridines	

Source: Severin, D. and David, T., *Petrol. Sci. Technol.*, 19(5–6), 469, 2001. With permission.

The chemistry and surface activities of heterocompounds found in different MDs toward neutral alumina are shown in Table 1.10. Paraffins are soluble in nonpolar paraffinic solvents and the least polar group, eluted with *n*-hexane, included only paraffins. Aromatic compounds are soluble in aromatic solvents and the more polar group, eluted with toluene, included all monoaromatic compounds, polycondensed aromatic compounds, and dibenzothiophene. A wide range of different heteroatom-containing molecules were found to adsorb onto neutral alumina, and, in terms of their surface activity, they were divided into four groups. The sulfur compounds identified in MDs were mainly dibenzothiophenes. A significantly more polar solvent, such as 10% THF in $CHCl_3$, was needed to remove all basic and nonbasic N compounds from neutral alumina. The nitrogen compounds were found to vary widely from low molecular weight compounds, such as anilines, pyridines, indoles, and quinolines, to high molecular weight compounds, such as carbazoles, acridines, and diphenylpyridines. The most surface active group of heterocompounds required a very polar solvent, such as 10% ethanol in THF, to remove diphenylpyridines and O-containing molecules. The oxygen compounds were mainly phenols and dibenzofurane but also benzaldehyde, 2-ethylhexanol, and p-octylanisol, which are a derivative of methyl phenyl ether (Severin and David, 2001). The surface activity of molecules toward neutral alumina was found to increase from nonpolar paraffins to aromatics and dibenzothiophenes, more polar basic and nonbasic N compounds to highly polar O-containing molecules that were found to be the most surface active.

The properties, including kinematic viscosity (KV), heteroatom content, and metal content of atmospheric and vacuum residua from Kuwaiti crude oil are shown in Table 1.11. The atmospheric residue, called long residue, and the vacuum residue, called short residue, from the same crude oil have different properties and composition. Crude oils and different petroleum fractions contain S, N, and O but only S and N are usually reported. Asphaltenes are defined as a solubility class that precipitates from oils by the addition of an excess of liquid paraffinic hydrocarbons. Defined as nonhydrocarbon components of crude oils, "true" asphaltenes are insoluble in low molecular weight alkanes, such as n-pentane (C_5 insolubles) and n-heptane (C_7

TABLE 1.11

Heteroatom and Metal Contents of Atmospheric and Vacuum Residua

	Kuwaiti Crude Oil Fraction		
	Atmospheric Residua	Vacuum Residua	VGO
Content of crude oil (vol%)	42	21	
Gravity (API)	13.9	5.5	22.4
KV at 100°C (cSt)	55	1900	
Pour point (°C)	18		
Sulfur (wt%)	4.4	5.5	3
Nitrogen (wt%)	0.3	0.4	0.1
Nickel (ppm)	14	32	<1
Vanadium (ppm)	50	102	<1
C_7 Insolubles (wt%)	2.4	7.1	0
CCR (wt%)	12.2	23.1	0.1

Source: Speight, J.G., *The Chemistry and Technology of Petroleum*, 4th ed., CRC Press, Boca Raton, FL, 2006. With permission.

TABLE 1.12

Crude Oil Distillation Products and Their Uses

Petroleum Products	Uses
Methane	Fuel and hydrogen production
Ethane	Fuel and petrochemical feedstock
Propane	Fuel and petrochemical feedstock
Butane	LPG and petrochemical feedstock
Gasoline	Blending stock and further processing
Naphtha	Blending stock and further processing
MDs	Kerosene, diesel, jet fuel, and heating oil
Light and heavy gas oils	Fuel oil and other products
Light and heavy VGOs	Other processing
Lube distillates	Lube oil base stock processing, wax
Vacuum residuum	Fuel oil, asphalt, coke, feedstock to fuel and lube oil manufacturing processes

Source: Reproduced from Sequeira, A. Jr., *Lubricant Base Oil and Wax Processing*, Marcel Dekker, Inc., New York, 1994. With permission.

insolubles), but soluble in benzene and toluene. In many cases, the literature uses the C_5 or C_7 insolubles as the asphaltene content (Pillon, 2007). Atmospheric and vacuum distillation separates crude oil into useful components which may be end products, such as stove oil or feedstocks for other refinery processes. The VGOs are derived directly from crude oils and are the feeds for lube oil refining. Refined crude oil is used to produce many different products, such as liquid fuels for use in transportation, heating of buildings, solvents, lubricants, and asphalt.

The crude oil distillation products and their uses are shown in Table 1.12. An increase in the demand for higher quality lube oil base stocks and more stringent environmental standards for some lubricants resulted in lube oil plant upgrades or closures. The global lubricant market in the past 10 years has also undergone dramatic changes due to industry consolidation. The total base oil global capacity was reported to be 1,024,706 barrels per day (BPD); however, refineries with lube oil plants with capacities under 3000 BPD, particularly in Europe and Asia, were not listed (Singh, 2002).

The global lube oil refining capacity, reported as BPD, including capacity of refineries in Africa is shown in Table 1.13. The lubricant industry was reported to face flat demand but also sharp shifts in consumption, increased competition, and greater pressure on profits. The world lubricants' demand has remained flat since 1991 despite some significant changes. The literature reports that the worldwide base oil demand was reported flat and was caused by an oversupply in some regions, such as western and eastern Europe, including Russia, but it is expected to continue to grow in Asia (Singh, 2002). The activities of the petroleum oil industry are aimed primarily at the production of fuel, and the proportion of crude oil that is refined into lubricant base oil was reported to be only 1% of the total. The literature reports that

TABLE 1.13

Global Lube Oil Refining Capacity

Region	Total Refineries	Total Capacity (BPD)	Country (BPD)
Asia	34	220,533	
Europe	24	177,444	
The United States	23	218,900	
Former Soviet Union	17	259,600	
The Middle East and Africa	15	49,245	
			Nigeria (3,900)
			South Africa (4,800)
			Algeria (2,400)
			Egypt (3,200)
			Morocco (2,100)
			Libya (640)
Latin America	13	59,805	
Canada	4	26,500	
Australia	3	12,679	

Sources: Lubricants World, 4th Annual Edition, 2002; Singh, H., *Science in Africa*, Umdaus Press, South Africa, 2002, http://www.scienceinafrica.co.za (accessed 2008).

it could be argued that the lube oils produced by a refinery are a by-product of the refining process. However, formulated lubricants represent a high technology and high-added value group of petroleum products (Betton, 2007). There are other valuable petroleum products, such as aromatic extracts and wax, which are by-products of conventional lube oil refining.

1.3 SOLVENT REFINING OF LUBE OILS

Most lube base oil plants are integrated with mainstream oil refineries, which produce a range of transportation and heating fuel products. Overall production capacity of lubricant base oils is relatively low compared to total refinery throughputs (Prince, 1997). The conventional refining of lube oils involves vacuum distillation, solvent extraction, solvent dewaxing, and finishing. The use of different crude oils affects the quality of VGO, used as feed to lube oil production, and not all crude oils are suitable for producing lube oils suitable for use in lubricants. The most important properties of lube oils that affect the lubricant performance are viscosity and viscosity index (VI). The viscosity of the lube oil base stocks is determined by the boiling range of their components, while the VI is related to their chemistry. VI describes the relationship of viscosity to temperature and high VI molecules tend to display less change in viscosity with temperature. The VI of lube oils is calculated based on their KV at 40°C and 100°C. For the hydrocarbons, having the same C number, their viscosity increases with an increase in aromatic ring number, which will affect the VI.

TABLE 1.14

Effect of Hydrocarbon Chemistry on Their KV and VI

C_{26} Hydrocarbons	KV at 40°C (cSt)	KV at 100°C (cSt)	VI
n-Paraffin			
n-Hexacosane	10.7	3.2	190
iso-Paraffin			
5,14-Di-*n*-butyloctadecane	11.3	2.8	83
4-Ring fused naphthenes (polycyclics)			
2-Octyl perhydrotriphenylene	103.3	8.7	25
2-Decyl perhydroindenoindene	48.1	7.0	100
n-Octyl perhydrochrysene	156.7	9.9	−18
4-Ring fused aromatics (polyaromatics)			
3-*n*-Hexylperylene	867.0	23.3	−60
3-*n*-Octylchrysene	375.3	14.5	−71

Source: Lynch, T.R., *Process Chemistry of Lubricant Base Stocks*, CRC Press, Boca Raton, FL, 2007. With permission.

The effect of chemistry on KV and VI of different hydrocarbons is shown in Table 1.14. The literature reports on VI of different hydrocarbon molecules, and *n*-paraffins and *iso*-paraffins have high VI, while naphthenes and aromatics are low VI components of crude oils (Pillon, 2007). It also reported on very high VI of long *n*-paraffins, in the range of 180–200, and a low or even negative VI of compact rigid molecules, such as multiring-fused naphthenes and aromatics (Lynch, 2007). The literature reports that typical crude oils used to manufacture lube oil base stocks are from the Middle East, North Sea, and U.S. mid-continent (Prince, 1997).

The variations in pour points and sulfur content of different crude oils are shown in Table 1.15. The Middle East, North Sea, and Indonesia crude oils used to produce lube oil base stocks were reported to have varying pour points from as low as −15°C to as high as 39°C and varying sulfur content in the range of 0.2–2.5 wt% (Prince, 1997). The literature reports on highly waxy Mukta crude oil from India having a low viscosity and a low sulfur content of 0.1 wt% but a high pour point of 27°C (Khan et al., 2000). The literature also reports on light Kuwaiti crude oil having a low pour point of −23°C but a high sulfur content of 2 wt% (Al-Kandary and Al-Jimaz, 2000).

An approved crude oil must undergo distillation to produce lube oil feedstocks. During the first distillation, under atmospheric pressure, naphtha, kerosene, gas oils, and the bottom product, called long residue, are produced. Long residue contains

TABLE 1.15

Variations in the Pour Points and the Sulfur Content of Different Crude Oils

	Crude Oil Origin				
	The Middle East	North Sea	Indonesia	India	Kuwait
KV at 40°C (cSt)	8	4	12	2.3	8.7
Pour point (°C)	−15	−3	39	27	−23
Sulfur (wt%)	2.5	0.3	0.2	0.1	2.0

Sources: Prince, R.J., Base oils from petroleum, in *Chemistry and Technology of Lubricants*, 2nd ed., Mortier, R.M. and Orszulik, S.T., Eds., Blackie Academic & Professional, London, U.K., 1997; Khan, H.U. et al., *Petrol. Sci. Technol.*, 18(7–8), 889, 2000; Al-Kandary, J.A.M. and Al-Jimaz, A.S.H., *Petrol. Sci. Technol.*, 18(7–8), 795, 2000.

useful molecules, which require vacuum distillation to prevent cracking and thermal degradation. Each distillate is recovered within a viscosity range corresponding to its base oil viscosity and normally stored in intermediate tanks before further processing. The literature reports on variations in VI, pour point, and S contents of lube oil distillates produced from different crude oils.

The properties of lube oil distillates from different crude oils are shown in Table 1.16. The literature reports on properties and composition of lube oil distillates produced from naphthenic and paraffinic type crude oils. The naphthenic lube oil distillate was reported to have a low VI of 17, a low pour point of −37°C, the aromatic

TABLE 1.16

Properties of Lube Oil Distillates from Different Crude Oils

	Lube Oil Distillates				
	Naphthenic	Paraffinic	Middle East	North Sea	Nigeria
KV at 100°F (SUS)	108	103			
KV at 40°C (cSt)			14	16	18
VI	17	86	70	92	42
Pour point (°C)	−37	24	19	25	18
Aromatics (wt%)	34	34	19	20	28
Sulfur (wt%)	0.1	1.1	2.6	0.3	0.3

Sources: Sequeira, A. Jr., *Lubricant Base Oil and Wax Processing*, Marcel Dekker, Inc., New York, 1994. With permission; Prince, R.J., Base oils from petroleum, in *Chemistry and Technology of Lubricants*, 2nd ed., Mortier, R.M. and Orszulik, S.T., Eds., Blackie Academic & Professional, London, U.K., 1997. With permission.

content of 34 wt%, and an S content of 0.1 wt%. The similar viscosity paraffinic lube oil distillate was reported to have a high VI of 86, a high pour point of 24°C, a similar aromatic content of 34 wt%, and a higher S content of 1.1 wt% (Sequeira, 1992). The distillates from the Middle East and North Sea crude oils were reported to have a relatively high VI, in the range of 70–92, and have a high pour point, in the range of 19°C–25°C, similar to lube oil distillate from paraffinic type crude oils. The lube oil distillate from Nigerian crude oil was found to have a relatively low VI of 42 and a relatively high pour point of 18°C indicating a variation in the properties of lube oil distillates (Prince, 1997). The literature also reports on the characterization of distillates from West Siberia crude oil for the production of high viscosity base stocks. Distillates were adsorbed on silica gel and different hydrocarbons were selectively removed using different solvents. Isooctane was used to dissolve and desorb paraffins and naphthenes. Toluene was used to dissolve and desorb aromatics. The resin fractions of about 3 wt% were removed by dissolving in acetone (Starkova et al., 2001).

The yields, VIs, and solid points of different hydrocarbons, separated from medium viscosity distillate produced from West Siberian crude oil are shown in Table 1.17. Medium viscosity distillate have a viscosity of 21.72 mm²/s at 50°C was reported to have a VI of 85 and a solid point of 14°C. After the hydrocarbon separation, the VI and solid points of different fractions were reported to vary. Medium viscosity distillate was reported to contain 47% of saturate hydrocarbons, methanonaphthenes, having a high VI of 114 and a high solid point of 30°C. Medium viscosity lube oil distillate was also found to contain 32% of light aromatics have a lower VI of 85 and a lower solid point of 23°C, and 14% of heavy aromatics having a lower VI of 41 and a lower solid point of 16°C. After the hydrocarbon separation, heavy aromatics were found to have a lower VI and a decrease in their solid point was reported. The literature reports on the VI and solid points of different hydrocarbons separated from heavy viscosity distillate obtained from West Siberian crude oil (Starkova et al., 2001).

The yields, VIs, and solid points of different hydrocarbons separated from high viscosity distillate produced from West Siberian crude oil are shown in Table 1.18.

TABLE 1.17

VI and Solid Points of Hydrocarbons from Medium Viscosity Distillate

West Siberia Crude Oil	Yield (wt%)	VI	Solid Point (°C)
Medium viscosity distillate	100	85	14
Methano-naphthenes	47.1	114	30
Aromatics light	31.7	85	23
Aromatics middle	4.6	72	18
Aromatics heavy	13.6	41	16

Source: Starkova, N.N. et al., *Chem. Technol. Fuels Oils*, 37(3), 191, 2001. With permission.

TABLE 1.18

VI and Solid Points of Hydrocarbons from High Viscosity Distillate

West Siberia Crude Oil	Yield (wt%)	VI	Solid Point (°C)
High viscosity distillate	100	64	12
Methano-naphthenes	38	114	36
Aromatics light	27	80	24
Aromatics middle	13	67	17
Aromatics heavy	19	36	14

Source: Starkova, N.N. et al., *Chem. Technol. Fuels Oils*, 37(3), 191, 2001. With permission.

High viscosity distillate with a viscosity of 53.59 mm²/s at 50°C was reported to have a lower VI of 64 and a solid point of 12°C. After the hydrocarbon separation, the VI and solid points of different fractions were reported to vary. High viscosity distillate was reported to contain 38 wt% of saturate hydrocarbons, methano-naphthenes, having not only a high VI of 114 but also higher solid point of 36°C. High viscosity lube oil distillate was also found to contain 27 wt% of light aromatics, have a lower VI of 80 and a lower solid point of 23°C, and 19 wt% of heavy aromatics, having a lower VI of 36 and a lower solid point of 14°C. After the hydrocarbon separation, heavy aromatics were found to have a lower VI and a decrease in their solid point was also reported (Starkova et al., 2001).

To remove low VI components of crude oils, conventional lube oil refining uses solvent extraction. Extraction solvents in commercial use include furfural, phenol, *N*-methylpyrrolidone (NMP), and sulfur dioxide. According to the literature, furfural is the most widely used solvent and the use of sulfur dioxide is rare. While the use of phenol solvent is declining, the use of NMP is increasing (Prince, 1997). Solvent refining uses furfural, phenol, or NMP to extract undesirable components, such as low VI aromatics, naphthenes, and some heteroatom-containing molecules. The resulting products are high VI raffinates and highly aromatic extracts. The amount of extraction solvent, such as furfural, ranges from two to five times that of the feed. If too little of the extraction solvent is used, the recovered raffinate will have a high yield and poor VI (Shell, 2001).

The effect of furfural solvent on VI of lube oil distillate from Venezuela crude oil is shown in Table 1.19. The VI of base stocks depends on the chemistry of hydrocarbons and will increase with a decrease in aromatic content of waxy raffinates. The literature reports on the higher VI of *n*-paraffins in a range of 175, *iso*-paraffins in a range of 155, 1-ring naphthenes in a range of 142 and a lower VI of polynaphthenes in a range of 70 and aromatics in a range of 50 (Sequeira, 1994).

The composition of different lube oil raffinates from Pennsylvania crude oil is shown in Table 1.20. Lube oil distillates need to have a high VI and low pour points to be used in lubricating oils. Some crude oils were reported to contain as much as 60 wt% of wax. The typical lube oil distillates contain 15–20 wt% of wax (Sequeira,

TABLE 1.19
Effect of Furfural Solvent on VI of
Lube Oil Distillate from Venezuela
Crude Oil

Furfural Extraction	Volume (%)	VI
Distillate feed	100	−17
Raffinate 1	78.2	14
Raffinate 2	64.8	36
Raffinate 3	59.3	43
Raffinate 4	54.1	48
Raffinate 5	53.2	57
Raffinate 6	40.0	68

Source: Lynch, T.R., *Process Chemistry of Lubricant Base Stocks*, CRC Press, Boca Raton, FL, 2007. With permission.

TABLE 1.20
Composition of Different Lube Oil Raffinates from Pennsylvania Crude Oil

	Lube Oil Fractions					
	Distillate —Feed	Raffinate —#1	Raffinate —#2	Raffinate —#3	Raffinate —#4	Raffinate —#5
KV at 40°C (cSt)	37.9	130.0	101.8	124.0	18.5	39.0
KV at 100°C (cSt)	5.8	8.7	9.1	11.4	3.9	6.4
VI	101	−5	58	83	111	125
Paraffins (%)	73	53	60	65	80	85
Naphthenes (%)	16	7	15	20	18	15
Aromatics (%)	9	40	25	15	2	0

Source: Lynch, T.R., *Process Chemistry of Lubricant Base Stocks*, CRC Press, Boca Raton, FL, 2007. With permission.

1994). The wax type molecules, undesirable as lube oil molecules due to their high pour points, are removed during the dewaxing step. Different solvent-based processes have been developed for the dewaxing of lube oils, but they are all based on three basic steps: first, the dilution and chilling of the feedstock with solvent (crystallization); second, the filtration of the wax; and third, the solvent recovery from the wax cake and the filtrate.

The effect of different crude oils on VI of solvent-dewaxed VGO is shown in Table 1.21.

TABLE 1.21

Effect of Different Crude Oils on VI of Dewaxed VGO

	Crude Oils	
Source	VI of Solvent-Dewaxed VGO (Dewaxed Pour Point of −12°C)	VGO Wax (Content %)
Alaska north slope	15	7
Arab light	60	9
Arab heavy	55	9
Brent	65	19
California light	8	5
Isthmus	40	10
Kuwait	50	7
Maya	40	16
Pennsylvania	100	0
West Texas intermediate	55	12.5

Source: Reproduced from Henderson, H.E., Chemically modified mineral oils, in *Synthetics, Mineral Oils, and Bio-Based Lubricants Chemistry and Technology*, Rudnick, L.R., Ed., CRC Press, Boca Raton, FL, 2005. With permission.

The most important properties of lube oil base stocks are VI and their KV. To meet the viscosity requirements of different lubricant products, VGO needs to have a narrow boiling range. The atmospheric residue (reduced crude) is fractionated in a vacuum distillation unit into fractions of the desired viscosity and flash point.

The boiling range of different viscosity lube oil distillates is shown in Table 1.22. The lube oils, which require the removal of aromatics usually by solvent extraction and solvent dewaxing are referred to as solvent neutrals (SN), where solvent means

TABLE 1.22

Boiling Range of Different Viscosity Lube Oil Distillates

Vacuum Fractions	Boiling Range (°C)	Lube Oil Distillates	Boiling Range (°C)
VGO	343–538	Light lube distillate	329–385
		Medium lube distillate	385–441
		Heavy lube distillate	441–566
Vacuum residue	>538–566		

Source: Sequeira, A. Jr., *Lubricant Base Oil and Wax Processing*, Marcel Dekker, Inc., New York, 1994. With permission.

TABLE 1.23

Properties of Different Viscosity Solvent Refined Paraffinic Base Stocks

	Paraffinic Base Stocks			
	100N	**150N**	**320N**	**850N**
KV at 100°F (SUS)	107	155	332	844
KV at 38°C (cSt)	22.3	33.1	70.9	191.3
VI	95	96	97	89
Pour point (°C)	−12	−15	−12	−9
Aromatics (wt%)	16.1	23.4	25.4	27.6
S (wt%)	0.1	0.3	0.3	0.4

Source: Sequeira, A., *Pre-Prints Div. Petrol. Chem.*, ACS, 37(4), 1286, 1992.

that the base stock was solvent refined and neutral means that the oil is of neutral pH (Prince, 1997). Often they are marked as neutral (N).

The properties and composition of different viscosity solvent-refined base stocks from paraffinic crude oil are shown in Table 1.23. The dewaxing solvents, in commercial use, include ketones, mixed solvents, and propane. The majority of refiners use ketone solvents or their mixtures. According to the literature, the use of mixed solvents helps to control the oil solubility and wax crystallization properties better than the use of single solvents (Prince, 1997). The literature reports that while solvent dewaxing is used to remove wax and lower the pour point of base stocks, a small decrease in VI is also observed (Gary and Handwerk, 2001). It also reported on the effect of different solvent dewaxing conditions on viscosity and VI of waxy raffinate feed obtained from a crude oil blend containing 78% Arabian Berri and 22% of Arabian Medium (Taylor, McCormack, and Nero, 1992).

The effect of different solvent dewaxing conditions on KV and VI of raffinate from different crude oil blends is shown in Table 1.24. The refined waxy distillate was reported to have a VI of 110 and a pour point of 27°C and it contained 21 wt% of aromatics, < 0.5 wt% of S, and 50 ppm of N. With a decrease in wax content of waxy raffinate, a decrease in dewaxed pour point leads to an increase in viscosity at 40°C and a decrease in the VI is observed. The typical conventional refining of lube oils to produce base stocks involves a vacuum distillation, a solvent extraction, a solvent dewaxing, and a finishing step (Pirro and Wessol, 2001). The early literature reports on the use of clay treatment as a finishing step. Clays are natural compounds of silica and alumina that also contain oxides of sodium, potassium, magnesium, calcium, and other alkaline earth metals (Sequeira, 1994).

Attapulgite clay is effective in decolorizing and neutralizing any petroleum oil. Porocel clay, activated by heat, was reported to be effective in decolorizing, reducing organic acidity, deodorizing, and also adsorbing aromatic type molecules and polar compounds containing sulfur and nitrogen. Attapulgite clay is less effective in

TABLE 1.24

Effect of Solvent Dewaxing Conditions on Viscosity and VI of Raffinate

Solvent Dewaxing	Pour Point (°C)	KV at 100°F (SUS)	VI
Waxy raffinate feed	27	93	110
Solvent dewaxed	16	106	106
Solvent dewaxed	4	105	105
Solvent dewaxed	−7	104	102
Solvent dewaxed	−18	106	100
Solvent dewaxed	−21	107	101

Source: Reprinted from Taylor, R.J. and McCormack, A.J., *Ind. Eng. Chem. Res.*, 31, 1731, 1992. With permission.

TABLE 1.25

Effect of Clay Treatment on the Composition of Different Lube Oil Base Stocks

Clay Treatment	Base Stocks			
	Saturates (wt%)	Aromatics (wt%)	Sulfur (wt%)	Nitrogen (ppm)
Oil #1 before	71.9	27.8	0.19	22
Oil #1 after	72.2	27.3	0.17	<1
Oil #2 before	69.3	30.4	0.14	17
Oil #2 after	69.7	30.0	0.13	<1
Oil #3 before	98.9	0.9	0	1
Oil #3 after	98.8	0.9	0	<1

Source: Lynch, T.R., *Process Chemistry of Lubricant Base Stocks*, CRC Press, Boca Raton, FL, 2007. With permission.

removing odorous compounds and metals and does not adsorb aromatics. The literature describes the Attapulgite clay as magnesium–aluminum silicate, while porocel is described as hydrated aluminum oxide known as bauxite (Sequeira, 1994).

The effect of clay treatment, used as the finishing step, on the composition of different lube oil base stocks is shown in Table 1.25. After clay treatment, a small increase in saturates with a small decrease in aromatics and S with no presence of N in mineral base stocks was reported. The processing severity will affect the properties and the trace polar content of mineral base stocks. Depending on the crude oil and the quality requirements, there are three to five processing steps required to produce lube oil base stocks. Naphthenic base stocks are made in small quantities from selected crude oils, having a low VI and a low or no wax content. The use of naphthenic type crude oils requires the solvent refining of vacuum distillate to increase VI, and dewaxing might not be needed but requires the finishing step to improve color and oxidation stability. The use of paraffinic type crude oils requires solvent refining of vacuum distillate, dewaxing to reduce pour point, and a finishing step. The selection of crude oil type can have a significant effect on the VI and the pour point of base stocks. The majority of lubricant products use paraffinic base stocks produced using the conventional refining of crude oils based on solvent extraction and solvent dewaxing of lube oil distillates.

1.4 PROCESSING OF CYLINDER OILS AND BRIGHT STOCKS

Crude oils are usually classified as paraffin base, naphthene base, mixed base, and asphalt base, and only some crude oils are suitable for producing lube oil base stocks. Paraffin base crude oils contain wax and little or no asphalt, and are suitable for producing solvent refined lube oil base stocks and wax products. Naphthene base crude oils, which contain little or no wax and contain no asphalt are suitable for producing naphthenic oils and specialty oils. Mixed base crude oils contain wax and asphalt and are used to produce low yield lube oil base stocks. Asphalt base crude oils contain high S and N content, and are used to produce base stocks and asphalts. If crude oils are not suitable for producing lube oil base stocks, they are used to produce asphalts.

The typical content and the use of different types of crude oils are shown in Table 1.26.

Typically crude oils are used to produce light distillates, used as fuels and heating oils, and lube oil distillates used to produce lube oil base stocks. The viscosity of lube oil distillates can vary from light distillates, known as spindle oils, to heavy distillates, known as machine oils. The vacuum residue contains a heavy residual oil, which is mixed with asphalt and resins and requires additional processing to obtain useful products.

TABLE 1.26
Typical Content and Use of Different Types of Crude Oils

Crude Oil Type	Typical Content	Typical Use
Paraffin base	Varying wax	Paraffinic base stocks
	Little or no asphalt	Wax products
Naphthene base	Little or no wax	Naphthenic base stocks
	Little or no asphalt	Specialty oils
Mixed base	Wax	Low yield base stocks
	Asphalt	
Asphalt base	High S and N	Base stocks
		Asphalt

Source: Reproduced from Sequeira, A. Jr., *Lubricant Base Oil and Wax Processing*, Marcel Dekker, Inc., New York, 1994. With permission.

TABLE 1.27
Typical Crude Oil Distillation Products

Atmospheric Distillation Products	Vacuum Distillation Products
Naphtha	Light distillate or spindle oil
Kerosene	Medium distillate or machine oil
Gas oil	Heavy distillate or medium machine oil
Long residue	Short residue

Source: Shell, http://www.shell-lubricants.com (accessed 2001).

The typical crude oil distillation products are shown in Table 1.27. The bottom product, called short residue, contains useful, very high viscosity molecules, which require extraction to prevent their thermal degradation and discoloration. To recover the heavy oils, most refiners use a propane deasphalting process. Propane is a gaseous paraffinic hydrocarbon present in natural gas and crude oil, and is also termed as, along with butane, liquefied petroleum gas (LPG). Asphaltenes are insoluble in LPG and propane is used commercially in the processing of residues. Liquid propane is used for its high selectivity and solubility in oil compared to the heavier material known as asphalt. The use of the same crude oil leads to VGO and residual oil having different compositions. The literature reports that the deasphalting reduces the aromatic, acidic, and basic contents of oils (Speight, 2006).

The composition of different VGOs and residual oils produced from Daqing crude oil is shown in Table 1.28. Light VGO was reported to contain about 70 wt%

TABLE 1.28
Composition of Different VGOs and Residual Oils from Daqing Crude Oil

	LVGO	HVGO	Residual Oil
	Composition (wt%)		
Total saturates	69.8	54.1	28
Naphthenics	41.4	35.8	18.2
1-Ring aromatics	8.6	14	19.2
2-Ring aromatics	3.9	7.5	10.9
3-Ring aromatics	2.3	4.2	4.9
4-Ring aromatics	1.2	1.7	2.6
5-Ring aromatics	0.8	1.5	1.1
Unidentified	3	5.4	1
Gum	9.3	9.9	27.9
Asphaltenes	0.1	0.3	0.5

Source: Reprinted from Kai-Fu, H. et al., *Petrol. Sci. Technol.*, 18(7–8), 815, 2000. With permission.

of saturates, 41 wt% of naphthenes, 17 wt% of aromatics, and some unidentified molecules. Heavy VGO was reported to contain about 54 wt% of saturates, 36 wt% of naphthenes, 29 wt% of aromatics, and some unidentified molecules.

With a decrease in the saturate and naphthenic contents and an increase in the aromatic content, a gum-forming content of 9.3–9.9 wt% was reported. From the same crude oil, residual oil was reported to contain only 28 wt% of saturates, only 18 wt% of naphthenes, and a high content of 39 wt% of aromatics. A further decrease in the saturate and naphthenic content of residual oil with an increase in aromatic content was reported. Residual oil was also found to significantly increase the gum-forming content from 9–10 to 27.9 wt% and a small increase in asphaltene content from 0.1–0.3 to 0.5 wt% was also reported.

The composition of deasphalted oil (DAO) and asphalt produced from Bombay High crude oil is shown in Table 1.29. The DAO from Bombay High crude oil was reported to contain 38.7 wt% of saturates and 48.9 wt% of aromatics. The DAO was also reported to contain 2.9 wt% of bases, 4.3 wt% of acids, and 0.8 wt% of C_5 insolubles. The asphalt from Bombay High crude oil was found to contain a lower saturate content of only 6.1 wt% but a similar aromatic content of 50.9 wt%. The asphalt was also found to contain a higher base content of 6.4 wt%, a higher acid content of 11.6 wt%, and a higher C_5 insoluble content of 2.4 wt% (Kohli, Bhatia, and Madhwal, 2001).

The composition of DAO and asphalt produced from Assam Mix crude oil is shown in Table 1.30. The DAO from Assam Mix crude oil was reported to contain 30.2 wt% of saturates and 56.2 wt% of aromatics. The DAO was also reported to contain 4.1 wt% of bases, 5.7 wt% of acids, and 1.1 wt% of C_5 insolubles. The asphalt from Assam Mix crude oil was found to contain no saturate content and a higher aromatic content of 62.2 wt%. The asphalt was also found to contain a higher base content of 9.4 wt%, a higher acid content of 13.8 wt%, and a higher C_5 insoluble

TABLE 1.29

Composition of DAO and Asphalt Produced from Bombay High Crude Oil

	Bombay High Crude Oil	
	DAO	Asphalt
Saturates (wt%)	38.7	6.1
Aromatics (wt%)	48.9	50.9
Bases (wt%)	2.9	6.4
Acids (wt%)	4.3	11.6
C_5 Insolubles (wt%)	0.8	2.4

Source: Kohli, B. et al., Petrol. Sci. Technol., 19(7–8), 885, 2001. With permission.

TABLE 1.30

Composition of DAO and Asphalt Produced from Assam Mix Crude Oil

	Assam Mix Crude Oil	
	DAO	Asphalt
Saturates (wt%)	30.2	0
Aromatics (wt%)	56.2	62.2
Bases (wt%)	4.1	9.4
Acids (wt%)	5.7	13.8
C_5 Insolubles (wt%)	1.1	11.5

Source: Kohli, B. et al., *Petrol. Sci. Technol.*, 19(7–8), 885, 2001. With permission.

TABLE 1.31

Properties of DAOs Suitable for Producing Heavy Lube Oils

Properties	DAOs		
	Ordovician Crude Oil	West Texas Crude Oil	Kuwaiti Crude Oil
Yield (% of crude oil)	3.6	9.9	12.2
KV at 210°F (SUS)	161	163	162
KV at 99°C (cSt)	34.5	35	34.7
VI	96	84	81
Pour point (°C)	27	38	32
Sulfur (wt%)	0.2	1.4	2.9
Carbon residue (wt%)	1.9	1.8	2.0

Source: Beuther, H. et al., *I&EC Prod. Res. Dev.*, 3(3), 174, 1964.

content of 11.5 wt% (Kohli, Bhatia, and Madhwal, 2001). The use of different crude oils affects the composition of DAOs and asphalts. The early literature reports on the properties of DAOs produced from crude oils, which were suitable for producing heavy lube oil base stocks (Beuther, Donaldson, and Henke, 1964).

The yields and properties of DAOs from different crude oils are shown in Table 1.31. The DAOs from Ordovician crude oil were reported to have a high VI of 96, a high pour point of 27°C, and a low S content of 0.2 wt%. The DAO from West Texas crude oil was reported to have a VI of 84, a pour point of 38°C, and a higher S content of 1.4 wt%. The DAO from Kuwaiti crude oil was reported to

TABLE 1.32

Properties and the Composition of Cylinder Oil and Bright Stock

Properties	Cylinder Oil	Bright Stock
KV at 100°F (SUS)	9440	2586
KV at 38°C (cSt)	2043.3	559.7
VI	70	95
Dewaxed pour point (°C)	−7	−12
Aromatics (wt%)	36.6	32.5
Sulfur (wt%)	0.7	0.5
Carbon residue (wt%)	2.9	0.7

Source: Sequeira, A., *Pre-Prints Div. Petrol. Chem., ACS,* 37(4), 1286, 1992.

have a VI of 81, a pour point of 32°C, and a high S content of 2.94 wt% (Beuther, Donaldson, and Henke, 1964). More recent literature reports on the properties of DAO from the Middle East crude oil having an API gravity of 19.8, a viscosity of 231 cSt at 100°C, a lower VI of 74, and a higher pour point of over 54°C (Sequeira, 1994). Deasphalting of vacuum residua followed by solvent refining, dewaxing, and finishing of DAOs are used to produce heavy oils. Cylinder oils and bright stocks are residual oils produced from paraffinic and naphthenic vacuum residue. The literature reports that cylinder oils are produced using propane deasphalting with solvent dewaxing rarely used to reduce their pour point. Bright stocks are usually manufactured using propane deasphalting, solvent extraction, and solvent dewaxing (Sequeira, 1994). Only in some cases, DAOs are solvent extracted to further increase their VIs.

The properties and composition of high viscosity cylinder oil and bright stock are shown in Table 1.32. High viscosity cylinder oil was reported to have a low VI of 70 and a dewaxed pour point of −7°C. The cylinder oil was also reported to have a high aromatic content of 36.6 wt%, an S content of 0.7 wt%, and a higher carbon residue of 2.9 wt%. High viscosity bright stock was reported to have a high VI of 95 and a dewaxed pour point of −12°C. It was reported to contain an aromatic content of 32.5 wt%, an S content of 0.5 wt%, and a carbon residue of 0.7 wt% (Sequeira, 1992). While typical lube oil distillates meet the viscosity requirements of many petroleum derived lubricants, some lubricant products require higher viscosity lube oil base stocks. Cylinder oils are used to blend only some high viscosity lubricant products. High viscosity bright stocks are blended with lower viscosity base stocks to meet the viscosity requirements of many different lubricant products.

1.5 COMPOSITION OF WAX PRODUCTS

A typical base oil manufacturing plant was reported to have a high vacuum unit, a propane deasphalting unit, a furfural extraction unit, and a methyl ethyl ketone

(MEK) dewaxing unit. The solvent dewaxing unit uses one solvent (e.g., MEK) or solvent mixture (e.g., MEK/MIBK) to remove high pour point waxy hydrocarbons by means of crystallization and filtration (Shell, 2001). The paraffinic VGOs are used to produce lube oil base stocks and wax. Processing of paraffinic and naphthenic crude oils leads to vacuum residue, containing high viscosity molecules, which are also suitable for producing heavy lube oil base stocks and wax. This class of base stocks is usually known as bright stock. The solvent dewaxing processes are suitable for dewaxing the entire range of different viscosity waxy refined oils, including bright stocks.

The feedstock entering the solvent dewaxing unit must be heated to ensure that the wax is in complete solution. The residence time at the required temperature is important in the case of heavy feedstocks. The boiling range and type of the wax affect the filtration rate, and solvent dewaxing aids are used to improve the filtration rates (Sequeira, 1994). The viscosity and wax content of the feed affect the solvent dilution ratio, the refrigeration requirements, the filtration rate, and the size of the solvent recovery facility. Residual wax is produced by a combination of solvent dilution and chilling (Sequeira, 1994).

The properties of solvent-dewaxed paraffinic base stocks from Arabian crude oils and having different viscosities are shown in Table 1.33. With an increase in the viscosity of solvent refined paraffinic base stocks from spindle oil to bright stock, a decrease in VI from 100–103 to 92 with an increase in dewaxed pour point from −15°C to −9°C were reported. Also, with an increase in the viscosity of solvent refined paraffinic base stocks, an increase in S content from 0.4 to 1.5 wt% was observed (Prince, 1997). The wax of refined oils, removed during the dewaxing step, is called slack wax. The solvent dewaxing of different viscosity refined oils produces different slack waxes, also known as crude waxes and raw waxes. Filtration rate is the lowest for the most viscous feedstock and also requires the highest solvent dilution. The waxes derived from solvent dewaxing of bright stock are called malcrystalline waxes because they filter poorly (Sequeira, 1994). The composition and

TABLE 1.33

Properties of Solvent-Dewaxed Paraffinic Base Stocks Having Different Viscosities

	Solvent Dewaxed—Viscosity Grade			
	Paraffinic— Spindle Oil	Paraffinic— 150N	Paraffinic— 500N	Paraffinic— Bright Stock
KV at 40°C (cSt)	12.7	27.3	95.5	550
VI	100	103	97	92
Dewaxed pour point (°C)	−15	−12	−9	−9
Sulfur (wt%)	0.4	0.9	1.1	1.5

Source: Prince, R.J., Base oils from petroleum, in *Chemistry and Technology of Lubricants*, 2nd ed., Mortier, R.M. and Orszulik, S.T., Eds., Blackie Academic & Professional, London, U.K., 1997. With permission.

TABLE 1.34

Effect of Boiling Range of Feed on Wax Properties

	Petroleum Fraction			
	LGO	LGO	Spindle Oil	Spindle Oil
Boiling range (°C)	210–320	240–330	300–400	414–480
Dewaxed oil				
Yield (wt%)	84	68	83	85
Pour point (°C)	−63	−46	−20	−12
Extracted wax				
n-Paraffins (wt%)	99	99	85	90
Melting point (°C)	6	12	35	55

Source: Lynch, T.R., *Process Chemistry of Lubricant Base Stocks*, CRC Press, Boca Raton, FL, 2007. With permission.

properties of waxes removed from different petroleum fractions can vary depending on their boiling range and viscosity.

The effect of boiling range on the properties of wax, removed using the urea dewaxing process, from different light gas oils (LGOs) and spindle oils is shown in Table 1.34. The wax removed from LGO, have a boiling range of 210°C–320°C, was reported to have a very high n-paraffin content of 99% and a melting point of 6°C. With an increase in the boiling range of LGO, the wax containing 99% of n-paraffins has a higher melting point of 12°C. The wax removed from spindle oil, having a boiling range of 300°C–400°C, was reported to have a lower n-paraffin content of 85% and a higher melting point of 35°C. The wax removed from another spindle oil, having a higher boiling range of 414°C–480°C, was reported to have an n-paraffin content of 90% and a higher melting point of 55°C. For the same crude oil, in the gasoline range, the paraffinic content was reported to be as high as 80 wt% but only 30 wt% in lube distillates (Speight, 2006).

The solvent dewaxing of paraffinic refined oils produces different composition slack waxes depending on whether light, intermediate, or heavy lube oil distillate is processed. The literature reports that the paraffin wax obtained from light paraffin distillate has a distinctive crystalline structure, and it is sold as pale raw wax. In higher boiling oils, waxy hydrocarbons occur as microcrystalline wax and they filter more slowly. The microcrystalline wax obtained from heavy paraffin distillate forms small crystals, and it is sold as dark raw wax (Speight, 2006). The wax from intermediate paraffin distillate contains waxes that are a mixture of paraffin and microcrystalline wax, and are known as semimicrocrystalline wax or "intermediate wax" (Shell, 2001). Semimicrocrystalline wax has certain characteristics of both paraffin and microcrystalline waxes.

The composition of wax obtained from different viscosity waxy raffinates is shown in Table 1.35. With an increase in the viscosity of waxy raffinate from 100N to 850N, the n-paraffinic content of wax was reported to decrease from

TABLE 1.35

Composition of Wax from Different Viscosity Waxy Raffinates

	100N	150N	320N	850N	Bright Stock
	Wax Composition				
n-Paraffins (%)	71	59	31	20	0
iso-Paraffins (%)	17	24	35	29	32
Naphthenes (%)	8	13	26	36	38
Aromatics (%)	4	4	8	10	30

Source: Taylor, R.J. et al., Pre-Prints Div. Petrol. Chem., ACS, 37(4), 1337, 1992.

71 to 20 wt%, and their iso-paraffinic content was reported to increase from 17 to 32 wt%. With an increase in the viscosity of waxy raffinate, their naphthenic content increased from 8% to 36%, and was reported to contain 1–3 ring naphthenic hydrocarbons. With an increase in the viscosity of waxy raffinate, their aromatic content also increased from 4 to 30 wt%, and was reported to contain alkylbenzenes. The bright stock slack wax was reported to contain 32 wt% of iso-paraffins, 37 wt% of naphthenics, and a high content of 30 wt% of aromatics. Essentially no n-paraffins were found in slack wax produced from dewaxing the bright stock (Taylor, McCormack, and Nero, 1992). The slack wax from dewaxing of the bright stock was reported to contain naphthenic and aromatic molecules having alkyl side-chains, which are long enough to give these molecules a high VI and pour point characteristics of n-paraffins (Sequeira, 1994). There are two types of waxes: macrocrystalline and microcrystalline. Paraffin wax is known as macrocrystalline because it forms large crystals.

The effect of feedstock viscosity on the type of petroleum wax is shown in Table 1.36. The slack wax removed from solvent refined light lube oil distillate,

TABLE 1.36

Effect of Feedstock Viscosity on Paraffin Content and Type of Petroleum Wax

Feedstock	Wax Type	Wax Name
Light distillate	Macrocrystalline	Paraffin
Medium distillate	Microcrystalline	Microcrystalline
Heavy distillate	Microcrystalline	Microcrystalline
Bright stock	Microcrystalline	Petrolatum

Source: Pillon, L.Z., Interfacial Properties of Petroleum Products, CRC Press, Boca Raton, FL, 2007. With permission.

produced from paraffinic crude oils, is used to produce paraffin wax. The solvent dewaxing of medium or heavy paraffinic refined oils is used to produce the microcrystalline wax. The wax derived from residual oils is also a microcrystalline type. Bright stock slack wax or bright stock crude wax is used to produce wax known as petrolatum. The variation in the composition and the oil content will affect the properties of wax. The literature reports that the wax from solvent dewaxing of 100N paraffinic base stock contained 8 wt% of oil in the wax (Taylor, McCormack, and Nero, 1992).

Slack waxes from solvent dewaxing of waxy raffinates can contain up to 50 wt% of oil. The literature reports that the slack wax containing 10–20 wt% of oil and 79–89 wt% of paraffins, has a melting point in the range of 48°C–55°C, and has a flash point of 150°C. Another slack wax was reported to have a max 15 wt% of oil and 89% of paraffins, a melting point of 54°C–55°C, and a flash point of 150°C. It was also reported to contain 0.5–0.7 wt% of sulfur and have a trace of water (Hinco, 2002). These slack waxes are fed to other refinery units or deoiled to produce wax.

The literature reports on the effect of deoiling on slack wax, containing 9%–20% oil, and its properties and composition (Billon et al., 1980). To produce wax products, the slack wax requires the use of a wax deoiling process to reduce the oil content below 1 wt%. Some industrial waxes can contain 1.8–3 wt% of oil. The commercial wax deoiling processes are based on the wax sweating, the recrystallization, the warm-up deoiling process, and the spray deoiling (Sequeira, 1994). The wax-sweating process is an old process, which can be used for the deoiling of the paraffin wax because it forms large crystals. The wax recrystallization process, also called wax fractionation, was developed to replace the wax-sweating process and can be used to deoil all types of waxes, including microcrystalline. The wax recrystallization process separates wax into fractions by making use of the different solubilities of the wax fractions in a solvent. To obtain low oil content, two or three filtrations are required. The warm-up deoiling process is more cost effective than the recrystallization process because energy requirements are lower. It is operated in conjunction with the solvent dewaxing unit. The spray deoiling process can be used to deoil paraffin wax containing up to 15 wt% of oil. The recrystallization or warm-up deoiling process is used to produce a hard wax and a soft wax known as foots oil (Sequeira, 1994). The low oil content waxes are difficult to produce.

The melting point and composition of deoiled macrocrystalline wax containing 0.2%–0.4% oil are shown in Table 1.37. The macrocrystalline waxes, containing 0.2–0.4 wt% oil, were reported to have an *n*-paraffin content varying in the range of 65–87 wt% and an S content varying in the range of 30–600 ppm. After deoiling, their melting point was reported to vary in the range of 50°C–65°C, and some variation in their penetration properties was also reported. The microcrystalline waxes, derived from medium and heavy distillates, differ from paraffin waxes in having higher viscosity, poorly defined crystalline structure, darker color, and higher melting points (Pillon, 2007). The ASTM D 87 is used to measure the melting point of petroleum wax, which is highly paraffinic or crystalline in nature. The ASTM D 127 is used to measure the drop melting point of microcrystalline wax, including petrolatum.

TABLE 1.37

Melting Point and the Composition of Deoiled Macrocrystalline Wax

Macrocrystalline Deoiled Wax	Oil Content			
	0.4%	0.2%	0.2%	0.2%
Melting point (°C)	63–65	58–60	52–54	50–52
n-Paraffins (wt%)	65	70	87	82
Sulfur (ppm)	300–600	100–400	50–200	30–50

Source: Lynch, T.R., *Process Chemistry of Lubricant Base Stocks*, CRC Press, Boca Raton, FL, 2007. With permission.

TABLE 1.38

Drop Melting Point and the Composition of Deoiled Microcrystalline Wax

Microcrystalline Deoiled Wax	Oil Content	
	2%–4%	0.5%–2%
Drop melting point (°C)	75–80	75–80
n-Paraffins (wt%)	20	20
Sulfur (ppm)	3000–8000	3000–8000

Source: Lynch, T.R., *Process Chemistry of Lubricant Base Stocks*, CRC Press, Boca Raton, FL, 2007. With permission.

The drop melting point and composition of deoiled microcrystalline wax containing 0.5%–4% oil are shown in Table 1.38. After deoiling, the drop melting point of microcrystalline wax was reported to vary in the range of 75°C–80°C. The microcrystalline waxes, containing 0.5–4 wt% of oil, were reported to have a lower n-paraffin content of 20 wt% and a higher S content varying in the range of 3000–8000 ppm. Some variation in their penetration properties was also reported (Lynch, 2007). Another method to deoil waxy raffinates is to produce dewaxed base stocks, and hard and soft wax during a combined processing step. Production of hard waxes was reported to be expensive and difficult due to stringent requirements for very low oil content. The solvent free wax and oil, called "foots oil," were reported to be used as feed to the catalytic cracker unit (Sequeira, 1994). According to the literature, the use of a combined dewaxing–deoiling process, while more economical, does not meet the quality requirements of some wax products. To meet a very low oil content of a maximum of 0.45 wt%, the literature reports that additional deoiling of the slack wax with a high purity solvent and a high dilution ratio is required (Kachlishvili and Filippova, 2003).

TABLE 1.39

Typical Properties and the Composition of Paraffin and Microcrystalline Waxes

Typical Properties	Paraffin Wax	Microcrystalline Wax
KV at 210°F (SUS)	40–50	60–120
KV at 99°C (cSt)	4–7	10–25
Average molecular weight	350–420	600–800
C atoms per molecule	20–36	30–75
n-Paraffins (%)	80–95	0–15
iso-Paraffins (%)	2–15	15–30
Naphthenes (%)	2–8	65–75
Melting point (°C)	45–65	60–90
Flash point (°C)	204	260
Crystallinity	Large crystals	Small crystals
Color	Light	Dark

Sources: Pedersen, K.S. et al., *Properties of Oils and Natural Gases*, Gulf Publishing Company, Houston, TX, 1989. With permission; Sequeira, A. Jr., *Lubricant Base Oil and Wax Processing*, Marcel Dekker, Inc., New York, 1994. With permission.

The typical composition and properties of paraffin and microcrystalline wax are shown in Table 1.39. The paraffin wax, containing 80–95 wt% of n-paraffins, 2–15 wt% of iso-paraffins, and 2–8 wt% of naphthenes, was reported to have a melting point of 45°C–65°C and a flash point of 204°C. The paraffin wax forms large crystals and has a light color (Pedersen, Fredenslund, and Thomassen, 1989). With an increase in the viscosity of feedstock, an increase in the polynaphthenes and aromatics leads to different wax type molecules affecting their physical properties. The microcrystalline wax, containing only 0–15 wt% of n-paraffins, 15–30 wt% of iso-paraffins, and 65–75 wt% of naphthenes, was reported to have a higher melting point in the range of 60°C–90°C (Pedersen, Fredenslund, and Thomassen, 1989). The literature reports that typical grades of microcrystalline waxes widely vary in their physical characteristics. Some microcrystalline waxes are ductile and others are brittle or crumble easily. The lower melting grades, 57°C–63°C, are flexible and adhesive (Pillon, 2007). The literature reports on the variations in oil content, congealing point, and n-paraffin content, measured by GC, of macrocrystalline wax (Sperber, Kaminsky, and Geissler, 2005).

The variation in composition and properties of deoiled petroleum wax is shown in Table 1.40. The ASTM D 938 is used to measure the congealing point of petroleum waxes, including petrolatum. The test measures the temperature at which a sample that is being cooled develops a resistance to flow and may be close to the solid state, or it may be semisolid, depending on the wax composition. After deoiling, the macrocrystalline wax, containing 3.1% of oil, was reported to have a congealing point of 50.5°C and contained 19.3% of n-alkanes as measured by GC.

TABLE 1.40
Variations in Composition and Properties of Deoiled Petroleum Wax

Macrocrystalline	Oil Content			
Deoiled Wax	3.1%	0.4%	0.3%	0.7%
Congealing point (°C)	50.5	62.5	64.5	66.0
n-Alkanes (GC) (%)	19.3	76.5	96	12.3

Source: Reprinted from Sperber, O. et al., *Petrol. Sci. Technol.*, 23(1), 47, 2005. With permission.

Another macrocrystalline wax, containing 0.7% of oil, was reported to have a higher congealing point of 66°C and contained only 12.3% of n-alkanes as measured by GC. The macrocrystalline waxes, containing only 0.3–0.4 wt% of oil, were reported to have a congealing point in the range of 63°C–65°C and a high n-alkane content varying in the range of 77%–96%. A significant variation in their penetration properties was also reported (Sperber, Kaminsky, and Geissler, 2005). The literature reports on the use and properties of the technical grade petrolatum, which can have color varying from amber, red to dark brown (Pillon, 2007).

The properties of different technical grade petrolatums are shown in Table 1.41. The literature reports that petrolatums are mixtures of pale to yellow colored semi-solid hydrocarbons and their melting point can vary in the range of 38°C–60°C (Sequeira, 1994). The amber colored petrolatum was reported to have a melting point of 57°C (Chevron, 2002). The dark brown colored petrolatum was reported to have a lower melting point in the range of 46°C–52°C (Penreco, 2002). Usually, paraffin and microcrystalline waxes are deoiled and, depending on their application, they might require the finishing step to remove trace contaminants. Due to its pharmacological properties, petrolatum was found to be an effective skin moisturizer and used by the cosmetic industry in skin-care products. Petrolatum and its manufacture were

TABLE 1.41
Properties of Technical Grade Petrolatum

Technical Grade	Petrolatum	Petrolatum
KV at 100°C (cSt)	15.1	13.0
Melting point (°C)	57	46–52
Consistency at 25°C	195	130–170
Color	Amber	Dark brown

Source: Chevron, http://www.chevron.com (accessed 2002); Penreco, http://www.penreco.com (accessed 2002).

patented in 1872 under the name "Vaseline" (Morrison, 2002). In highly refined form, finished petroleum wax, including petrolatum, can be white or snow white.

The literature reports that a treatment of molten wax with activated clay improves color, and reduces odor and taste of the finished wax. In conventional wax refining, the decoloring operation, known as percolation, is a batch process. It also removes traces of possibly harmful polycyclic aromatic hydrocarbons that are considered potential carcinogens. The clay is regenerated before reuse by passing it through a multiple hearth furnace to remove the adsorbed color bodies (Sequeira, 1994). In the past, the wax in the residual oil was removed by cold settling. After the addition of a large volume of naphtha to a residual oil, the mixture was allowed to stand as long as necessary in a tank exposed to low temperatures. The waxy components would congeal and settle to the bottom of the tank. After the removal of naphtha, the steam-refined stock was filtered through charcoal to improve its color and it was called bright stock. The wax product that settled to the bottom of the cold settling tank was crude petrolatum (Speight, 2006). Further treatments with sulfuric acid and clay produced white grades of petrolatum.

REFERENCES

Al-Kandary, J.A.M. and Al-Jimaz, A.S.H., *Petroleum Science and Technology*, 18(7&8), 795, (2000).

Al-Sabagh, A.M. et al., *Polymers for Advanced Technologies*, 13, 346 (2002).

Bagci, S., Kok, M.V., and Turksoy, U., *Petroleum Science and Technology*, 19(3&4), 359 (2001).

Betton, C.I., Lubricants, in *Environmental Technology in the Oil Industry*, Orszulik, S.T., Ed., Springer, Dordrecht, The Netherlands, 2007.

Beuther, H., Donaldson, R.E., and Henke, A.M., *I&EC Product Research and Development*, 3(3), 174 (1964).

Billon, A. et al., *Proceedings of the American Petroleum Institute, Division of Refining*, 59, 168–177 (1980).

Chevron, http://www.chevron.com (accessed 2002).

Gary, J.H. and Handwerk, G.E., *Petroleum Refining, Technology and Economics*, 4th ed., Marcel Dekker, Inc., New York, 2001.

Goldstein, R.F., *The Petroleum Chemicals Industry*, 2nd ed., E. & F.N. SPON, London, 1958.

Henderson, H.E., Chemically modified mineral oils, in *Synthetics, Mineral Oils, and Bio-Based Lubricants Chemistry and Technology*, Rudnick, L.R., Ed., CRC Press, Boca Raton, FL, 2005

Hinco Group, http://www.hinco.com (accessed 2002).

Kachlishvili, I.N. and Filippova, T.F., *Chemistry and Technology of Fuels and Oils*, 39(5), 233 (2003).

Kai-Fu, H. et al., *Petroleum Science and Technology*, 18(7&8), 815 (2000).

Khan, H.U. et al., *Petroleum Science and Technology*, 18(7&8), 889 (2000).

Khidr, T.T. and Mohamed, M.S., *Petroleum Science and Technology*, 19(5&6), 547 (2001).

Kohli, B., Bhatia, B.M.L., and Madhwal, D.C., *Petroleum Science and Technology*, 19(7&8), 885 (2001).

Kumar, S., Gupta, A.K., and Agrawal, K.M., *Petroleum Science and Technology*, 21(7&8), 1253 (2003).

Ling, T.F., Lee, H.K., and Shah, D.O., Surfactants in enhanced oil recovery, in *Industrial Applications of Surfactants*, Karsa, D.R., Ed., The Royal Society of Chemistry, Burlington House, London, U.K., 1986.

Liu, G., Xu, X., and Gao, J., *Energy and Fuels*, 18, 918 (2004).

Lubricants World, 4th Annual Edition, 2002.

Lynch, T.R., *Process Chemistry of Lubricant Base Stocks*, CRC Press, Boca Raton, FL, 2007.

Morrison, D.S., Petrolatum, http://www.penreco.com (accessed 2002).

Pedersen, K.S., Fredenslund, Aa., and Thomassen, P., *Properties of Oils and Natural Gases*, Gulf Publishing Company, Houston, TX, 1989.

Penreco, http://www.penreco.com (accessed 2002).

Pillon, L.Z., *Interfacial Properties of Petroleum Products*, CRC Press, Boca Raton, FL, 2007.

Pirro, D.M. and Wessol, A.A., *Lubrication Fundamentals*, 2nd ed., Marcel Dekker Inc., New York, 2001.

Prince, R.J., Base oils from petroleum, in *Chemistry and Technology of Lubricants*, 2nd ed., Mortier, R.M. and Orszulik, S.T., Eds., Blackie Academic & Professional, London, U.K., 1997.

Sequeira, A., *Pre-Prints Division of Petroleum Chemistry, ACS*, 37(4), 1286 (1992).

Sequeira, A. Jr., *Lubricant Base Oil and Wax Processing*, Marcel Dekker, Inc., New York, 1994.

Severin, D. and David, T., *Petroleum Science and Technology*, 19(5&6), 469 (2001).

Shell, http://www.shell-lubricants.com (accessed 2001).

Singh, H., *Science in Africa*, 2002, http://www.scienceinafrica.co.za (accessed 2008).

Speight, J.G., *The Chemistry and Technology of Petroleum*, 4th ed., CRC Press, Boca Raton, FL, 2006.

Sperber, O., Kaminsky, W., and Geissler, A., *Petroleum Science and Technology*, 23(1), 47 (2005).

Srivastava, S.P. et al., *Petroleum Science and Technology*, 20(3&4), 269 (2002).

Starkova, N.N. et al., *Chemistry and Technology of Fuels and Oils*, 37(3), 191 (2001).

Taylor, R.J., McCormack, A.J., and Nero, V.P., *Pre-Prints Division of Petroleum Chemistry, ACS*, 37(4), 1337 (1992).

2 Nonconventional Processing of Base Stocks

2.1 HYDROFINISHING AND HYDROTREATMENT

Conventional processing, based on solvent extraction and solvent dewaxing of lube oil distillates, is sufficient for many lubricant products and needed to produce wax. Traditional ways of lube oil finishing include adsorbent clay, which removes some undesirable molecules, such as residual solvents. In recent years, finishing of solvent-refined base stocks has been replaced by a nonconventional catalytic process, such as hydrofinishing or hydrofining (HF). The commercial HF processes are based on heating the feedstock in a furnace and passing it with hydrogen through a reactor filled with catalyst. After passing through the reactor, the treated oil is cooled, separated from the excess hydrogen, and pumped to a stripper tower, where hydrogen sulfide is removed by steam, vacuum, or flue gas. The finished product leaves the bottom of the stripper tower and has improved color, odor, and lower sulfur content (Speight, 2006). The HF process is used to finish naphthas, gas oils, and lube oil base stocks. Under mild HF conditions, olefins are saturated and the aromatic contents of finished products are usually not affected. The nitrogen-, sulfur-, and oxygen-containing compounds undergo hydrogenolysis to split out ammonia, hydrogen sulfide, and water, respectively (Speight, 2006).

The effect of clay treatment and HF on the properties of solvent-refined base stock produced from Western Canadian crude oil and having a low kinematic viscosity (KV) is shown in Table 2.1. Both clay treatment and HF were reported to increase the dewaxed pour point from −15°C to −12°C with no change in their cloud points. However, HF was found to be more effective in reducing the S content of base stock (Lynch, 2007). Hydrogen-finishing processes are used in the place of more costly acid and clay finishing processes for the purpose of improving color, odor, and thermal and oxidative stability of mineral base stocks. The literature reports that most HF operations are operated at a severity set by the color improvement needed (Sequeira, 1994).

The effect of clay treatment and HF on the properties of solvent-refined base stock produced from Western Canadian crude oil and having a high KV is shown in Table 2.2. Both clay treatment and HF were reported to decrease the viscosity and increase the cloud point from −2°C to 1°C. However, HF leads to a higher yield and is more effective in decreasing the S content of the base stock. While HF is more effective in decreasing the S content, the literature reports that the clay treatment is more effective in decreasing the N content of solvent-refined base stocks (Lynch, 2007). Mineral base stocks are traditionally classified according to Saybolt Universal Seconds (SUS) and a 100N (neutral) base stock has a viscosity of 100 SUS at 100°F

TABLE 2.1

**Effect of Clay Treatment and HF on the Properties of
Low Viscosity Oil**

Western Canadian Crude Oil Base Stock	Solvent-Refined No Finishing	Solvent-Refined Clay Treatment	Solvent-Refined HF
KV at 100°F (SUS)	157	157	157
KV at 38°C (cSt)	33.5	33.5	33.5
VI	91	91	91
Cloud point (°C)	−10	−10	−10
Pour point (°C)	−15	−12	−12
Sulfur (wt%)	0.15	0.12	0.06
Flash point (°C)	202	207	204

Source: Lynch, T.R., *Process Chemistry of Lubricant Base Stocks*, CRC Press, Boca Raton, FL, 2007. With permission.

TABLE 2.2

**Effect of Clay Treatment and HF on the Properties of
Higher Viscosity Oil**

Western Canadian Crude Oil Base Stock	Solvent-Refined No Finishing	Solvent-Refined Clay Treatment	Solvent-Refined HF
KV at 100°F (SUS)	643	624	632
KV at 38°C (cSt)	139.2	135.1	136.8
VI	91	91	91
Cloud point (°C)	−2	1	1
Pour point (°C)	−4	−4	−4
Sulfur (wt%)	0.18	0.15	0.07
Flash point (°C)	266	263	266
Carbon residue (wt%)	0.05	0.03	0.02

Source: Lynch, T.R., *Process Chemistry of Lubricant Base Stocks*, CRC Press, Boca Raton, FL, 2007. With permission.

(Sequeira, 1994). At the present time, the KV of base stocks is reported in cSt at 40°C and 100°C (Pillon, 2007).

The effect of solvent refining (SR) and HF on the composition of 100N paraffinic base stock is shown in Table 2.3. The lube oil distillate, from paraffinic crude oil, was reported to have an aromatic content of 34 wt% and an S content of 1.1 wt%. After solvent extraction, a decrease in the aromatic content to 17 wt% and a decrease in the S content to 0.7 wt% were reported. With a decrease in the aromatic content, the VI increased from 86 to 110, and the pour point increased from 24°C to 29°C.

TABLE 2.3
Effect of SR and HF on the Composition of 100N
Paraffinic Base Stock

Processing 100N Paraffinic Oil	Lube Oil Distillate	Lube Oil Raffinate	After Dewaxing and HF
KV at 100°F (SUS)	103	85	100
KV at 38°C (cSt)	21.3	17	20.6
VI	86	110	95
Pour point (°C)	24	29	−18
Aromatics (wt%)	34	17	16
Sulfur (wt%)	1.1	0.7	0.2
ASTM color	4.5	1.5	L0.5

Source: Sequeira, A., *Pre-Prints Div. Petrol. Chem., ACS,* 37(4), 1286, 1992. With permission.

The solvent extraction process was also reported to improve the color of paraffinic raffinates. After solvent dewaxing to a −18°C pour point and HF, a decrease in the VI to 95 is observed with a small decrease in the aromatic content to 16 wt% and the S content to 0.2 wt%. A further improvement in color was also reported. The literature reports that naphthene pale oils (NPOs) are vacuum distilled and solvent extracted to remove aromatics and, in many cases, HF is used as the last processing step (Sequeira, 1992).

The effect of solvent extraction and HF on the properties and composition of 100N NPO is shown in Table 2.4. The lube oil distillate, from naphthenic crude oil containing no wax, was reported to have an aromatic content of 34 wt%, and an S content of 0.1 wt%. After solvent extraction, a decrease in the aromatic content

TABLE 2.4
Effect of Solvent Extraction and HF on the
Composition of 100N Naphthenic Oil

Processing 100N NPO	Lube Oil Distillate	Lube Oil Raffinate	After HF
KV at 100°F (SUS)	108	100	100
KV at 38°C (cSt)	22.5	20.6	20.6
VI	17	61	61
Natural pour point (°C)	−37	−29	−29
Aromatics (wt%)	34	24	24
Sulfur (wt%)	0.1	<0.1	<0.1
ASTM color	7	1.5	L0.5

Source: Sequeira, A., *Pre-Prints Div. Petrol. Chem., ACS,* 37(4), 1286, 1992. With permission.

to 24 wt% and a decrease in the S content were reported. A low VI of only 17 increased to 61, and a low natural pour point of –37°C increased to –29°C. The solvent extraction process also improves the color of naphthenic raffinates. After HF, there is no change in the VI, pour point, and aromatic content; however, a further decrease in the S content with an improvement in color is observed. The composition of crude oils affects the properties of vacuum gas oil (VGO) and the processing severity. A low VI of VGO leads to high extraction losses in solvent extraction. A high wax content of VGO increases the cost of the dewaxing. A high sulfur content of VGO requires hydrotreating (HT) of finished base stocks to eliminate sulfur (Mang, 2001). The paraffinic content of base stocks includes normal *n*-paraffins and branched *iso*-paraffins, and the early literature reports on the effect of the HT process on their hydrocarbon composition and VI (Beuther, Donaldson, and Henke, 1964).

The effect of the HT process on the VI and composition of base stocks produced from different crude oils is shown in Table 2.5. Solvent refined base stock, having a VI of 108, was reported to contain 3.5 wt% of paraffins, 85.4 wt% of naphthenes, and 11.1 wt% of aromatics. After the HT of the base stock, from Kuwait crude oil, its VI increased to 125 with a small increase in the paraffinic content and a drastic decrease in aromatic content. After the HT of the base stock, from West Texas crude oil, its VI increased to 117 with no presence of paraffinic content and a drastic decrease in aromatic content. After the HT of the base stock, from Ordovician crude oil, its VI increased to 120, with the presence of paraffinic content and a drastic decrease in aromatic content. A recent literature reports on the effect of a two-stage HT process on the VI and the composition of base stocks, including the N content (Lynch, 2007).

The effect of a two-stage HT process on the heteroatom content of different viscosity base stocks produced from paraffinic crude oil is shown in Table 2.6. After the first stage HT of different viscosity base stocks, with a decrease in the

---

TABLE 2.5
Effect of SR and HT on the VI and the Composition of Different Base Stocks

Crude Oils Base Stock Processing	Solvent Refined	Kuwait HT	West Texas HT	Ordovician HT
VI	108	125	117	120
Paraffins (%)	3.5	6.9	0	1.4
Mononaphthenes (%)	58.3	68.2	71.2	69.0
Condensed naphthenes (%)	27.1	23.7	27.5	27.8
Monoaromatics (%)	10.0	0.9	1.0	1.6
Condensed aromatics (%)	1.1	0.3	0.3	0.2

Source: Reprinted from Beuther, H. et al., *I&EC Prod. Res. Dev.*, 3(3), 174, 1964. With permission.

TABLE 2.6

Effect of Two-Stage HT Process on VI and the Composition of Paraffinic Base Stocks

	80N	160N	650N	Bright Stock
		First Stage HT		
KV at 40°C (cSt)	14.7	34.7	117.0	440
VI	91	110	101	105
Aromatics (wt%)	13.1	12.5	3.1	14.7
Sulfur (ppm)	<2	<2	7	7
Nitrogen (ppm)	2	4	3	11
		Second Stage HT		
KV at 40°C (cSt)	14.9	35.1	116.8	439
VI	92	111	101	105
Aromatics (wt%)	0.5	0.2	0.2	0.6
Sulfur (ppm)	<2	<2	<2	<2
Nitrogen (ppm)	1	1	2	2

Source: Lynch, T.R., *Process Chemistry of Lubricant Base Stocks*, CRC Press, Boca Raton, FL, 2007. With permission.

aromatic content, a decrease in the S content below 7 ppm, and an N content varying in 2–11 ppm was reported. After the second stage HT of the same base stocks, a drastic decrease in the aromatic content from 3.1 to 14.7 wt% to 0.2–0.6 wt% with no significant increase in the VI was observed. However, an S content was further reduced to below 2 ppm, and the N content was further reduced to 2 ppm or less.

The effect of HT process on properties and S content of different viscosity naphthenic oils is shown in Table 2.7. Variation in crude oils and finishing conditions

TABLE 2.7

Effect of HT on the Properties and the S Content of Different Naphthenic Oils

Properties	HT Naphthenic Oil	HT Naphthenic Oil
KV at 40°C (cSt)	7.5	29.8
KV at 100°C (cSt)	2.1	4.6
VI	56	35
Pour point (°C)	−54	−39
Sulfur (wt%)	<0.1	0.1

Source: Reproduced from Lynch, T.R., *Process Chemistry of Lubricant Base Stocks*, CRC Press, Boca Raton, FL, 2007. With permission.

can lead to variation in properties and composition of lube oil base stocks. The refined paraffinic oils have high VI and contain waxes, which crystallize out at low temperatures. Base stocks, produced from paraffinic crude oils, require dewaxing to lower their pour points. Naphthenic oils are made in small quantities from selected crude oils, which have low wax content, and they generally do not require dewaxing. After the HT of different viscosity naphthenic oils, their VIs of 35–56 are still low, but they have very low natural pour points of −54°C to −39°C. Their S contents were reported to be of 0.1 wt% or lower and flash points of 151°C–157°C. Some low viscosity naphthenic oils are used in lubricant products, which do not require a high VI but require very good low temperature fluidity.

2.2 HYDROREFINING OF WHITE OILS

Most organic molecules and functional groups are transparent in the portions of the electromagnetic spectrum that we call the ultraviolet (UV) region, however, some molecules containing O, N, S and aromatics can adsorb, depending on their chemistry (Pavia, Lampman, and Kriz, 1996). Compounds that are highly colored have adsorption in the visible (VIS) region. Aliphatic ketones, aromatic esters, and aromatic acids, such as benzoic acid, were reported to adsorb in a similar UV range of 270–280 nm. Naphthalene, which is a 2-ring condensed aromatic compound, was reported to adsorb in the UV range of 220–320 nm. In the UV range of 280–350 nm, some aliphatic ketones and many different chemistry aromatic molecules can adsorb (Pavia, Lampman, and Kriz, 1996). Polynuclear aromatics include two or more aromatic rings and can be fused, such as naphthalene and phenantrene, or can be separate as biphenyl. With an increase in condensed ring number, the UV absorbance increases to 250–500 nm. Anthracene, which is a 3-ring condensed aromatic compound, was reported to adsorb in the UV range of 255–380 nm (Pavia, Lampman, and Kriz, 1996). The literature reports on the variation in the UV absorbance of large polynuclear hydrocarbons containing 9–11 condensed aromatic rings (Fetzer, 1988). Coronene, containing six condensed aromatic rings, was found to adsorb in a wide range of 250–400 nm (Pillon, 2007).

The chemistry and content of different polyaromatic hydrocarbons found in mineral oils are shown in Table 2.8. In conventional refining, lube oil distillates are solvent extracted and solvent dewaxed, and HF is used as the last processing step. At low pressures and low temperatures, the hydrogen-finishing process does not saturate aromatics or break the carbon–carbon bonds. The use of higher pressure might saturate some aromatics, while the use of higher temperature might lead to some cracking (Sequeira, 1994). The typical HF process, used to finish lube oil base stocks, requires a reactor temperature of 260°C–315°C and a reactor pressure of 500–1000 psig. The commercial hydrogen-finishing catalysts consist of cobalt–molybdenum on alumina, nickel–molybdenum on alumina, iron–cobalt–molybdenum on alumina, and nickel–tungsten on alumina or silica–alumina. Promoters, such as fluorides or phosphorus, are sometimes added to enhance the catalyst performance (Sequeira, 1994). The early literature reports on the effect of temperature on the thermal stability of base stocks. Their decomposition, in nitrogen atmosphere, was measured as the amount of material boiling below 345°C (Beuther, Donaldson, and Henke, 1964).

TABLE 2.8
Chemistry and Content of Polyaromatic Hydrocarbons Found in Mineral Base Stocks

Polyaromatic Hydrocarbons	Content (mg/kg)
Fluoranthene	0.02
Pyrene	0.42
Benzonaphtho-thiophene	0.34
Benzofluoranthenes	0.3
Benzophenanthrene	<0.01
Cyclopentapyrene	<0.01
Benzanthracene	0.08
Chrysen + triphenylene	2.1
Benzopyrenes	0.4
Indenopyrene	0.06
Dibenzoanthracene	<0.01
Benzoperylene	0.5
Anthanthrene	0.03
Coronene	0.13

Source: Betton, C.J., Lubricants and their environmental impact, in *Chemistry and Technology of Lubricants*, 2nd ed., Mortier, R.M. and Orszulik, S.T., Eds., Blackie Academic & Professional, London, U.K., 1997.

TABLE 2.9
Effect of Temperature on the Thermal Stability of Different Base Stocks

	Base Stocks	
	Solvent Refined	Hydrotreated
Thermal Treatment	Decomposition (%)	
370°C/3 h/nitrogen	0	0
385°C/3 h/nitrogen	13	5
400°C/3 h/nitrogen	26	19

Source: Beuther, H. et al., *I&EC Prod. Res. Dev.*, 3(3), 174, 1964. With permission.

The effect of temperature on the thermal stability in nitrogen atmosphere of solvent refined and hydrotreated base stocks is shown in Table 2.9. The early literature reports that at increased temperatures, above 370°C, and, in nitrogen atmosphere, hydrotreated base stocks were more thermally stable and less decomposition was observed. However, even in the case of hydrotreated base stocks, some decomposition was taking place when exposed to high temperatures for prolonged time indicating cracking of molecules. Hydrogen-finishing processes, known as HF, are mild hydrogenation processes. The selective hydrogenation process, known as hydrorefining, is a high-pressure HF process. The single stage hydrorefining process is used to produce the technical grade white oils. In the hydrorefining process, the sulfur and the nitrogen contents of the technical white oils are removed and the aromatic content is reduced to a very low level (Sequeira, 1994).

The typical HF and hydrorefining process conditions are shown in Table 2.10. At low pressures and low temperatures, the hydrogen-finishing process does not saturate aromatics or break the carbon–carbon bonds. The use of higher pressure might saturate some aromatics, while the use of higher temperature might lead to some cracking (Sequeira, 1994). The typical hydrorefining process requires a reactor temperature of 260°C–315°C and a reactor pressure of 1500–3000 psig. Some lubricant applications require the use of high purity oils that contain no aromatic hydrocarbons. At the present time, hydrorefining is used to produce white oils from solvent neutral mineral oils or high quality naphthenic distillates.

The effect of hydrorefining severity on properties of low viscosity white oils is shown in Table 2.11 The use of hydrorefining leads to the saturation of aromatics and the removal of the sulfur present in low viscosity feed. A decrease in the viscosity with no change in the flash point is observed. White oils are produced by high-pressure hydrogenation used to saturate aromatics and remove essentially any sulfur and nitrogen from conventional mineral base stocks. The single stage hydrorefining process is used to produce the technical grade white oils. The second stage hydrorefining treatment is used to saturate the last traces of aromatic compounds to produce the pharmaceutical and food grade white oils (Sequeira, 1994). The use of hydrorefining can lead to technical white oil, which contains practically no aromatics and a low sulfur content of 3 ppm, or food grade and medicinal grade white oil that contains no aromatics and no S content.

The effect of hydrorefining severity on properties of high viscosity white oils oil is shown in Table 2.12. The use of hydrorefining leads to a total saturation of aromatics

---

TABLE 2.10
Typical HF and Hydrorefining Process Conditions

Process Variables	HF	Hydrorefining
Pressure (psig)	500–1000	1500–3000
Temperature (°C)	260–315	260–315
Space velocity ($V_o/V_c/h$)	1.0–1.5	0.5–1.0
Lube oil yield (vol%)	98+	95–98
Catalyst life (years)	1–2	1–2

Source: Sequeira, A. Jr., *Lubricant Base Oil and Wax Processing*,
          Marcel Dekker, Inc., New York, 1994. With permission.

TABLE 2.11

Effect of Hydrorefining Severity on the Properties of Low Viscosity White Oils

Hydrorefining Process	Neutral Mineral Oil Feed	White Oil Technical Grade	White Oil Food Grade
KV at 20°C (cSt)	319	220	221
Aromatics (wt%)	12	Trace	Trace
Sulfur (ppm)	1500	3	0
Flash point (°C)	258	258	256

Source: Lynch, T.R., *Process Chemistry of Lubricant Base Stocks*, CRC Press, Boca Raton, FL, 2007. With permission.

TABLE 2.12

Effect of Hydrorefining on the Properties of High Viscosity White Oils

Hydrorefining Process	Heavy Machine Oil Feed	White Oil Technical Grade	White Oil Food Grade
KV at 20°C (cSt)	940	542	545
Aromatics (wt%)	28	Trace	Trace
Sulfur (ppm)	12,000	4	0
Flash point (°C)	262	250	262

Source: Lynch, T.R., *Process Chemistry of Lubricant Base Stocks*, CRC Press, Boca Raton, FL, 2007. With permission.

and removal of the S present in heavy machine oil feed. A drastic decrease in the viscosity with some variation in the flash point is observed. The use of hydrorefining can lead to technical white oil, which contains practically no aromatics and a low sulfur content of 4 ppm, or food grade and medicinal grade white oil that contains no aromatics and no S content. Lubricants, which require food grade or medicinal grade mineral oils, need to meet severe U.S. FDA requirements. The level of severity of refining is tested by following the UV absorbance of dimethyl sulfoxide (DMSO) extract for different grade white oils in the range from 260 to 420 nm.

The U.S. FDA UV absorbance limits for technical and food/medicinal grade white oils are shown in Table 2.13. The use of hydroprocessing is effective in removing aromatics, S, and N molecules from any petroleum product ranging in viscosity from light oils to waxes. HF is used to finish base stocks and wax. The use of a single stage hydrorefining process is needed to meet the UV absorbance requirements of DMSO extracts for technical white oils. The technical white oils are used in cosmetics, textile lubrication, insecticide, vehicles, and paper impregnation. The technical white oils are also used in products, such as waxes, agricultural spray oils,

TABLE 2.13
U.S. FDA UV Absorbance Limits for White Oils

White Oils UV Wavelength Range (nm)	Technical Grade UV Absorbance (Max)	Food and Medicinal Grade UV Absorbance (Max)
280–289	4.0	
290–299	3.3	
300–329	2.3	
330–350	0.8	
260–350		0.1

Source: Lynch, T.R., *Process Chemistry of Lubricant Base Stocks*, CRC Press, Boca Raton, FL, 2007. With permission.

and mineral seal oils (Speight, 2006). The use of a two-stage hydrorefining process is required to produce medicinal white oils, which are colorless, odorless, and taste-less. Medicinal white oils need to be free of polycyclic aromatic hydrocarbons to meet the U.S. Pharmacopoeia (USP) requirements for use in medicinal and cosmetic formulations. Some applications require food or medicinal grade petroleum wax and, at the present time, hydrorefining has replaced clay and percolation processes as the process of choice for the manufacture of wax, which must meet the purity specifications prescribed by the Government. The level of severity of refining is tested following the UV absorbance in the range from 280 to 400 nm for petroleum wax and petrolatum. Variation in feed and hydrorefining conditions can lead to vari-ation in the composition of white oils and wax.

2.3   HYDROCRACKING PROCESS

The literature reports that in hydroconversion units, using hydrogen in the presence of a catalyst under high temperatures and pressures, aromatic molecules can be converted into useful base stock molecules. The feed can vary from vacuum distil-late, deasphalted oil (DAO) to raffinate. The final product has only a few aromatics, a higher VI, and contains a significant amount of light materials that are removed in a re-distilling unit where the viscosity of the final product is controlled (Shell, 2001). While the use of conventional refining leads to mineral base stocks with a VI ranging from 85 to 95, use of hydroprocessing can yield a VI over 100 from almost any crude oil. Instead of unwanted, low VI molecules being solvent extracted, they are chemically changed into high VI molecules, usually of lower molecular weights. This enables an increase in the output of light lubricating oils for which there is a growing demand.

The boiling point and carbon number range of different petroleum fractions, used to produce lube oils, are shown in Table 2.14. The literature reports that carbon number ranges are referred to by the boiling points of the nearest *n*-paraffins. For example, the carbon number range of a 650°F–850°F fraction is C_{20}–C_{30}, which has a boiling point range of 651°F–843°F (Lynch, 2007). An increase in the boiling point

TABLE 2.14

Boiling Point and Carbon Number Range of Different Lube Oil Fractions

Petroleum Fractions	Boiling Range (°C)	Carbon Number
Light VGO (LVGO)	316–482	18–34
Heavy VGO (HVGO)	427–593	28–53
DAO	510+	38+

Source: Reproduced from Lynch, T.R., *Process Chemistry of Lubricant Base Stocks*, CRC Press, Boca Raton, FL, 2007. With permission.

range of VGO from 316°C–482°C to 427°C–593°C leads an increase in the carbon range from C_{18-34} to C_{28-53}. With a further increase in the boiling point of DAO, above 510°C, a further increase in the carbon number range, above C_{38}, was reported.

The effect of furfural extraction and hydrocracking (HC) on product distribution from heavy distillate is shown in Table 2.15. After furfural extraction, the viscosity of heavy distillate decreased from 14.1 to 10.5 cSt at 100°C, and the VI increased from 53 to 95 leading to 60 wt% of 400N lube oil base stocks. After HC, the viscosity of heavy distillate further decreased to 8.7 cSt at 100°C, and the VI increased to 97 also leading to 60 wt% of 400N base stocks. However, after HC, an additional 18 wt% of 200N and 5 wt% of 100N lube oil base stocks were also obtained, thus leading to an increase in yields of low viscosity base stocks.

The properties and composition of VGO, from conventional refining of crude oil, and, HC of DAO, are shown in Table 2.16. LVGO (Light VGO), from conventional refining of crude oil, was reported to have a high pour point of 29°C, and contain an S content of 1.2 wt% and an N content of 0.1 wt%. After the dewaxing and the

TABLE 2.15

Effect of Furfural Extraction and HC on Product Distribution from Heavy Distillate

Heavy Distillate Feed	Furfural Extraction	HC
KV at 99°C (cSt)	11	9
Dewaxed VI	95	97
Flash point (°C)	245	225
Product distribution		
400N (wt%)	60.0	60.4
200N (wt%)	—	18.0
100N (wt%)	—	5.1

Source: Lynch, T.R., *Process Chemistry of Lubricant Base Stocks*, CRC Press, Boca Raton, FL, 2007. With permission.

TABLE 2.16

Properties of VGO from Conventional Refining of Crude Oil and HC of DAO

Processing Petroleum Fractions	From Crude Oil LVGO	From Crude Oil HVGO	From HC DAO HVGO
KV at 100°C (cSt)	5.8	14.5	16.7
Pour point (°C)	29	41	43
Sulfur (wt%)	1.2	1.3	0.1
Nitrogen (wt%)	0.1	0.2	0.1
Dewaxed oil			
KV at 100°C (cSt)	6.2	16.8	17.9
VI	32	18	58
Pour point (°C)	−9	−15	−15
Wax (wt%)	6.2	6.3	6.3

Source: Lynch, T.R., *Process Chemistry of Lubricant Base Stocks*, CRC Press, Boca Raton, FL, 2007. With permission.

removal of 6.2 wt% of wax, dewaxed LVGO was reported to have a VI of 32 and a dewaxed pour point of −9°C. HVGO (Heavy VGO), from conventional refining of crude oil, was reported to have a higher pour point of 41°C and a similar S content of 1.2 wt% but a higher N content of 0.2 wt%. After the dewaxing and the removal of 6.3 wt% of wax, dewaxed HVGO was reported to have a low VI of 18 and a lower dewaxed pour point of −15°C. Another HVGO, from HC of DAO, was reported to have a similar high pour point of 43°C and a lower S of only 0.1 wt% but a similar N content of 0.1 wt%. After the dewaxing and the removal of 6.3 wt% of wax, dewaxed HVGO, from HC of DAO, was reported to have a higher VI of 58 and also a dewaxed pour point of −15°C.

The effect of furfural extraction and HC on product distribution from DAO is shown in Table 2.17. Furfural extraction and dewaxing of DAO leads to 61 wt% of high viscosity >600N lube oil base stock having a high VI of 98 and a flash point of 280°C. After the HC and the dewaxing of DAO, lube oil products also have a high VI of 98 but a lower flash point of 240°C. Products distribution indicates a decrease in the volume of high viscosity >600N oil but an increase in the volume of lower viscosity <600N oils. The feed and processing conditions can affect the viscosity, VI, and volume of lube oil products.

The effect of hydroprocessing severity on VI and dewaxed oil yield produced from DAO is shown in Table 2.18. With an increase in hydroprocessing severity, a decrease in dewaxed oil yield, a decrease in viscosity, and an increase in VI from 74 to 103 are observed. The severity of the HC reaction is measured by the degree of conversion of the feed to lighter products. Conversion is defined as the volume percent of the feed, which disappears to form products boiling below the desired product end point (Gary and Handwerk, 2001).

TABLE 2.17
Effect of Furfural Extraction and HC on
Product Distribution from DAO

DAO	Furfural Extraction	HC
KV at 99°C (cSt)	34	16
Dewaxed VI	98	98
Flash point (°C)	280	240
Product distribution		
>600N (wt%)	61.0	57.9
600N (wt%)	—	10.1
350N (wt%)	—	9.0
150N (wt%)	—	9.0

Source: Lynch, T.R., *Process Chemistry of Lubricant Base Stocks*, CRC Press, Boca Raton, FL, 2007. With permission.

TABLE 2.18
Effect of Hydroprocessing Severity on VI and Dewaxed Oil Yield

Middle East Crude Oil Properties	Deasphalted Oil DAO Feed	Hydroprocessing Low Severity	Hydroprocessing High Severity
Dewaxed oil yield (vol%)		77.7	64.4
Gravity (API)	19.8	24.9	28.6
KV at 210°F (SUS)	231	149.9	75.5
VI	74	84	103
Pour point (°C)	54	−23	−26

Source: Reproduced from Sequeira, A. Jr., *Lubricant Base Oil and Wax Processing*, Marcel Dekker, Inc., New York, 1994. With permission.

The effect of SR and HC on yield and VI of base stocks is shown in Table 2.19. The use of conventional SR leads to higher viscosity base stocks including bright stock, having a VI of 95. The use of HC and dewaxing leads to higher volume of lower viscosity base stocks having a higher VI of 115. The early literature reports on the hydrotreatment of mixtures of distillates and DAOs leading to a VI of 120–125 (Beuther, Donaldson, and Henke, 1964). The recent literature reports on the "VI droop," which refers to a decrease in VI, after dewaxing, with an increase in HC temperature (Lynch, 2007).

The effect of HC temperature on VI of dewaxed oil is shown in Table 2.20. With an increase in HC temperature from 392°C to 427°C, the VI of dewaxed base stocks increases from 88 to 133; however, after a further increase in HC temperature to 433°C the VI decreases to 127. Lube oil base stocks need to have a high VI and a low

TABLE 2.19

Effect of SR and HC on Yield and VI of Base Stocks

Base Stocks Dewaxed to –12°C	SR Yield (%)	SR VI	HC Yield (%)	HC VI
150N	9.9	95	10.8	115
260N	4.1	95	12.1	115
600N	7.4	95	8.7	115
750N	—	—	5.1	115
Bright stock	8	95	—	—
Total yield	29.4		36.6	

Source: Reproduced from Lynch, T.R., *Process Chemistry of Lubricant Base Stocks*, CRC Press, Boca Raton, FL, 2007. With permission.

TABLE 2.20

Effect of HC Temperature on VI of Dewaxed Oil

	HC Reactor Temperature (°C)				
	392	**406**	**417**	**429**	**433**
After dewaxing (343°C+)					
KV at 38°C (cSt)	154.1	72.6	42.4	23.7	16.5
KV at 99°C (cSt)	13.2	9.2	6.8	4.8	3.7
VI	88	112	125	133	127

Source: Reproduced from Lynch, T.R., *Process Chemistry of Lubricant Base Stocks*, CRC Press, Boca Raton, FL, 2007. With permission.

pour point to be used in lubricating oils. In addition, distillation is needed to adjust the viscosity, flash point, and volatility (Shell, 2001). Different processing techniques are used to meet the volatility targets of base stocks at the max yields. The volatility can be improved by removing the low boiling front end, known as topping, that increases the viscosity of the oils. Another route to improve the volatility is to remove the high boiling and low boiling ends, known as heart-cut, which maintains a constant viscosity (Cody et al., 2002).

The effect of different processing techniques on volatilities and yields of 100N HC base stock is shown in Table 2.21. The volatility can be measured or estimated using the gas chromatography (GC) technique. The volatility of base stocks is usually measured as the percent loss of material at 250°C in the standard Noack volatility test (ASTM D 5800). With a decrease in volatility from 38.5% to 18.7% off, a decrease in yield from 99.9% to 39.6% is also observed. While HC is less dependent on feedstock than SR, it can have a significant impact on the product volatility and yields.

TABLE 2.21

Effect of Different Processing Techniques on Volatility and Yield of 100N HC Base Stock

HC Base Stock Processing	KV of 3.9 cSt at 100°C Volatility (% Off)	KV of 3.9 cSt at 100°C Yield (%)
None	38.5	99.9
Topping	21.1	76.2
Heart-cut	20.9	73.8
Heart-cut	19.2	52.2
Heart-cut	18.7	39.6

Source: Cody, I.A. et al., Raffinate hydroconversion process, CA Patent 2,429,500, 2002.

TABLE 2.22

Composition and Volatility of Solvent Refined and Different HC Base Stocks

Base Stocks Properties	Solvent Refined Conventional VI	HC Conventional VI	HC High VI
KV at 40°C (cSt)	30.0	42.0	39.5
KV at 100°C (cSt)	5.1	6.3	6.7
VI	95	95	125
Pour point (°C)	−15	−15	−18
Aromatics (wt%)	20.5	<1	<1
Sulfur	0.45 wt%	15 ppm	6 ppm
Nitrogen (ppm)	50	<1	<1
Volatility (% off)	20	13	5

Source: Lynch, T.R., *Process Chemistry of Lubricant Base Stocks*, CRC Press, Boca Raton, FL, 2007. With permission.

The composition and volatility of solvent refined and different HC base stocks are shown in Table 2.22. The use of conventional SR leads to base stocks having a VI of 95, a high aromatic content of 20.5 wt%, and a high volatility of 20% off. The use of HC can lead to base stocks having a conventional VI of 95 but a low aromatic content, below 1 wt%, only 15 ppm of S, and practically no N. A decrease in volatility was also reported. The use of HC can also lead to base stocks having an increased VI of 125 and a low aromatic content, below 1 wt%, only 6 ppm of S, and practically no N. A further decrease in volatility was reported. Heavy polynuclear aromatics were reported to form in small amounts from HC reactions and, when the fractionator bottoms are recycled, can build up to concentrations that cause the fouling of heat exchanger surfaces and equipment (Gary and Handwerk, 2001). The typically low quality feedstocks are used in HC, and the consequent severe conditions required to achieve the VI and volatility might result in the formation of toxic polynuclear aromatic molecules (Cody et al., 2002). Variation in HC conditions can lead to a variation in the composition of base stocks.

2.4 CATALYTIC DEWAXING

Lubricants require the use of base stocks that have low pour points, and the solvent dewaxing process is used to produce lube oil base stock and petroleum wax. In solvent dewaxing, the oil is diluted with a solvent that has a high affinity for oil, chilled to precipitate the wax, filtered to remove the wax, stripped of solvent, and dried. The solvent dewaxing is a crystallization–filtration process. The process uses refrigeration to crystallize the wax, and solvent to dilute the oil to allow for a rapid filtration and wax separation from the oil. As the temperature of solvent dewaxing decreases, the highest melting point components crystallize first. Higher melting *iso*-alkane waxes predominate in heavier petroleum fractions and the composition of waxes from solvent dewaxing of different viscosity feedstocks varies (Prince, 1997). There are two types of dewaxing processes in use today, which are solvent dewaxing and catalytic dewaxing. In catalytic dewaxing, the oil is contacted with hydrogen at increased temperature and pressure, and in the presence of catalyst that selectively cracks *n*-paraffins to methane, ethane, and propane. The oil is steam stripped and dried (Exxon, 2002). The literature reports that the catalytic dewaxing process reduces the pour point by cracking *n*-paraffins and similar molecules to gasoline and other lighter products, and is not affected by the S and N contents in the feed (Lynch, 2007). When a selective catalytic process, called catalytic dewaxing, is used as an alternative to solvent dewaxing for the removal of wax from lube oil base stocks, no wax is produced.

The effect of solvent dewaxing and catalytic dewaxing on VI and wax content of waxy raffinate having a boiling range of 360°C–470°C is shown in Table 2.23. Low viscosity waxy raffinate having a boiling range of 360°C–470°C was reported to have

TABLE 2.23
Effect of Solvent Dewaxing and Catalytic Dewaxing on VI and Wax Content of Raffinate

Raffinate Dewaxing	Waxy Feed 360°C–470°C	After Solvent Dewaxing	After Catalytic Dewaxing
VI	118	98	85
Pour point (°C)	32	−15	−18
Wax content (wt%)	18	2	1
Sulfur (wt%)	0.9	0.9	1.0
Nitrogen (ppm)	57	—	94

Source: Lynch, T.R., *Process Chemistry of Lubricant Base Stocks*, CRC Press, Boca Raton, FL, 2007. With permission.

TABLE 2.24

Effect of Dewaxing on VI and Wax Content of Higher Viscosity Raffinate

Raffinate Dewaxing	Waxy Feed 430°C–540°C	Solvent Dewaxing	Catalytic Dewaxing
VI	93	79	62
Pour point (°C)	46	−12	−7
Wax (wt%)	22	4	3
Sulfur (wt%)	1.8	2.1	2.2

Source: Lynch, T.R., *Process Chemistry of Lubricant Base Stocks*, CRC Press, Boca Raton, FL, 2007. With permission.

a VI of 118, a pour point of 32°C, and a wax content of 18 wt%. After solvent dewaxing to a pour point of −15°C, a decrease in the wax content to 2 wt% and a decrease in the VI to 98 were reported. After the catalytic dewaxing of the same raffinate to a pour point of −18°C, a decrease in the wax content to 1 wt% and a further decrease in the VI to 95 were observed. After solvent dewaxing and catalytic dewaxing, no decrease in the S content but a small increase in the N content was reported.

The effect of solvent dewaxing and catalytic dewaxing on VI and wax content of waxy raffinate having a higher boiling range of 430°C–540°C is shown in Table 2.24. Higher viscosity waxy raffinate having a boiling range of 430°C–540°C was reported to have a lower VI of 93, a higher pour point of 46°C, and a higher wax content of 22 wt%. After solvent dewaxing to a pour point of −12°C, a decrease in the wax content to 4 wt% and a decrease in the VI to 79 were reported. After the catalytic dewaxing of the same raffinate to a pour point of −7°C, a decrease in the wax content to 3 wt% and a further decrease in the VI to 62 were observed. After solvent dewaxing and catalytic dewaxing, a small increase in the S content was observed.

The effect of solvent dewaxing and catalytic dewaxing on VI and hydrocarbon composition of light neutral base stock dewaxed to the same pour point is shown in Table 2.25. Light waxy raffinate was reported to contain 15 wt% of *n*-paraffins, 22 wt% of *iso*-paraffins, 15 wt% of mononaphthenes, 25 wt% of polynaphthenes, and 23 wt% of aromatics. After solvent dewaxing and reduction in the pour point to −6°C, the *n*-paraffinic content was reduced to 1 wt% and the solvent-dewaxed light neutral base stock was reported to have a VI of 108. After catalytic dewaxing and reduction in the pour point to −6°C, the *n*-paraffinic content was totally eliminated and a small decrease in the *iso*-paraffinic content was observed. The reduction in the pour point of catalytically dewaxed base stocks results from the cracking of the *n*-paraffins, and the catalytically dewaxed light neutral base stock was reported to have a lower VI of 98. Depending on the dewaxing methods, some light viscosity base stocks produced from the same waxy raffinate and dewaxed to the same pour point might have varying VI.

The effect of solvent and catalytic dewaxing on VI and hydrocarbon composition of heavy neutral base stock dewaxed to the same pour point is shown in Table 2.26.

TABLE 2.25

Effect of Dewaxing on VI and Hydrocarbon Composition of Light Base Stock

Light Neutral Base Stock	Waxy Raffinate Feed	After Solvent Dewaxing	After Catalytic Dewaxing
Pour point (°C)	—	−6	−6
VI	—	108	98
n-Paraffins (wt%)	15	1	0
iso-Paraffins (wt%)	22	24	21
Mononaphthenes (wt%)	15	15	16
Polynaphthenes (wt%)	25	34	35
Aromatics (wt%)	23	26	28

Source: Reproduced from Sequeira, A. Jr., *Lubricant Base Oil and Wax Processing*, Marcel Dekker, Inc., New York, 1994. With permission.

TABLE 2.26

Effect of Dewaxing on VI and Hydrocarbon Composition of Heavy Base Stock

Heavy Neutral Base Stock	Waxy Raffinate Feed	After Solvent Dewaxing	After Catalytic Dewaxing
Pour point (°C)	—	−6	−6
VI	—	95	89
n-Paraffins (wt%)	3	0	0
iso-Paraffins (wt%)	20	18	13
Mononaphthenes (wt%)	15	15	17
Polynaphthenes (wt%)	24	24	27
Aromatics (wt%)	38	43	43

Source: Reproduced from Sequeira, A. Jr., *Lubricant Base Oil and Wax Processing*, Marcel Dekker, Inc., New York, 1994. With permission.

Heavy waxy raffinate was reported to contain only 3 wt% of n-paraffins, 20 wt% of iso-paraffins, 15 wt% of mononaphthenes, 24 wt% of polynaphthenes, and 38 wt% of aromatics. After solvent dewaxing and reduction in the pour point to −6°C, the n-paraffinic content was totally eliminated, and a small decrease in the iso-paraffinic content was observed. The solvent-dewaxed heavy neutral base stock was reported to have a VI of 95. After catalytic dewaxing and reduction in the pour point to −6°C, the n-paraffinic content was also totally eliminated and a significant decrease in the iso-paraffinic content from 20 to 13 wt% was also reported. The catalytically dewaxed heavy neutral base stock was reported to have a lower VI of 89. Depending on the dewaxing methods, some heavy neutral base stocks produced from the same waxy raffinate and dewaxed to the same pour point might have also varying VI.

The effect of solvent dewaxing and catalytic dewaxing on the hydrocarbon composition of bright stock dewaxed to the same pour point is shown in Table 2.27. The n-paraffin content of mineral waxy raffinates decreases with an increase in their viscosity. Bright stock raffinate was reported to contain no n-paraffins, 16 wt% of iso-paraffins, 14 wt% of mononaphthenes, 23 wt% of polynaphthenes, and a high content of 47 wt% of aromatics. After solvent dewaxing and reduction in the pour point to −3°C, a small decrease in the iso-paraffinic and mononaphthenic contents was observed. The solvent-dewaxed bright stock was reported to have a VI of 95. After catalytic dewaxing and reduction in the pour point to −3°C, a small decrease in the iso-paraffinic and mononaphthenic contents was also observed. The catalytically dewaxed bright stock was reported to have the same VI of 95. The bright stocks contain no n-paraffins, and their pour point reduction was reported to result from the cracking of the paraffinic side chains on the naphthenic and aromatic molecules (Sequeira, 1994). The literature reports that the catalytic dewaxing, known as isodewaxing, is a less severe cracking process used to crack the wax molecules to light hydrocarbons (Lynch, 2007).

The effect of solvent dewaxing and isodewaxing on the properties of different viscosity neutral base stocks is shown in Table 2.28. The ASTM D 2500 test method for the cloud point of petroleum products is used to measure the temperature when the wax crystals first appear upon cooling, usually visible as milky cloud. The cloud point is the first sign of the wax formation, usually related to the presence of n-paraffins, while the pour point is the lowest temperature at which oil will flow under specific testing conditions. The literature reports that the presence of water in base stocks might lead to the haze formation and interfere with the cloud point measurement (Prince, 1997). After solvent dewaxing, with an increase in the viscosity of

---

TABLE 2.27

Effect of Different Dewaxing on the Hydrocarbon Composition of Bright Stock

Very High Viscosity Oil Bright Stock	Waxy Raffinate Feed	After Solvent Dewaxing	After Catalytic Dewaxing
Pour point (°C)	—	−3	−3
VI	—	95	95
n-Paraffins (wt%)	0	0	0
iso-Paraffins (wt%)	16	14	13
Mononaphthenes (wt%)	14	12	14
Polynaphthenes (wt%)	23	24	26
Aromatics (wt%)	47	50	47

Source: Reproduced from Sequeira, A. Jr., *Lubricant Base Oil and Wax Processing*, Marcel Dekker, Inc., New York, 1994. With permission.

TABLE 2.28

Effect of Solvent Dewaxing and Isodewaxing on Properties of Base Stocks

	Base Stocks			
	100N	**240N**	**500N**	**Bright Stock**
After Solvent Dewaxing				
KV at 100°C (cSt)	4.2	7.7	12.1	30.9
VI	87	93	94	104
Cloud point (°C)	−11	−12	−9	—
Pour point (°C)	−12	−15	−12	−15
After Isodewaxing				
KV at 100°C (cSt)	4.0	6.9	11.0	29.4
VI	94	102	106	116
Cloud point (°C)	−9	−8	−7	—
Pour point (°C)	−12	−12	−12	−13

Sources: Sequeira, A. Jr., *Lubricant Base Oil and Wax Processing*, Marcel Dekker, Inc., New York, 1994. With permission; Lynch, T.R., *Process Chemistry of Lubricant Base Stocks*, CRC Press, Boca Raton, FL, 2007. With permission.

base stocks, from 100N to bright stock, their VI was reported to increase, from 87 to 104. After isodewaxing, with an increase in their viscosity, their VI was reported to increase, from 94 to 116; however, their dewaxed pour points and cloud points were also found to increase indicating an increase in waxy hydrocarbons known to have a high VI. The literature reports that the isodewaxing process was commercialized to lower processing severity in the hydrocracker unit and to increase the lube oil base stock yields (Lynch, 2007).

The effect of solvent dewaxing and isodewaxing on yield and hydrocarbon composition of HC oil is shown in Table 2.29. HC oil, with a high pour point of 42°C, was reported to contain 33.7 wt% of paraffins, 65.7 wt% of 1–6 ring naphthenes, and 0.6 wt% of residual aromatics. After the solvent dewaxing of HC oil to −15°C, lube oil yield of 83.1 vol% having a VI of 120 was obtained. The paraffinic content of solvent-dewaxed HC oil was reported to be 29.6%. After the isodewaxing of HC oil to −15°C, a higher lube oil yield of 89.8 vol% having an increased VI of 121 was produced. A higher paraffinic content of 34.4%, after isodewaxing of HC oil, was reported. The hydrocarbon composition of HC oils indicated that the use of isodewaxing increased the paraffinic content from 29.6 to 34.4 wt%, which would lead to an increase in the lube oil yield and an increase in the VI. Solvent dewaxing and catalytic dewaxing are used to decrease the wax content of paraffinic base stocks and residual oils. No wax is produced from catalytic dewaxing, unless the wax is removed by solvent dewaxing prior to catalytic dewaxing. The variation in dewaxing conditions can lead to a variation in hydrocarbon composition and VI of base stocks.

TABLE 2.29
**Effect of Solvent Dewaxing and Isodewaxing on
Yields and Composition**

Processing	HC Oil	After Solvent Dewaxing	After Isodewaxing
Lube oil yield (vol%)		83.1	89.8
KV at 100°C (cSt)		5.8	5.8
VI		120	121
Pour point (°C)	42	−15	−15
Paraffins (%)	33.7	29.6	34.4
1-Ring naphthenes (%)	34.1	34.5	35.4
2-Ring naphthenes (%)	16.3	16.7	18.4
3-Ring naphthenes (%)	6.6	6.5	6.7
4-Ring naphthenes (%)	3.1	3.3	3.0
Monoaromatics (%)	0.6	0.6	0.3

Source: Reproduced from Lynch, T.R., *Process Chemistry of Lubricant Base Stocks*, CRC Press, Boca Raton, FL, 2007. With permission.

2.5 HYDROISOMERIZATION OF SLACK WAX

The worldwide demand for high quality lubricating oils is increasing, and some of the demands cannot be met with conventional processing of crude oils, such as high VI and low volatility.

In recent years, solvent extraction and solvent dewaxing of lube oil distillates have been replaced by nonconventional catalytic processing, such as HT, HC, or hydroisomerization, to meet the ever increasing demand for higher quality lubricants. Lower oil consumption demands lower volatility base stocks, and many engine oils require the use of base stocks, having a high VI, low volatility, and good low temperature fluidity. For the same viscosity base stocks, an increase in the VI leads to lower volatility. There are different feeds and methods to produce nonconventional base stocks having a VI above 140. The literature reports that lube oil base stocks can be produced by hydroisomerization of medium and heavy slack waxes, derived from solvent refined oils, over Pt catalyst supported on silicone–alumina oxides in a continuous down flow trickle bed (Calemma et al., 1999). Catalytic dewaxing is used to produce low pour point base stocks from raffinates, and HC or slack wax hydroisomerized oils (Shell, 2001).

The VI and pour points of dewaxed oils and liquid wax separated from different viscosity distillates are shown in Table 2.30. After the urea dewaxing of different viscosity distillates, their VI decreased from 66–78 to 61–69, and their pour points decreased from as high as −18°C to −2°C to as low as −48°C to −37°C. The analysis of separated liquid wax indicated very high VI of 139–167 and pour points of 16°C–27°C.

TABLE 2.30

VIs and Pour Points of Liquid Wax Separated from Different Viscosity Distillates

Urea Dewaxing	50N Viscosity Oil	80N Viscosity Oil	220N Viscosity Oil
Distillate feeds			
VI	78	75	66
Pour point (°C)	−9	2	−18
Dewaxed oil			
VI	69	61	64
Pour point (°C)	−48	−40	−37
Liquid wax			
VI	139	158	167
Pour point (°C)	24	16	27

Source: Lynch, T.R., *Process Chemistry of Lubricant Base Stocks*, CRC Press, Boca Raton, FL, 2007. With permission.

TABLE 2.31

Effect of Distillate Viscosity on the *n*-Paraffin Content of Separated Wax

Distillate Feeds	Light Neutral	Medium Neutral	Heavy Neutral
KV at 100°C (cSt)	3.7	5.6	10.5
Pour point (°C)	27	38	57
Solvent-dewaxed oil			
Pour point (°C)	−15	−12	−12
Separated wax (wt%)	12.3	14.4	25.3
n-Paraffins in wax (%)	71.3	42.3	24.0

Source: Lynch, T.R., *Process Chemistry of Lubricant Base Stocks*, CRC Press, Boca Raton, FL, 2007. With permission.

The effect of distillate viscosity on *n*-paraffin content of separated wax is shown in Table 2.31. After the solvent dewaxing of different distillates, their wax content increases with an increase in their viscosities; however, the *n*-paraffin content of wax decreases with an increase in the distillate viscosity. Fully refined wax needs to be deoiled to have an oil content below 1 wt% and needs to be finished. HF or hydrorefining is used to finish wax products by removing trace contaminants. Different analytical techniques, such as NMR and GC, are used to measure the branching of wax, including urea adduct.

TABLE 2.32
Estimated Branching of Wax by Different Analytical Techniques

Analysis	Fully Refined Wax	Rubber Compounding Wax
Average chain length by GC	C_{26}	C_{26-27}
Estimated branching (%)		
By urea adduct	20	21
By NMR	5.4	7.9
By GC and NMR	5.1	6.3

Source: Lynch, T.R., *Process Chemistry of Lubricant Base Stocks*, CRC Press, Boca Raton, FL, 2007. With permission.

The estimated branching of wax by different analytical techniques is shown in Table 2.32. The fully refined wax, having an average chain length of C_{26}, was reported to have an estimated branching of 20% by urea adduct but only 5% by GC and NMR. Another wax, called rubber compounding wax, having a similar average chain length of C_{26-27}, was reported to have an estimated branching of 21% by urea adduct but only 6%–8% by GC and NMR.

The effect of isodewaxing of slack wax from solvent dewaxing of waxy raffinates on yields and properties of lube oils is shown in Table 2.33. The literature reports on isodewaxing of deoiled slack wax, which produced lube oil base stocks having an extra high VI (XHVI) in the range of 145–155 and a pour point of −15°C. The catalytic dewaxing, when combined with HF, is known as hydrodewaxing (Sequeira, 1994).

TABLE 2.33
Effect of Isodewaxing of Slack Wax on Yields and Properties of Lube Oils

Isodewaxing	Slack Wax Feed #1	Lube Oil #1	Lube Oil #2
Lube oil yield (LV %)		60	66
KV at 100°C (cSt)		6.3	3.9
KV at 50°C (cSt)	6.8		
KV at 40°C (cSt)	13.2	32.8	—
VI	172	145	155
Cloud point (°C)		−11	—
Pour point (°C)		−15	−15

Source: Sequeira, A. Jr., *Lubricant Base Oil and Wax Processing*, Marcel Dekker, Inc., New York, 1994. With permission; Lynch, T.R., *Process Chemistry of Lubricant Base Stocks*, CRC Press, Boca Raton, FL, 2007. With permission.

TABLE 2.34

Effect of Wax Hydrodewaxing on the Yields and the Properties of Lube Oils

Wax Hydrodewaxing	Base Stock #1	Base Stock #2	Base Stock #3
Lube oil yield (%)	65	50	35
KV at 100°C (cSt)	6.8	6.5	6.3
VI	168	158	146
Cloud point (°C)	8	11	33
Pour point (°C)	−21	−39	−65
Volatility (% off)	10.7	13	15

Source: Lynch, T.R., *Process Chemistry of Lubricant Base Stocks*, CRC Press, Boca Raton, FL, 2007. With permission.

A selective HC process can be used as an alternative to solvent dewaxing and hydrogen finishing for the removal of wax, and finishing the lube oil base stock.

The effect of wax hydrodewaxing on yields and properties of lube oils is shown in Table 2.34. The use of wax hydrodewaxing can lead to 65 vol% of lube oil #1, having a high VI of 168, a pour point of −21°C, and a low volatility of 10.7% off. With a decrease in the yield (from 65 to 50 vol%), a decrease in the VI (from 168 to 158), a decrease in the pour point (from −21°C to −39°C), and an increase in the volatility (from 10.7% to 13% off) was reported. With a further decrease in yield (from 50 to 35 vol%), a further decrease in the VI (from 158 to 146), a further decrease in the pour point (from −39°C to −65°C), and an additional increase in the volatility (from 13% to 15% off) was reported. With a decrease in the pour points of lube oils, obtained by wax hydrodewaxing, their pour points were decreasing from −21°C to −65°C, while their cloud points were increasing from 8°C to 33°C. The high cloud point of petroleum oils indicates the crystallization of *n*-paraffins or the haze formation caused by water contamination or the presence of other molecules having a poor solubility in highly saturated oils (Prince, 1997). The literature reports on the slack wax hydroisomerization process, which involved HT of slack wax feed, hydroisomerization, vacuum stripping, and solvent dewaxing (Sequeira, 1994). The literature reported on commercial products, known as XHVI and Exxsyn, were reported to have a VI above 140 and a low volatility (Phillipps, 1999).

The properties and composition of different viscosity solvent-dewaxed XHVIs and Exxsyn base stocks produced by slack wax hydroisomerization are shown in Table 2.35. During the HC process, the aromatic content can be reduced to low levels through many different reactions such as heteroatom removal, aromatic ring saturation, dealkylation of the aromatic rings, ring opening, straight chain and side chain crackings, and wax isomerization (Cody et al., 2002). The main objective of the hydroisomerization reactions is to convert the naphthenic and aromatic ring structures into straight chain and branched chain compounds. It is necessary to

TABLE 2.35

Properties and Composition of Solvent-Dewaxed XHVI and Exxsyn Base Stocks

Slack Wax Hydroisomerization	XHVI 4 Dewaxed	XHVI 5 Dewaxed	Exxsyn 4 Dewaxed	Exxsyn 6 Dewaxed
KV at 100°C (cSt)	4	5.1	4.0	5.8
VI	144	146	140	142
Pour point (°C)	−18	−18	−18	−21
Saturates (wt%)	100	100	—	100
Aromatics (wt%)	0	0	—	0
Sulfur (ppm)	20	20	—	10
Nitrogen (ppm)	2	2	—	0
Volatility (% off)	16.4	11	15.2	9

Sources: Phillipps, R.A., Highly refined mineral oils, in *Synthetic Lubricants and High-Performance Functional Fluids*, 2nd ed., Rudnick, L.R. and Shubkin, R.L., Eds., Marcel Dekker Inc., New York, 1999. With permission; Lynch, T.R., *Process Chemistry of Lubricant Base Stocks*, CRC Press, Boca Raton, FL, 2007. With permission.

crack the side chains on these ring structures to reduce the chain length and to saturate the molecules thus formed. Not all ring structures will open (Phillipps, 1999). The literature reports on the properties and the composition of different slack wax isomerate (SWI) base stocks having XHVI, above 140 (Pillon, 2007).

The effect of hydroprocessing severity on VIs and solvency properties of very high VI base stocks is shown in Table 2.36.

With an increase in the VI of petroleum derived base stocks, a decrease in the aromatics and the naphthenic hydrocarbons with an increase in the paraffinic and

TABLE 2.36

Effect of Hydroprocessing Severity on the Solvency Properties of VHVI Base Stocks

Processing	HF	Hydrotreatment	HC	Hydroisomerization
Severity	Low	Moderate	High	High
Feedstock	Solvent refined	Solvent refined	Distillates	Slack wax
Purpose	Saturate olefins	Saturate olefins	Saturate olefins	Saturate olefins
	Remove S and N	Remove S and N	Remove S and N	Remove S and N
		Saturate aromatic	Ring opening	Ring opening
VI	90–105	>105	95–130	>140
Solvency	Excellent	Very good	Moderate	Poor

Source: Reproduced from Pirro, D.M. and Wessol, A.A., *Lubrication Fundamentals*, 2nd ed., Marcel Dekker Inc., New York, 2001. With permission.

iso-paraffinic contents will decrease their solvency properties. The use of high VI base stocks in lubricating oils and a decrease in their solvency properties will lead to a decrease in the solubility of additives, which might affect the lubricant performance. The recent literature reports on an updated slack wax hydroisomerization process based on HT of wax feed, hydrodewaxing, and HF (Lynch, 2007).

The original and updated wax hydroisomerization processes are shown in Table 2.37. The original slack wax hydroisomerization process requires solvent dewaxing to reduce pour point and HF, as the last processing step, to improve the color and to increase the daylight stability of SWI 6 base stocks (Pillon, 2007). The recent literature reports that a modified wax conversion process, studied using different catalysts, can lead to lube oils, having a VI above 140, but a solvent dewaxing step is still necessary to reduce their pour points (Lynch, 2007). The Fischer–Tropsch (F–T) wax, produced by converting gas to liquids, was reported to be over 99% paraffinic and can be HC, hydroisomerized, and isodewaxed to make high quality base stocks (Mang, 2001). Base stocks, produced by converting gases to liquids and by hydroisomerization of F–T wax, are known as GTL products. They were reported to have a high VI above 140; however, they also require dewaxing. The MS analysis of GTL base stocks indicated the presence of *iso*-paraffins and the absence of *n*-paraffins (Henderson, 2005b).

The properties and compositions of different viscosity GTL base stocks produced by hydroisomerization of F–T wax are shown in Table 2.38. The use of additives is not sufficient to meet the performance requirements of some lubricant products, and an increase in the saturate content of lube oil base stocks is needed to increase their quality in areas, such as oxidation stability, volatility, and fuel economy (Henderson, 2005a). The literature reports that the F–T process, used to produce synthetic wax followed by hydroisomerization, produced XHVI base stocks, having a high saturate content and an excellent oxidation stability (Phillipps, 1999). The use of hydroprocessing increases the VI of petroleum derived base stocks by decreasing their aromatic and naphthenic contents. Different chemistry hydrocarbons are used commercially as solvents. Naphthenic hydrocarbons have excellent solvency properties to dissolve organic molecules followed by aromatic hydrocarbons. Modern refineries

---

TABLE 2.37
Original and Updated Wax Hydroisomerization Process

Original SWI Process	Updated Wax Hydroisomerization Process
Slack wax feed	Light and heavy neutral wax feed
HT of feed to reduce catalyst poison	HT of feed to reduce S
Two-stage hydroisomerization	Hydrodewaxing to reduce pour point
Vacuum stripping to remove volatiles	HF to improve color and stability
Solvent dewaxing to reduce pour point	

Source: Lynch, T.R., *Process Chemistry of Lubricant Base Stocks*, CRC Press, Boca Raton, FL, 2007. With permission.

TABLE 2.38
Properties and Composition of Different Viscosity GTL Base Stocks

F–T Wax Feed Hydroisomerization	GTL 2 Dewaxed	GTL 3 Dewaxed	GTL 5 Dewaxed	GTL 7 Dewaxed
KV at 40°C (cSt)	5.0	10.4	20.1	38.4
KV at 100°C (cSt)	1.7	2.8	4.5	7.0
VI	—	122	144	147
Pour point (°C)	−59	−45	−21	−21
Saturates (wt%)	100	100	100	100
Flash point (°C)	159	207	238	260
Volatility (% off)	—	28.8	7.8	2.0

Source: Reproduced from Henderson, H.E., Gas to liquids, in *Synthetics, Mineral Oils, and Bio-Based Lubricants Chemistry and Technology*, Rudnick, L.R., Ed., CRC Press, Boca Raton, FL, 2005. With permission.

use different combinations of conventional processing and catalytic processing to produce lube oil base stocks. With an increase in paraffinic hydrocarbons, known to have poor solvency properties, a decrease in the solvency properties of base stocks might affect the solubility of some additives and the lubricant performance.

REFERENCES

Betton, C.J., Lubricants and their environmental impact, in *Chemistry and Technology of Lubricants*, 2nd ed., Mortier, R.M. and Orszulik, S.T., Eds., Blackie Academic & Professional, London, U.K., 1997.

Beuther, H., Donaldson, R.E., and Henke, A.M., *I&EC Product Research and Development*, 3(3), 174 (1964).

Calemma, V. et al., *Preprints-ACS, Division of Petroleum Chemistry*, 44(3), 241 (1999).

Cody, I.A. et al., Raffinate hydroconversion process, CA Patent 2,429,500, 2002.

Exxon, http://www.exxon.com (accessed 2002).

Fetzer, J.C., Correlations between the spatial configuration and behaviour of large polynuclear aromatic hydrocarbons, in *Polynuclear Aromatic Compounds*, Ebert, L.B., Ed., American Chemical Society, Washington, DC, 1988.

Gary, J.H. and Handwerk, G.E., *Petroleum Refining, Technology and Economics*, 4th ed., Marcel Dekker, Inc., New York, 2001.

Henderson, H.E., Chemically modified mineral oils, in *Synthetics, Mineral Oils, and Bio-Based Lubricants Chemistry and Technology*, Rudnick, L.R., Ed., CRC Press, Boca Raton, FL, 2005a.

Henderson, H.E., Gas to liquids, in *Synthetics, Mineral Oils, and Bio-Based Lubricants Chemistry and Technology*, Rudnick, L.R., Ed., CRC Press, Boca Raton, FL, 2005b.

Lynch, T.R., *Process Chemistry of Lubricant Base Stocks*, CRC Press, Boca Raton, FL, 2007.

Mang, T., Base oils, in *Lubricants and Lubrications*, Mang, T. and Dresel, W., Eds., Wiley-VCH GmbH, Weinheim, 2001.

Pavia, D.L., Lampman, G.M., and Kriz, G.S., *Introduction to Spectroscopy*, Harcourt Brace College Publishers, Orlando, FL, 1996.

Phillipps, R.A., Highly refined mineral oils, in *Synthetic Lubricants and High-Performance Functional Fluids*, 2nd ed., Rudnick, L.R. and Shubkin, R.L., Eds., Marcel Dekker Inc., New York, 1999.

Pillon, L.Z., *Interfacial Properties of Petroleum Products*, CRC Press, Boca Raton, FL, 2007.

Pirro D.M. and Wessol A.A., *Lubrication Fundamentals*, 2nd ed., Marcel Dekker Inc., New York, 2001.

Prince, R.J., Base oils from petroleum, in *Chemistry and Technology of Lubricants*, 2nd ed., Mortier, R.M. and Orszulik, S.T., Eds., Blackie Academic & Professional, London, U.K., 1997.

Sequeira, A., *Pre-Prints Division of Petroleum Chemistry*, *ACS*, 37(4), 1286 (1992).

Sequeira, A. Jr., *Lubricant Base Oil and Wax Processing*, Marcel Dekker, Inc., New York, 1994.

Shell, http://www.shell-lubricants.com (accessed 2001).

Speight, J.G., *The Chemistry and Technology of Petroleum*, 4th ed., CRC Press, Boca Raton, FL, 2006.

3 Low-Temperature Fluidity of Base Stocks

3.1 EFFECT OF ISOMERIZATION AND ESTERIFICATION ON MELTING POINTS OF MOLECULES

In cold conditions, the low temperature fluidity of some lubricants is an important requirement. The straight chain n-paraffins have a high viscosity index (VI) but they are undesirable as lube oil molecules due to their high melting points. The preferred molecules for the lube oil manufacture are iso-paraffins having a high VI and low melting points. The literature reports that the melting point of iso-paraffins decreases with an increase in the degree of branching, but the size of the decrease depends on the position and the length of the branching. Branching positions in the middle of the chain have a greater effect than those near the end of the chain, and longer side chains lead to a greater decrease of melting point (Calemma et al., 2004). The highly branched C_{30} hydrocarbon, known as squalane, was reported to have a VI of 117 (McCabe, Cui, and Cummings, 2001).

The literature reports that the normal hydrocarbon, n-$C_{13}H_{28}$ (known as n-tridecane), has a low melting point from $-6°C$ to $-4°C$ but a boiling point of 234°C, which is too low for lubricant products. The hydrocarbons meeting the boiling range requirement of lube oil base stocks need to have the carbon number in the range of C_{20}–C_{40}. The normal hydrocarbon n-$C_{20}H_{42}$, known as n-eicosane, has a boiling point of 343°C but a melting point of 35°C–37°C. The higher molecular weight normal hydrocarbon n-$C_{28}H_{58}$, known as n-octacosane, has a higher boiling point of 432°C and a higher melting point of 57°C–62°C. Another high molecular weight normal hydrocarbon n-$C_{32}H_{66}$, known as n-dotriacontane, has a very high boiling point of 467°C and a very high melting point of 65°C–70°C (Sigma-Aldrich, 2007–2008). The literature also reports that the C_{30} hydrocarbon, known as squalane, cosbiol, or robane, is actually 2,6,10,15,19,23-hexamethyltetracosane having a very low melting point of $-38°C$ (Sigma-Aldrich, 2007–2008).

The effects of molecular weight and isomerization on boiling point (Bp), melting point (Mp), and VI of different C_{26} and C_{30} hydrocarbons are shown in Table 3.1. With an increase in the molecular weight of normal hydrocarbons, their boiling point, melting point, and VI increase (Lynch, 2007). With an increase from n-$C_{26}H_{54}$ to n-$C_{30}H_{62}$, the boiling point increased from 420°C to 454°C, the melting point increased from 55°C–58°C to 64°C–67°C, and the VI increased from 190 to 195. The literature reports, based on the gas chromatography (GC)/MS analyses of n-C_{16} isomerization products, that the iso/n ratio of the cracking products was found to increase with an increase in the conversion. At low conversion levels, the class of mono-methyl isomers forms more than 90% of the iso-C_{16} lump. At higher conversion levels,

TABLE 3.1

Effects of Molecular Weight and Isomerization on the Boiling Point, the Melting Point, and the Viscosity Index of Hydrocarbons

Formula	Paraffin Name	Bp (°C)	Mp (°C)	VI
$C_{26}H_{54}$	n-Hexacosane $CH_3(CH_2)_{24}CH_3$	420	55–58	190
$C_{26}H_{54}$	5-Butyldocosane	440	21	136
$C_{26}H_{54}$	11-Butyldocosane	440	0	128
$C_{26}H_{54}$	5,14-Di-n-butyloctadecane	—	—	83
$C_{30}H_{62}$	n-Triacontane $CH_3(CH_2)_{28}CH_3$	454	64–67	195
$C_{30}H_{62}$	9-Octyldocosane	—	9	144
$C_{30}H_{62}$	2,6,10,15,19,23-Hexamethyltetracosane	350	−38	117

Sources: McCabe, C. et al., *Fluid Phase Equilib.*, 183, 363, 2001. With permission; Lynch, T.R., *Process Chemistry of Lubricant Base Stocks*, CRC Press, Boca Raton, FL, 2007. With permission; Sigma-Aldrich, *Handbook of Fine Chemicals*, St. Louis, MO, 2007–2008. With permission.

a decrease in the monobranched isomers and an increase in the dibranched isomers were observed (Calemma, Peratello, and Perego, 2000). Isomerized hydrocarbons have lower melting points and lower VI indicating a need to optimize their branching density to assure low melting points but still sufficiently high VI.

The polyalfaolefin (PAO) fluids are highly branched synthetic hydrocarbons, produced from 1-decene monomer as a starting material, and their final properties are determined by the processing conditions and the selective distillation of the light oligomers (Rudnick and Shubkin, 1999). The literature reports that the PAO product mixture contains C_{10} monomer, C_{20} dimer, C_{30} trimer, and C_{40} tetramer after polymerization and saturation. After distillation, different viscosity PAO fluids ranging from 2 cSt to as high as 100 cSt at 100°C are produced (Dresel, 2001). The lowest temperature at which the movement of the specimen is observed is recorded as the pour point. The ASTM D 97 test method is often used to test the pour point of petroleum oils. After the preliminary heating, the sample is cooled at a specified rate and examined at intervals of 3°C for flow characteristics.

The VI, pour point, and kinematic viscosity at −40°C of different synthetic PAO fluids are shown in Table 3.2. With an increase in the viscosity of different PAO fluids, their VI was reported to increase from 124 to 137, while their natural pour points increased from as low as −72°C to −53°C. At the low temperature of −40°C, the viscosity of PAO fluids increase; however, they remain liquids due to very low natural pour points. At high temperatures, different PAO fluids have a low volatility of 1.8%–11.8% off. The literature reports on the chemistry of synthetic PAO fluid

TABLE 3.2
VI, Pour Points, and KV at −40°C of Different Synthetic PAO Fluids

Synthetic iso-Hydrocarbon	PAO 4	PAO 6	PAO 8	PAO 10
KV at 40°C (cSt)	16.68	30.89	46.30	64.50
KV at 100°C (cSt)	3.84	5.98	7.74	9.87
VI	124	134	136	137
Pour point (°C)	−72	−64	−57	−53
KV at −40°C (cSt)	2,390	7,830	18,200	34,600
Volatility (% off)	11.8	6.1	3.1	1.8

Source: Reproduced from Rudnick, L.R. and Shubkin, R.L., Poly(alfa-olefins), in *Synthetic Lubricants and High-Performance Functional Fluids*, 2nd ed., Marcel Dekker Inc., New York, 1999. With permission.

TABLE 3.3
Effects of Molecular Weight and Esterification on Boiling Point and Melting Point

Formula	Hydrocarbon Chemistry	Bp (°C)	Mp (°C)
$C_{16}H_{34}$	*n*-Hexadecane	287	18
$C_{16}H_{32}O_2$	Myristic acid ethyl ester	178–180	11–12
$C_{16}H_{32}O_2$	Hexadecanoic acid (palmitic acid)	352	61–63
$C_{21}H_{42}O_2$	Ester methyl eicosonoate	215–216	45–48

Sources: Sigma-Aldrich, *Handbook of Fine Chemicals*, 2007–2008. With permission; Chemical Dictionary, http://www.chemicaldictionary.cn. With permission.

that is used to formulate lubricant products. The GC analysis of PAO base stock indicated the presence of one main component that was reported to be the C_{30} trimer (Rudnick and Shubkin, 1999).

The effects of molecular weight and esterification on the boiling point and the melting point of organic molecules are shown in Table 3.3. The *n*-hexadecane, which has a formula of $C_{16}H_{34}$, has a boiling point of 287°C and a melting point of 18°C. Myristic acid ester, which has a formula of $C_{16}H_{32}O_2$, has a lower boiling point of 178°C–180°C and a lower melting point of 11°C–12°C. The hexadecanoic acid, known as palmitic acid, which has the same formula of $C_{16}H_{32}O_2$, has a drastically higher boiling point of 352°C and a drastically higher melting point of 61°C–63°C. The oxidation of hydrocarbons can lead to an increase in their viscosity due to an increase in their boiling points and melting points. Another higher molecular weight ester $C_{21}H_{42}O_2$, known as ester methyl eicosonoate, was reported to have a Bp of 215°C–216°C and a

TABLE 3.4

VI, Pour Point, and KV at −40°C of Synthetic Polyol Esters and Diesters

Synthetic Esters	Polyol Ester		Diester	
KV at 40°C (cSt)	18.1	32.4	17.5	33.8
KV at 100°C (cSt)	4.1	5.6	4.3	6.4
VI	130	114	161	145
Pour point (°C)	−69	−54	−65	−57
KV at −40°C (cSt)	3,750	30,000	3,500	18,700
Volatility (% off)	3.8	3	7.7	4.7

Source: Lakes, S.C., Automotive crankcase oils, in *Synthetic Lubricants and High-Performance Functional Fluids*, 2nd ed., Marcel Dekker Inc., New York, 1999. With permission.

Mp of 45°C–48°C. The esterification of hydrocarbons can also lead to molecules having lower melting points. Commercial synthetic polyol ester and diester fluids, used in some lubricant products, have very low pour points and very high VIs.

The VI, pour point, and kinematic viscosity at −40°C of synthetic polyol esters and diesters are shown in Table 3.4. With an increase in the viscosity of different polyol esters, their VIs were reported to decrease from 130 to 114, and their natural pour point was reported to increase from as low as −69°C to −54°C. At the low temperature of −40°C, the kinematic viscosity of polyol esters increased; however, they remained liquids due to very low natural points. At high temperatures, different polyol esters have a very low volatility of 3%–4% off. With an increase in the viscosity of different diesters, their VIs were also reported to decrease from 161 to 145, and their natural pour point was reported to increase from as low as −65°C to −57°C. At the low temperature of −40°C, the kinematic viscosity of diesters increased; however, they remained liquids due to very low natural points. At high temperatures, different diesters have also a low volatility of 5%–8% off. The severe low temperature fluidity specifications of some lubricants require the use of synthetic fluids.

3.2 EFFECTS OF CRUDE OILS AND DEWAXING ON POUR POINTS OF BASE STOCKS

The solvent-refined (SR) paraffinic base stocks require dewaxing to lower their pour points. The literature reports about the effect of dewaxing ratio on the low temperature performance of lube oil base stock (Yutai, Huipeng, and Shuren, 2001). The literature also reports on the composition of some solubilized waxes and their effect on the flow properties of mineral lube oil base stocks (Anwar et al., 1999). The presence of n-paraffins will increase the pour points of base stocks while the presence of high molecular weight hydrocarbons will decrease the volatility and increase the low temperature kinematic viscosity (Rudnick and Shubkin, 1999). Naphthenic oils are made in small quantities from selected crude oils, having low

TABLE 3.5

VI, Natural Pour Point, and the Composition of 100N NPOs

	100N NPO Oils			
	Oil #1	Oil #2	Oil #3	Oil #4
KV at 100°F (SUS)	100	100	100	100
KV at 38°C (cSt)	20.6	20.6	20.6	20.6
VI	−7	20	34	61
Natural pour point (°C)	−51	−45	−40	−29
Aromatics (wt%)	31.9	25.6	31.8	24.0
Sulfur (wt%)	0.02	0.06	0.07	0.02

Source: Reproduced from Sequeira, A. Jr., *Lubricant Base Oil and Wax Processing*, Marcel Dekker, Inc., New York, 1994. With permission.

or no wax content, and do not require dewaxing. The literature reports on processing and the variation in VI and the natural pour of different naphthene pale oils (NPOs) (Sequeira, 1994). Mineral base stocks are traditionally classified according to Saybolt Universal Seconds (SUS) and a 100N NPOs were reported to have a viscosity of 100 SUS at 100°F. At present, the viscosity of base stocks is reported in cSt at 40°C and 100°C.

The VI, the natural pour point, and the composition of different 100N NPOs are shown in Table 3.5. Some NPOs, known as naphthenic oils, such as oil #1 was reported to have a negative VI of −7 and a very low natural pour point of −51°C. For the same viscosity 100N naphthenic oils, with an increase in their VI, from −7 to 61, an increase in their natural pour point, from −51°C to −29°C, was reported. Their aromatic content was found to vary in the range of 24–32 wt%, and their S content was found to vary in the range of 0.02–0.07 wt% (Sequeira, 1994). The straight chain *n*-paraffins, found in paraffinic crude oils, have a high VI but they are undesirable as lube oil molecules due to their high pour points and are removed during the dewaxing step. The use of hydroprocessing, such as, hydrotreatment (HT), hydrocracking (HC), and slack wax hydroisomerization (SWI), leads to lube oil base stocks, also known as extra high VI (XHVI), which also have high pour points and require dewaxing.

The VI and the low temperature properties of different solvent-dewaxed base stocks having a kinematic viscosity of 4–5 cSt at 100°C are shown in Table 3.6. After solvent dewaxing, different 100N SR base stocks were reported to have a typical VI of 89–98 and a typical dewaxed pour point of −12°C to −15°C. At the temperature of −40°C, below their dewaxed pour points, they were "solids." At high temperatures, their volatility was high in the range of 30%–37% off. After solvent dewaxing, VHVI base stock was reported to have a higher VI of 121, and the dewaxed pour point of −21°C. At the temperature of −40°C, below its dewaxed pour point, it was a "solid." At high temperature, VHVI base stock was reported to have a lower volatility of 22.2% off. After solvent dewaxing, XHVI base stock was reported to have an XHVI of 149 and a dewaxed pour point of −15°C. However, with a decrease in the temperature to −40°C, below its dewaxed pour point, it was also

TABLE 3.6
Low Temperature Properties of Lower Viscosity Solvent-Dewaxed Base Stocks

	Solvent-Dewaxed Base Stocks				
	100N			**VHVI**	**XHVI**
KV at 40°C (cSt)	18.6	20.2	20.1	16.2	24.1
KV at 100°C (cSt)	3.81	4.06	4.02	3.75	5.14
VI	89	98	94	121	149
Dewaxed pour point (°C)	−15	−12	−15	−21	−15
KV at −40°C (cSt)	Solid	Solid	Solid	Solid	Solid
Volatility (% off)	37.2	30.0	29.5	22.2	8.8

Source: Rudnick, L.R. and Shubkin, R.L., Poly(alfa-olefins), in *Synthetic Lubricants and High-Performance Functional Fluids*, 2nd ed., Marcel Dekker Inc., New York, 1999. With permission.

TABLE 3.7
Low Temperature Properties of Higher Viscosity Solvent-Dewaxed Base Stocks

	Solvent-Dewaxed Base Stocks				
	HT 160N	**200SN**	**240N**	**325SN**	**325N**
KV at 40°C (cSt)	33.1	40.8	47.4	63.7	58.0
KV at 100°C (cSt)	5.77	6.31	6.98	8.30	8.20
VI	116	102	103	99	110
Dewaxed pour point (°C)	−15	−6	−12	−12	−12
KV at −40°C (cSt)	Solid	Solid	Solid	Solid	Solid
Volatility (% off)	16.6	18.8	10.3	7.2	5.1

Source: Rudnick, L.R. and Shubkin, R.L., Poly(alfa-olefins), in *Synthetic Lubricants and High-Performance Functional Fluids*, 2nd ed., Marcel Dekker Inc., New York, 1999. With permission.

a "solid." At high temperature, XHVI base stock was reported to have a very low volatility of 8.8% off.

The VI and the low temperature properties of different solvent-dewaxed base stocks having a kinematic viscosity of 6–8 cSt at 100°C are shown in Table 3.7. After solvent dewaxing, HT 160N base stock was reported to have a VI of 116 and a typical dewaxed pour point of −15°C. At the temperature of −40°C, below its dewaxed pour point, it was "solid." At high temperature, the volatility was 16.6% off. After solvent dewaxing, 200N SR base stock was reported to have a VI of 102 and a high-dewaxed pour point of −6°C. At the temperature of −40°C, below its dewaxed pour point, it was "solid." At high temperature, the volatility was 18.8% off. After solvent dewaxing, another 240N SR base stock was reported to have a VI of 103 and a dewaxed pour point of −12°C. At the temperature of −40°C, below its dewaxed pour point, it was "solid." At

TABLE 3.8

Effects of Solvent and Catalytic Dewaxing on the Composition of 100N Base Stock

	100N Base Stock		
	Raffinate Feed	**Solvent Dewaxing**	**Catalytic Dewaxing**
KV at 100°F (SUS)	93	106	117
KV at 38°C (cSt)	18.9	22.0	24.6
VI	116	100	89
Pour point (°C)	27	−18	−15
Saturate fraction			
n-Paraffins (%)	21	—	—
iso-Paraffins (%)	13	24	21
Naphthenes (%)	33	37	39
Other saturates (%)	12	13	13
Aromatic fraction			
Alkylbenzenes (%)	6	8	8
Other aromatics (%)	15	18	19

Source: Taylor, R.J. et al., *Pre-Prints Div. Petrol. Chem., ACS*, 37(4), 1337, 1992.

high temperature, with an increase in the viscosity, a lower volatility of 10.3% off was reported. After solvent dewaxing, different 325N SR base stocks were reported to have a VI of 99–110 and dewaxed pour points of −12°C. At the temperature of −40°C, below its dewaxed pour point, they were "solids." At high temperature, with a further increase in the viscosity, a further decrease in the volatility to 5%–7% off was reported.

The effects of solvent and catalytic dewaxing on low temperature properties and the composition of 100N paraffinic base stock are shown in Table 3.8. After solvent dewaxing of 100N paraffinic waxy raffinate, with a decrease in the pour point, from 27°C to −18°C, the VI also decreases from 116 to 100. The literature reports that no *n*-paraffins are present in solvent-dewaxed 100N base stock. After catalytic dewaxing of 100N paraffinic waxy raffinate, with a decrease in the pour point, from 27°C to −15°C, the VI further decreases from 116 to 89. The literature also reports that no *n*-paraffins are present in catalytically dewaxed 100N base stock. In conventional solvent dewaxing and nonconventional catalytic dewaxing, the *n*-paraffin content of 100N waxy raffinate is totally removed and only *iso*-paraffins are present. According to the literature, the cracking of some *iso*-paraffins further decreases the VI of catalytically dewaxed 100N dewaxed base stock. The use of the MS technique, to analyze the hydrocarbon composition of base stocks, binds *n*-paraffins and *iso*-paraffins together.

The MS analysis of hydrocarbon composition of different solvent-dewaxed 100N base stocks is shown in Table 3.9. After solvent dewaxing, 100N SR base stocks were reported to have 25.7%–29.0% of paraffins, 20.8%–25.0% of monocycloparaffins, 27.9%–31.7% of polycycloparaffins, 14.2%–24.9% of monoaromatics, and 0.1%–0.7% of thiophenes. After solvent dewaxing, HC 100N base stock was reported to have 23.7% of paraffins, 30.8% of monocycloparaffins, 39.1% of polycycloparaffins,

TABLE 3.9

MS Analysis of Different Solvent-Dewaxed 100N Base Stocks

100N Base Stocks	Solvent Refined	Solvent Refined	HC	Severe HC
Dewaxing	Solvent	Solvent	Solvent	Solvent
MS analysis				
n- and iso-Paraffins (%)	25.7	29.0	23.7	32.6
Monocycloparaffins (%)	20.8	25.0	30.8	34.2
Polycycloparaffins (%)	27.9	31.7	39.1	32.9
Aromatics (%)	24.9	14.2	6.4	0.6
Thiophenes (%)	0.7	0.1	0	0

Source: Henderson, H.E., Chemically modified mineral oils, in *Synthetics, Mineral Oils, and Bio-Based Lubricants Chemistry and Technology*, Rudnick, L.R., Ed., CRC Press, Boca Raton, FL, 2005. With permission.

6.4% of monoaromatics, and no thiophenes. A decrease in the paraffinic and the aromatic contents with an increase in the naphthenic content is observed. Since *n*-paraffins crack easily, the use of HC and solvent dewaxing will lead to base stocks containing no *n*-paraffins.

The use of a more severe HC leads to an increase in the paraffinic and the monocycloparaffin contents with a decrease in the polycycloparaffins and a further decrease in the aromatics. The MS analysis of different isodewaxed HC 100N base stocks is shown in Table 3.10. Hydrotreating can lead to highly refined mineral base stocks having a low aromatic content, a low sulfur content, and a low nitrogen content. In most cases, HC is used to decrease the aromatic content and the S content of base stocks. Hydrotreated and hydrocracked base stocks might have conventional or high VI and require dewaxing to lower their pour point. With an increase in the demand for higher quality base stocks, isodewaxing is used to produce higher yields of lube

TABLE 3.10

MS Analysis of Different Isodewaxed HC 100N Base Stocks

100N Base Stocks	HC	Severe HC	Severe HC
Dewaxing	Isodewaxing	Isodewaxing	Isodewaxing
MS analysis			
n- and iso-Paraffins (%)	30.2	51.4	76.1
Monocycloparaffins (%)	30.5	24.4	14.7
Polycycloparaffins (%)	35.3	23.9	9.2
Aromatics (%)	4.0	0.3	0
Thiophenes (%)	0	0	0

Source: Henderson, H.E., Chemically modified mineral oils, in *Synthetics, Mineral Oils, and Bio-Based Lubricants Chemistry and Technology*, Rudnick, L.R., Ed., CRC Press, Boca Raton, FL, 2005. With permission.

oils having high VI (Lynch, 2007). After isodewaxing, with an increase in the HC severity the paraffin content increases from 30.2% to 51.4%–76.1%, monocyclopar-affin content decreases from 30.5% to 14.7%–24.4%, polycycloparaffin content decreases from 35.3% to 9.2%–23.9%, aromatic content further decreases from 4% to 0.3% or less, and no presence of thiophenes is observed. The reactor temperature is the primary means of conversion control and, with an increase in the temperature, an increase in cracking of high molecular weight hydrocarbons will take place. Since *n*-paraffins crack easily, the use of HC and isodewaxing will also lead to base stocks containing no *n*-paraffins. The literature reports that the cracking of *n*-paraffins is easy while the cracking of *iso*-paraffins is more difficult.

3.3 BROOKFIELD VISCOSITIES AND COLD CRANKING STIMULATOR VISCOSITIES OF BASE STOCKS

The pour point is the lowest temperature at which the petroleum oil can be made to flow under gravity. The "solidified" oil does not pour under gravity but it can be moved if sufficient force is applied, for example, applying torque to a rotor suspended in the oil (Prince, 1997). The literature reports that cooling rates affect the wax crystallization and the oil, which was cooled rapidly, might still flow below its pour point. The mechanical agitation can break up the crystal structure and oil might flow below its pour point (Pirro and Wessol, 2001). The low temperature and the low-shear-rate viscosity of the lubricants can be measured by a Brookfield viscometer where the bath-controlled temperature is selected within the range from +5°C to −40°C. A lubricant sample is preheated, allowed to stabilize at RT, and then poured into a glass cell with a special spindle. The glass cell is then placed into a precooled cold cabinet, at the test temperature ranging from 5°C to −40°C for 16 h. Then the viscometer is used to rotate the spindle at the speed giving a max torque reading on the viscometer and the result is reported as the Brookfield viscosity. Another test, Cold Cranking Stimulator (CCS), measures the low temperature viscosity of lubricating oils under high shear conditions and the result is reported as the CCS viscosity.

The CCS viscosity at −20°C of different petroleum-derived base stocks and synthetic PAO fluid is shown in Table 3.11. The SR 100N base stock, having a VI of 92 and a dewaxed pour point of −18°C, was reported to have a CCS viscosity of 670 cP measured at −20°C. 100N mineral base stock was reported to have a typical saturate content of 81.1 wt% and a high volatility of 32% off. Severely HT 80N base stock, having a VI of 99 and a dewaxed pour point of −21°C, was reported to have a lower CCS viscosity of 520 cP measured at −20°C. A lower viscosity 80N base stock, having a 100 wt% saturate content, was reported to have a higher volatility of 40% off. HC base stock, having a higher VI of 130 and a lower dewaxed pour point of −27°C, was reported to have a lower CCS viscosity of 459 cP measured at −20°C. A similar viscosity HC oil, containing 92.3 wt% of saturates was reported to have a lower volatility of 15.3% off. Synthetic PAO base stock, having a VI of 124 and a very low natural pour point of −54°C, was reported to have a lower CCS viscosity of 413 cP measured at −20°C. PAO base stock, containing 100% saturate content, was also reported to have a lower volatility of 13% off.

TABLE 3.11
CCS Viscosity at −20°C of Different Petroleum-Derived Base Stocks and PAO

Process	Solvent Refined	Severe HT	HC	Synthetic
Base Stocks	100N	80N	Oil	PAO
KV at 100°C (cSt)	3.8	3.5	4.0	3.9
VI	92	99	130	124
Pour point (°C)	−18	−21	−27	−54
CCS at −20°C (cP)	670	520	459	413
Saturates (wt%)	81.8	100	92.3	100
Aromatics (wt%)	18.2	0	7.7	0
Sulfur (wt%)	0.5	0	0.1	0
Nitrogen (ppm)	17	5	12	0
Volatility (% off)	32	40	15.3	13

Source: Reproduced from Lynch, T.R., *Process Chemistry of Lubricant Base Stocks*, CRC Press, Boca Raton, FL, 2007. With permission.

TABLE 3.12
CCS Viscosity at −20°C of Different Petroleum-Derived 150N Base Stocks and PAO

150N Base Stocks	Solvent Refined	HC	VHVI	PAO 6
KV at 40°C (cSt)	30.1	29.6	32.5	31.3
KV at 100°C (cSt)	5.1	5.1	6.0	5.9
VI	95	99	133	135
Pour point (°C)	−12	−12	−15	−60
CCS viscosity at −20°C (cP)	2100	2000	1230	900
Volatility (% off)	17	16.5	7.8	7.0

Source: Henderson, H.E., Chemically modified mineral oils, in *Synthetics, Mineral Oils, and Bio-Based Lubricants Chemistry and Technology*, Rudnick, L.R., Ed., CRC Press, Boca Raton, FL, 2005. With permission.

The CCS viscosity at −20°C of different petroleum-derived 150N base stocks and synthetic PAO 6 fluid is shown in Table 3.12. The higher viscosity 150N SR base stock, having a VI of 95 and dewaxed pour point of −12°C, was reported to have a higher CCS viscosity of 2100 cP measured at −20°C. The 150N mineral base stock was also reported to have a lower volatility of 17% off. Higher viscosity HC base stock, having a VI of 99 and a dewaxed pour point of −12°C, was reported to have a higher CCS viscosity of 2000 cP at −20°C. It was reported to have a volatility of

16.5% off. Another higher viscosity VHVI base stock, having a higher VI of 133 and a dewaxed pour point of −15°C, was reported to have a lower CCS viscosity of 1230 cP at −20°C. It was reported to have a low volatility of 7.8% off. Similar viscosity synthetic PAO 6 base stock, having a VI of 135 and a very low natural pour point of −60°C, was reported to have a lower CCS viscosity of 900 cP at −20°C. It was reported to have a low volatility of 7% off. The CCS viscosity, measured at −20°C, of different 150N base stocks was found to decrease with a decrease in their pour points. The Brookfield viscosity test measures the low temperature viscosity of lubricating oils under low shear conditions.

The Brookfield viscosity and the CCS viscosity of different petroleum-derived 100N base stocks and PAO 4 fluid measured at −25°C are shown in Table 3.13. At −25°C, 100N SR base stock, having a dewaxed pour point of −21°C, is a "solid" under low sheer conditions but has a CCS viscosity of 1280 cP under high sheer conditions. At −25°C, HVI base stock, having a higher dewaxed pour point of −12°C, is also a "solid" under low sheer conditions and has a higher CCS viscosity of 1350 cP under high sheer conditions. At −25°C, VHVI base stock, having a lower dewaxed pour point of −27°C, has a Brookfield viscosity of 1160 cP under low sheer conditions and has a lower CCS viscosity of 580 cP under high sheer conditions. At −25°C synthetic PAO 4 base stock having a very low natural pour point of −64°C has a lower Brookfield viscosity of 600 cP under low sheer conditions and has a lower CCS viscosity of 490 cP under high sheer conditions. For petroleum-derived 100N base stocks, their Brookfield viscosity and their CCS viscosity decrease with a decrease in their pour points. Under low sheer conditions, 100N base stocks are solids when tested below their dewaxed pour points.

The Brookfield viscosity and the CCS viscosity of different petroleum-derived 150N base stocks and PAO 6 fluid measured at −25°C are shown in Table 3.14. At −25°C, 150N SR base stock, having a dewaxed pour point of −12°C, is a "solid" under

TABLE 3.13

Brookfield Viscosity and CCS Viscosity at −25°C of Different 100N Base Stocks and PAO 4

	100N Base Stocks			
	SR	HVI	VHVI	PAO 4
KV at 100°C (cSt)	3.79	4.50	3.79	3.9
Pour point (°C)	−21	−12	−27	−64
Brookfield viscosity at −25°C (cP)	Solid	Solid	1160	600
CCS viscosity at −25°C (cP)	1280	1350	580	490

Source: Reproduced from Rudnick, L.R. and Shubkin, R.L., Poly(alfa-olefins), in *Synthetic Lubricants and High-Performance Functional Fluids*, 2nd ed., Marcel Dekker Inc., New York, 1999. With permission.

TABLE 3.14

Brookfield Viscosity and CCS Viscosity at −25°C of Different 150N Base Stocks and PAO 6

	150N Base Stocks			PAO 6
	SR	HVI	VHVI	
KV at 100°C (cSt)	5.17	5.79	5.38	5.86
Pour point (°C)	−12	−9	−9	−58
Brookfield viscosity at −25°C (cP)	Solid	Solid	Solid	1550
CCS viscosity at −25°C (cP)	4600	2740	1530	1300

Source: Reproduced from Rudnick, L.R. and Shubkin, R.L., Poly(alfa-olefins), in *Synthetic Lubricants and High-Performance Functional Fluids*, 2nd ed., Marcel Dekker Inc., New York, 1999. With permission.

low sheer conditions but has a CCS viscosity of 4600 cP under high sheer conditions. At −25°C, HVI base stock, having a higher dewaxed pour point of −9°C, is also a "solid" under low sheer conditions but has a lower CCS viscosity of 2740 cP under high sheer conditions. At −25°C, VHVI base stock, having a dewaxed pour point of −9°C, is also a "solid" under low sheer conditions but has a lower CCS viscosity of 1530 cP under high sheer conditions. At −25°C, synthetic PAO 6 base stock, having a very low natural pour point of −58°C, has a Brookfield viscosity of 1550 cP under low sheer conditions and has a lower CCS viscosity of 1300 cP under high sheer conditions. Under low sheer conditions, higher viscosity petroleum-derived 150N base stocks are also solids when tested below their dewaxed pour points. Under high sheer conditions, their CCS viscosity measured at −25°C decrease with an increase in their VI. Gas to liquid (GTL) processing of Fischer–Tropsch wax leads to different viscosity GTL base stocks, having a high VI and low volatility, which require dewaxing.

The CCS viscosity of different viscosity GTL and synthetic PAO base stocks are shown in Table 3.15. At −30°C, lower viscosity GTL 5 base stock, having a VI of 144, a dewaxed pour point of −21°C, and a volatility of 8% off, was reported to have a CCS viscosity of 1320 cP. At −30°C, synthetic PAO 5 fluid, having a lower VI of 132, a lower natural pour point of −67°C, and a higher volatility of 13% off, was reported to have a lower CCS viscosity of 1250 cP. At a higher temperature of −25°C, higher viscosity GTL 7 base stock, having a VI of 147, a dewaxed pour point of −21°C, and a low volatility of 2% off, was reported to have a CCS viscosity of 2660 cP. At a higher temperature of −25°C, synthetic PAO 7 fluid, having a lower VI of 134, a lower natural pour point of −54°C, and a higher volatility of 5% off, was reported to have a lower CCS viscosity of 2340 cP. Despite a higher VI, GTL base stocks were found to have a higher CCS viscosity, when compared to similar viscosity synthetic PAO fluids, due to higher pour points and lower volatility, which might indicate the presence of higher molecular weight hydrocarbons causing an increase in their low temperature viscosity.

TABLE 3.15

CCS Viscosity of Different Viscosity GTL and PAO Base Stocks

	Base Stocks			
	GTL 5	PAO 5	GTL 7	PAO 7
KV at 100°C (cSt)	4.5	4.6	7.0	7.0
VI	144	132	147	134
Pour point (°C)	−21	−67	−21	−54
CCS at −25°C (cP)	—	—	2660	2340
CCS at −30°C (cP)	1320	1250	—	—
Volatility (% off)	8	13	2	5

Source: Henderson, H.E., Gas to liquids, in *Synthetics, Mineral Oils, and Bio-Based Lubricants Chemistry and Technology*, Rudnick, L.R., Ed., CRC Press, Boca Raton, FL, 2005. With permission.

3.4 USE OF POUR POINT DEPRESSANTS AND VI IMPROVERS

The dewaxing of petroleum-derived base stocks is not sufficient to assure good flow properties of lubricating oils and pour point depressants are used to prevent congealing of wax at low temperatures and offer an economically attractive alternate to entirely removing wax by dewaxing. Certain high molecular weight polymers inhibit the formation of a wax crystal structure. There are various theories on the mechanism of pour point depression and they can be combined into three general categories. The cocrystallization theory, which proposes that the wax-like components of the additive are incorporated into the wax crystal lattice and thus modify the morphology of the growing crystal. The adsorption theory, which proposes that the wax crystals are covered by an adsorbed layer of the additive and that this protective layer then inhibits further crystal growth. The nucleation theory, which proposes that the additive heterogeneously nucleates the wax crystals and this results in crystals of a relatively smaller size. Different chemistry polymeric additives are used as pour point depressants also called wax inhibitors. According to the literature, alkyaromatic polymers adsorb on the wax crystals and prevent them from growing and adhering to each other. Polymethacrylates (PMAs) cocrystalize with wax to prevent crystal growth (Pirro and Wessol, 2001). The literature reports on the use of DSC to study the wax precipitation from different petroleum fractions (Srivastava et al., 2002).

The DSC onset wax crystallization temperature and a Brookfield viscosity at −40°C of different dewaxed base stocks containing PMA pour point depressant are shown in Table 3.16. The literature reports on the effect of the additional removal of wax on the DSC onset crystallization temperature and the low sheer Brookfield viscosity at −40°C of different base stocks, containing PMA pour point depressant (Sharma and Stipanovic, 2003). Some dewaxed oils #1–2, having a similar low viscosity, a similar VI of 120–121, and a similar onset wax crystallization temperatures of −18 to −17°C, were found to have a Brookfield viscosity at −40°C varying from

TABLE 3.16

DSC Onset Wax Crystallization Temperature and Brookfield Viscosity at −40°C

	Dewaxed Base Stocks (PMA)			
	Oil #1	Oil #2	Oil #3	Oil #4
KV at 40°C (cSt)	12.94	12.83	12.99	12.8
KV at 100°C (cSt)	3.25	3.24	3.24	3.25
VI	120	121	117	122
DSC analysis				
Onset wax crystallization temp. (°C)	−17.3	−17.7	−33.8	−35.6
ASTM D 2983				
Brookfield viscosity at −40°C (cP)	7500	3650	5010	3250

Source: Reprinted from From Sharma, B.K. and Stipanovic, A.J., *Ind. Eng. Chem. Res.*, 42, 1522, 2003. With permission.

3650 to 7500 cP. Other dewaxed oils #3–4, having a similar low viscosity and a VI of 117–122, were found to have a much lower onset wax crystallization temperatures of −36°C to −34°C and a lower Brookfield viscosity at −40°C varying from 3250 to 5010 cP. The use of hydroprocessing produces base stocks that have a higher VI but still require dewaxing and the use of pour point depressants in lubricant products.

The variation in depressed pour points of hydroprocessed and dewaxed base stocks containing PMA pour point depressant is shown in Table 3.17. After dewaxing, different hydroprocessed base stocks, containing the PMA pour point depressant, were found to have depressed pour points varying in the range of −30°C to −18°C. The performance of PMA pour point depressant can be affected by many factors such

TABLE 3.17

Variation in Depressed Pour Points of Hydroprocessed and Dewaxed Base Stocks

Hydroprocessed Base Stocks (PMA)	KV at 40°C (cSt)	VI	Depressed Pour Point (°C)
Dewaxed oil	20.2	99	−18
Dewaxed oil	20.3	110	−27
Dewaxed oil	20.9	118	−24
Dewaxed oil	22.1	125	−30
Dewaxed oil	18.7	132	−18
Dewaxed oil	17.2	133	−21

Source: Sharma, B.K. et al., *Energ. Fuels*, 18, 952, 2004.

as the dewaxed pour point and the amount of residual wax. Moreover, high molecular weight polymers used as pour point depressants might have poor solubility in some highly saturated base stocks, containing low aromatic content and low naphthenic contents. The literature reports that the polyalkyl methylacrylate pour point depressants have a comb-like structure and contain C_{12}–C_{24} paraffinic side chains. Instead of the needle-like paraffin crystals which cause gelation, densely packed round crystals are formed which have a less effect on the flowing properties even below the pour point (Braun and Omeis, 2001).

The effect of PMA pour point depressant on the pour point depression of GTL base stocks dewaxed to −21°C is shown in Table 3.18. GTL base stocks require pour point depressants to improve the low temperature fluidity of lubricant products. The addition of 0.1 wt% of PMA type pour point depressant was reported effective in decreasing the dewaxed pour point of GTL 5 and GTL 7, from −21°C to −39°C.

According to the literature, GTL base stocks are similar to synthetic PAO base stocks and their only shortfall was reported to be their depressed pour point (Henderson, 2005b). While there is no presence of *n*-paraffins in dewaxed GTL base stocks, the use of polymeric pour point depressants in base stocks, having a high 100% saturate content, can decrease their solubility and affect their performance. The PMA polymers, the most widely used pour point depressants in lube oil base stocks, have usually a low molecular weight of 7,000–10,000 to ensure their solubility in oils. Some commercial products were reported to contain a paraffinic side chain of at least C_{12}, mixed alkyl side chains or can be branched (Prince, 1997).

The effect of different PMA polymers on pour point depression of SWI 6 base stock having a VI of 143 and solvent dewaxed to −21°C is shown in Table 3.19. The use of PMA polymer, having a low molecular weight of <10,000 and a side chain length of C_{13}, was found effective in depressing the pour point of SWI 6 base stock by 15°C. The use of another PMA polymer, having a higher molecular weight of 43,000 and a side chain length of C_{12}, was found effective in depressing the pour point of SWI

TABLE 3.18

Effect of PMA on the Pour Point Depression of Dewaxed GTL Base Stocks

Hydroprocessing of F–T Wax	GTL 5	GTL 7
KV at 40°C (cSt)	20.1	38.4
KV at 100°C (cSt)	4.5	7.0
VI	144	147
Dewaxed pour point (°C)	−21	−21
PMA pour point depressant (wt%)	0.1	0.1
Depressed pour point (°C)	−39	−39

Source: Henderson, H.E., Gas to liquids, in *Synthetics, Mineral Oils, and Bio-Based Lubricants Chemistry and Technology*, Rudnick, L.R., Ed., CRC Press, Boca Raton, FL, 2005. With permission.

TABLE 3.19

Effect of Different PMA Polymers on the Pour Point Depression of SWI 6 Base Stock

SWI 6 (VI 143/−21°C) PMA (wt%)	PMA (MWt)	PMA Side Chain	Depressed Pour Point (°C)	Pour Point Difference (°C)
0.5	<10,000	C_{13}	−36	15
0.5	43,000	C_{12}	−30	9
0.5	92,000	C_{13}	−30	9
0.5	98,000	C_{15}	−33	12
0.5	350,000	C_{12}	−30	9

Source: Pillon, L.Z., *Petrol. Sci. Technol.*, 20(1–2), 101, 2002. With permission.

6 base stock by only 9°C. The use of PMA polymer, having a molecular weight of 92,000 and a side chain length of C_{13}, was also found effective in depressing the pour point of SWI 6 base stock by only 9°C. However, the use of another PMA polymer, having a higher molecular weight of 98,000 and a side chain length of C_{15}, was found effective in depressing the pour point of SWI 6 base stock by 12°C. The use of PMA polymer, having a high molecular weight of 350,000 and a side chain length of C_{12}, was found effective in depressing the pour point of SWI 6 base stock by only 9°C. The variation in molecular weight and a side chain length was found to affect the performance of PMA polymer in solvent-dewaxed SWI 6 base stock.

The effect of treat rate of PMA pour point depressant on pour point depression of SWI 6 base stock having a VI of 146 and solvent dewaxed to −21°C is shown in Table 3.20. Lube oil base stocks usually require 0.5 wt% of pour point depressant. At a treat rate of 0.5 wt%, the addition of PMA polymer, having a molecular weight of 92,000 and a side chain length of C_{14}, decreased the pour point of SWI 6 base stock by 12°C. At the higher treat rate of 1 wt% and even 3 wt%, no further reduction in the depressed pour point of SWI 6 base stock was observed. The literature reports

TABLE 3.20

Effect of Treat Rate of PMA on the Pour Point Depression of SWI 6 Base Stock

SWI 6 (VI 146/−21°C) PMA (wt%)	PMA (MWt)	PMA Side Chain	Depressed Pour Point (°C)
0.5	92,000	C_{14}	−33
1.0	92,000	C_{14}	−33
3.0	92,000	C_{14}	−33

Source: MacAlpine, G.A. et al., Wax isomerate having a reduced pour point, CA Patent 2,054,096, 1991.

TABLE 3.21
Effect of Different FVA Polymers on the Pour Point Depression of SWI 6 Base Stock

SWI 6 (VI 143/−21°C) FVA (wt%)	FVA (MWt)	FVA Side Chain	Depressed Pour Point (°C)	Pour Point Difference (°C)
0.5	<10,000	C_{13}	−30	9
0.5	<10,000	C_{14}	−27	6
0.5	60,000	C_{12}	−30	9

Source: Pillon, L.Z., *Petrol. Sci. Technol.*, 20(1–2), 101, 2002. With permission.

that the efficiency of polymers to depress the pour point of different oils is usually controlled by their concentration (Braun and Omeis, 2001). The PMA polymer, having a molecular weight of 92,000 and C_{14} side chain, is not effective when used above 0.5 wt% treat rate. While special polyalkyl methylacrylates are normally used in mineral based lubricating oils, ethylene vinyl acetate and polyacrylate pour point depressants are usually used in crude oils. The fumarate-vinylacetate (FVA) polymers, used in crude oils, also have different molecular weights and varying side chains.

The effect of different FVA polymers on pour point depression of SWI 6 base stock having a VI of 143 and solvent dewaxed to −21°C is shown in Table 3.21. The use of FVA polymer, having a low molecular weight of <10,000 and a side chain length of C_{13}, was found effective in depressing the pour point of SWI 6 base stock by 9°C. At the same treat rate of 0.5 wt%, the use of another FVA polymer, having a low molecular weight of <10,000 and a side chain length of C_{14}, was found effective in depressing the pour point of SWI 6 base stock by only 6°C. The use of FVA polymer, having a higher molecular weight of about 60,000 and a side chain length of C_{12}, was found effective in depressing the pour point of SWI 6 base stock by 9°C. A variation in molecular weight and a side chain length was also found to affect the performance of FVA polymers in depressing the pour point of the solvent-dewaxed SWI 6 base stock. A variety of other polymeric additives, such as VI modifiers, were also found to modify the wax crystal structure and the wax precipitation from lubricant products. Early literature reports that small amounts of rubber dissolved in mineral oils were effective in increasing their VI. Similar behavior was later observed for other polymers, such as PMAs and polyisobutylene (Stambaugh, 1997).

The chemistry and the molecular weight of some polymers used as VI modifiers, also known as VI improvers, are shown in Table 3.22. Although the viscosity of base stocks decreases with an increase in the temperature, the decrease in the viscosity is smaller in the presence of the VI improver. The VI improvers are long chain, high molecular weight polymers that function by causing the relative viscosity of an oil to increase more at high temperature than at low temperature. It is postulated that in cold oil, the molecules of the polymer adopt a coiled form so that their effect on the viscosity is minimized. At high temperatures, the polymer tends to straighten out and produces a thickening effect (Pirro and Wessol, 2001). Some VI improvers are multifunctional and can act as VI improvers and pour point depressants.

TABLE 3.22

Chemistry and Molecular Weight of Some Polymers Used as VI Improvers

Polymer Chemistry	Name	Molecular Weight
Polyalkylmethacrylates	PMA	10,000–600,000
Ethylene–propylene copolymer	OCP	60,000
Styrene–butadiene copolymer	HSB	140,000
Styrene–isoprene copolymer	SIP	197,000
PMA–olefin copolymer	PMA/OCP	121,000

Source: MacAlpine, G.A. et al., Wax isomerate having a reduced pour point,
CA Patent 2,054,096, 1991.

TABLE 3.23

Effect of PMA and VI Improvers on the Depressed Pour Point of SWI 6 Base Stock

SWI 6 (VI 146/−21°C)	Oil #1	Oil #2	Oil #3	Oil #4	Oil #5	Oil #6
PMA (92,000) (wt%)	0.5	0.5	0.5	0.5	0.5	0.5
PMA (43,000) (wt%)	2					
PMA (325,000) (wt%)		2				
OCP (60,000) (wt%)			2			
HSD (140,000) (wt%)				2		
SIP (197,000) (wt%)					2	
PMA/OCP (121,000) (wt%)						2
Depressed pour point (°C)	−42	−36	−33	−33	−33	−33

Source: MacAlpine, G.A. et al., Wax isomerate having a reduced pour point, CA Patent 2,054,096,
1991.

The effect of different polymer mixtures on the depressed pour point of SWI 6 base stock having a VI of 146 and solvent dewaxed to −21°C is shown in Table 3.23. At the treat rate of 0.5 wt%, the use of PMA polymer having an average molecular weight of 92,000 was found effective in depressing the pour point of SWI 6 base stock from −21°C to −33°C only. The addition of another PMA polymer having a lower molecular weight of 43,000 was found effective in depressing the pour point of SWI 6 base stock, already containing the PMA pour point depressant to −42°C. The addition of another PMA polymer having a higher molecular weight of 325,000 was found effective in depressing the pour point of SWI 6 base stock, already containing the PMA pour point depressant to −36°C. At the same treat rate of 2 vol%, the addition of OCP polymer, HSD polymer, SIP polymer, and PMA/OCP copolymer was found not effective in depressing the pour point of SWI 6 base stock, containing the PMA type pour point depressant below −33°C indicating no effect. The patent

TABLE 3.24

Effect of Polymer Mixtures on the Pour Point Depression of SWI 6 Base Stock

SWI 6 (VI 143/−21°C)	Oil #1	Oil #2	Oil #3	Oil #4	Oil #5
FVA (60,000) (wt%)	0.5	0.5			
PMA (92,000) (wt%)			0.5	0.5	1.5
PMA (43,000) (wt%)		0.5		0.5	0.5
Depressed pour point (°C)	−30	−33	−33	−36	−42

Source: Pillon, L.Z., *Petrol. Sci. Technol.*, 20(1–2), 101, 2002.

literature reports that the use of the mixture of different PMA polymers, having a molecular weight of 92,000 and 43,000 was the most effective in depressing the pour point of solvent-dewaxed SWI 6 base stock (MacAlpine, Halls, and Asselin, 1991).

The effect of different polymer mixtures on the pour point depression of SWI 6 base stock having a VI of 143 and solvent dewaxed to −21°C is shown in Table 3.24. The FVA chemistry pour point depressants were found not as effective as the PMA type polymers in depressing the pour point of solvent-dewaxed SWI 6 base stock. The addition of 0.5 wt% of PMA, having a low molecular weight of 43,000, was found effective in improving the performance of FVA pour point depressant by decreasing the depressed pour point by additional 3°C to −33°C. The use of a mixture of different PMA polymers, at a total treat rate of 1 wt%, was found to also be effective in depressing the pour point of SWI base stock by additional 3°C to −36°C. However, the use of a mixture containing 1.5 wt% of PMA polymer, having a molecular weight of 92,000 and 0.5 wt% of PMA polymer, having a molecular weight of 43,000, was reported effective in depressing the pour point of SWI 6 base stock by additional 6°C to −42°C. The pour point depressants are usually used in lubricating oils at low treat rates of 0.1–0.5 wt%, however, some lubricant products might contain 1 wt%. The selection of polymeric additives, such as pour point depressants and VI improvers, require that their chemistry and molecular weight be optimized to assure good solubility in lubricating oils.

3.5 PROCESSING EFFECT

The pour point of most SR base stocks is related to the crystallization of wax. The use of pour point depressants was reported to be less effective in lubricants containing high viscosity mineral base stocks. With an increase in the viscosity of mineral base stocks, their *n*-paraffinic content decreases, and heavy mineral base stocks and bright stocks contain no *n*-paraffins. The pour point depressants were developed to interfere with the growth of *n*-paraffinic type wax and do not entirely prevent wax crystal growth. They only lower the temperature at which a rigid wax structure is formed (Pirro and Wessol, 2001). XHVI and SWI base stocks were reported to be the best modified mineral base stocks on the market. For the base stocks, having a similar kinematic viscosity at 100°C, the volatility of the slack wax isomerate base stocks is lower than the volatility of the mineral and the hydrocracked base stocks (Phillipps, 1999). The GC of XHVI base stocks indicated the presence of different molecular weight hydrocarbons whereas only one type molecular weight hydrocarbon was found in PAO fluid (Rudnick and Shubkin, 1999). The literature also reports that the poor low temperature properties of SWI base stocks are due to the solvent dewaxing step that is not efficient in removing the trace *n*-paraffin residual waxes (Henderson, 2005a). The slack wax isomerate product is fractionated to obtain a lube oil cut, boiling above 330°C, and dewaxed to a pour point of −15°C to −24°C (Pillon and Asselin, 1993). The solvent-dewaxed SWI base stocks have a very high VI and a low volatility.

The cloud point and the MS analysis of hydrocarbon composition of different dewaxed XHVI and SWI base stocks are shown in Table 3.25. With a decrease in the temperature, the viscosity of oil increases until it reaches the pour point and it stops to flow. The cloud point is the temperature at which the first signs of wax formation can be detected. The presence of cloud point might be related to the presence of

TABLE 3.25
Cloud Point and Hydrocarbon Composition of Dewaxed XHVI and SWI Oils

	Base Stocks		
	XHVI 6	XHVI 5	SWI 6
KV at 40°C (cSt)	28.82	24.44	29.46
KV at 100°C (cSt)	5.77	5.13	5.84
VI	147	144	146
Cloud point (°C)	−12	—	—
Dewaxed pour point (°C)	−15	−21	−21
Aromatics (wt%)	2.2	0.5	0.3
Sulfur (ppm)	48	3	<1
Nitrogen (ppm)	2	2	<1
MS of saturate fraction (vol%)			
n-Paraffins and *iso*-paraffins	75.7	82.8	87.1
1-Ring naphthenes	12.3	9.3	8.0
2-Ring naphthenes	5.9	4.5	2.9
3-Ring naphthenes	1.8	1.1	0.4
4-Ring naphthenes	1.0	0.7	0.4
5-Ring naphthenes	0.6	0.6	0.5
6-Ring naphthenes	0.3	0.5	0.4
Residual monoaromatics	0.2	0	0
Average ring number per mole	0.4	0.3	0.2
Volatility (% off)	14–15	11	9

Sources: Phillipps, R.A., Highly refined mineral oils, in *Synthetic Lubricants and High-Performance Functional Fluids*, 2nd ed., Rudnick, L.R. and Shubkin, R.L., Eds., Marcel Dekker Inc., New York, 1999. With permission; Pillon, L.Z., *Petrol. Sci. Technol.*, 20(1–2), 223, 2002. With permission.

n-paraffins but also high molecular weight *iso*-paraffins or aromatic hydrocarbons having high melting points.

While a low cloud point or no cloud point indicates that the petroleum oils will remain bright and clear, base stocks that do not have clearly defined cloud points might indicate the viscosity limited pour points. XHVI 6 base stock, having a high VI of 147, a volatility of 14%–15% off and dewaxed to −15°C, was found to have a cloud point at −12°C. The MS analysis of the saturate fraction indicated the presence of 76 vol% of total *n*-paraffins/*iso*-paraffins and an average ring number per mole of 0.4. Lower viscosity XHVI 5 base stock, having a VI of 144, a lower volatility of 11% off and dewaxed to a pour point of −21°C, did not have a visible cloud point. The MS analysis of the saturate fraction indicated the presence of 83 vol% of total *n*-paraffins/*iso*-paraffins and an average ring number per mole of 0.2. SWI 6 base stock having a VI of 146, a low volatility of 9% off, and solvent dewaxed to a pour point of −21°C, also did not have a visible cloud point. The MS analysis of the saturate fraction indicated the presence of 87 vol% of total *n*-paraffins/*iso*-paraffins and an average ring number per mole of 0.2. As *n*-paraffins crack easily, dewaxed XHVI and SWI 6 base stocks will contain only *iso*-paraffins.

The DSC wax content of XHVI 5 and SWI 6 base stocks produced from different slack wax feed and hydroisomerization temperature is shown in Table 3.26. The

TABLE 3.26

DSC Wax Content of XHVI 5 and Different SWI 6 Base Stocks

	Dewaxed Base Stocks		
	XHVI 5	**SWI 6 #1**	**SWI 6 #2**
KV at 40°C (cSt)	24.44	28.96	29.47
KV at 100°C (cSt)	5.13	5.77	5.72
VI	144	146	139
Dewaxed pour point (°C)	−21	−21	−21
DSC wax (wt%)			
−100°C to −40°C	22.45	24.38	17.68
−40°C to −35°C	2.39	2.68	2.05
−35°C to −30°C	3.03	3.38	2.51
−30°C to −25°C	2.67	2.81	2.06
−25°C to −20°C	3.29	3.07	2.21
−20°C to −15°C	2.52	1.65	1.28
−15°C to −10°C	0.83	0.51	0.55
−10°C to −5°C	0.21	0.21	0.24
−5°C to 0°C	0.13	0.14	0.17
0°C to 2°C	0.04	0.05	0.05
2°C to 5°C	0.03	0.03	0.04
Total wax (wt%)	37.59	38.91	28.84

Source: Reprinted from Pillon, L.Z., *Petrol. Sci. Technol.*, 20(1–2), 101, 2002. With permission.

ASTM D 4419 test measures the transition temperatures of petroleum waxes by the DSC technique. The initial DSC analyses for the wax content of different slack wax isomerate base stocks were not consistent and indicated the presence of one wide crystallization peak. After the test procedure was modified to include the annealing of the slack wax isomerates at −84°C for 30 min, the DSC analysis showed two crystallization peaks in the temperature range from −20°C to −40°C, and at about −75°C (Pillon, 2002a). After the annealing of dewaxed XHVI 5 oil, the DSC analysis showed the total wax content, measured in the temperature range, from −100°C to 5°C, of 37.59 wt%. XHVI 5 was found to have some soluble wax of 3.76 wt% present above the temperature of −20°C. After the annealing of dewaxed SWI 6 oil #1, the DSC analysis showed the total wax content measured in the temperature range from −100°C to 5°C, of 38.91 wt%. SWI 6 #1 was also found to have a soluble wax of 2.57 wt% present above the temperature of −20°C. After the annealing of another dewaxed SWI 6 oil #2, the DSC analysis showed a lower total wax content, measured in the temperature range from −100°C to 5°C, of 28.84 wt%. SWI 6 #2 was also found to have less soluble wax of 2.18 wt% present above the temperature of −20°C.

The effect of PMA polymers on the pour point depression of XHVI 5 and SWI 6 base stocks containing different wax content is shown in Table 3.27. The use of 0.5 wt% of PMA pour point depressant, having a molecular weight of 92,000, depressed the pour point of both XHVI 5 and SWI #1 base stocks, having a higher wax content, to −33°C. The use of 1 wt% of PMA mixture did not depress their pour points any further. The use of 0.5 wt% of same PMA pour point depressant, depressed the pour point of SWI 6 #2 base stock, having a lower wax content to −39°C indicates a significant improvement. The use of 1 wt% of the PMA mixture depressed the pour point of SWI #2 base stock by additional 3°C to −42°C thus further confirming the significance of reducing the wax content. The use of different slack wax feed, in terms of oil content and hydroisomerization temperature was found to affect the VI and the wax content of different solvent-dewaxed SWI 6 base stocks, and also their

TABLE 3.27

Effect of Wax Content on the Pour Point Depression of XHVI 5 and SWI 6 Base Stocks

	Dewaxed Base Stocks		
	XHVI 5	**SWI 6 #1**	**SWI 6 #2**
VI	144	146	139
Dewaxed pour point (°C)	−21	−21	−21
DSC wax (−100°C to 5°C) (wt%)	37.59	38.91	28.84
PMA (92,000) (0.5 wt%)			
Depressed pour point (°C)	−33	−33	−39
PMA mixture (1.0 wt%)			
Depressed pour point (°C)	−33	−33	−42

Source: Reprinted from Pillon, L.Z., *Petrol. Sci. Technol.*, 20(1–2), 101, 2002. With permission.

TABLE 3.28
^{13}C NMR Analysis of Branching Density of XHVI 5 and Different SWI Oils

	Dewaxed Base Stocks			
	XHVI 5	SWI 6 #1	SWI 6 #2	SWI 6 #3
KV at 40°C (cSt)	24.44	28.96	29.2	29.49
KV at 100°C (cSt)	5.13	5.77	5.68	5.74
VI	144	146	138	140
Dewaxed pour point (°C)	−21	−21	−24	−21
^{13}C NMR analysis				
Methyl per 100 carbons	5.8	5.8	6.2	6.8
Ethyl per 100 carbons	1.9	1.7	2	2
Methyl per 1000 bulk (−CH$_2$−)	210	205	255	277
Ethyl per 1000 bulk (−CH$_2$−)	67	60	80	79

Source: Pillon, L.Z., *Petrol. Sci. Technol.*, 20(3–4), 357, 2002. With permission.

branching density (Pillon, 2002c). The use of the NMR technique can determine the number of each of the distinct types of hydrogen nuclei as well as to obtain information on the nature of their environment. The ^{13}C NMR provides information about the carbon nuclei and their environment (Pavia, Lampman, and Kriz, 1996).

The ^{13}C NMR analysis of the branching density of XHVI 5 and the different SWI 6 base stocks is shown in Table 3.28. The ^{13}C NMR spectra of the saturate fractions separated from different SWI base stocks indicated the presence of some important components of the hydrocarbon structure. The peaks, such as branch-ethyl (12 ppm), terminal-methyl (14 ppm), branch-methyl (20 ppm), and bulk methylene (30 ppm), were identified (Pillon, 2002c). The ^{13}C NMR analysis of XHVI 5 oil, having a VI of 144 and a high DSC wax content, indicated a branching density of 5.8 methyl and 1.9 ethyl side chain groups per 100 carbons. The methyl-branching of 210 and the ethyl-branching of 67 per 1000 bulk (−CH$_2$−) was calculated. The ^{13}C NMR analysis of SWI 6 oil #1, having a VI of 146 and a high DSC wax content, indicated 5.8 methyl and 1.7 ethyl side chain groups per 100 carbons. The methyl-branching of 205 and the ethyl-branching of 60 per 1000 bulk (−CH$_2$−) was calculated. The ^{13}C NMR analysis of SWI 6 oil #2, having a lower VI of 138 and a lower DSC wax content, indicated the presence of 6.2 methyl and 2 ethyl side chain groups per 100 carbons. A significantly higher methyl-branching of 255 and the ethyl-branching of 80 per 1000 bulk (−CH$_2$−) was calculated. The ^{13}C NMR analysis of another SWI 6 oil #3, also having a lower VI of 140, indicated 6.8 methyl and 2 ethyl side chain groups per 100 carbons. Moreover, a higher methyl-branching of 277 and a higher ethyl-branching of 79 per 1000 bulk (−CH$_2$−) was calculated. The branching density values calculated with respect to the bulk methylene (−CH$_2$−) signal were reported to provide the most sensitive measure of the total branching in slack wax isomerate base stocks (Pillon, 2002c).

TABLE 3.29

Effect of Branching Density on the Pour Point Depression of XHVI 5 and SWI 6 Oils

	Dewaxed Base Stocks			
	XHVI 5	**SWI 6 #1**	**SWI 6 #2**	**SWI 6 #3**
VI	144	146	138	140
Dewaxed pour point (°C)	−21	−21	−24	−21
Methyl per 1000 bulk ($-CH_2-$)	210	205	255	277
PMA (92,000) (wt%)	0.5	0.5	0.5	0.5
Depressed pour point (°C)	−33	−33	−39	−39
Pour point difference (°C)	12	12	15	18

Source: Reprinted from Pillon, L.Z., *Petrol. Sci. Technol.*, 20(3–4), 357, 2002. With permission.

The effect of PMA polymers on pour point depression of XHVI 5 and different SWI 6 base stocks having a different branching density is shown in Table 3.29. With an increase in the branching density, an improvement in pour point depression of solvent-dewaxed SWI 6 base stocks is observed. Solvent-dewaxed SWI base stocks with an increased branching density have a lower wax content. The literature also reports that the low temperature properties of the solvent-dewaxed slack wax isomerate base stocks are not as good as the low temperature properties of synthetic PAO base stocks (Prince, 1997). The SWI process was developed to produce base stocks having a very high VI and a low volatility. The high VI of base stocks related to low branching density *iso*-paraffins and a low volatility related to high molecular weight hydrocarbons will lead to poor low temperature properties. Some oils, essentially wax-free, were reported to have viscosity-limited pour points. The pour point depressants, developed to interfere with the growth of the crystal structure, are less effective in depressing the viscosity-limited pour points (Mang, 2001).

The literature reports that solvent dewaxing is less effective in removing the *n*-paraffin residual waxes whereas hydrodewaxing can isomerize them and/or convert them into shorter fuel type products (Henderson, 2005a). Hydrodewaxing is a generic term where hydrogen is used in the presence of a catalyst to convert the waxy components in the feed. The recent literature reports that hydrodewaxing of different viscosity wax, from solvent dewaxing of neutral oils, leads to lube oil base stocks, having a high VI, and known as Visom 4 and Visom 6. However, solvent dewaxing is still required to lower their pour points (Lynch, 2007). Despite the absence of *n*-paraffins, many *iso*-paraffins, having a low branching density, will have high pour points and lead to wax formation. At temperatures below their dewaxed pour points, SR and hydroprocessed base stocks contain varying degrees of waxy type molecules, not related to *n*-paraffins, which will crystallize and prevent the free flow of oil.

REFERENCES

Anwar, M. et al., *Petroleum Science and Technology*, 17(5&6), 491 (1999).

Braun, J. and Omeis, J., Additives, in *Lubricants and Lubrications*, Mang, T. and Dresel, W., Eds., Wiley-VCH, Weinheim, 2001.

Calemma, V., Peratello, S., and Perego, C., *Applied Catalysis A: General,* 190, 207 (2000).

Calemma, V. et al., *Industrial and Engineering Chemistry Research,* 43, 934 (2004).

Chemical Dictionary, http://www.chemicaldictionary.cn (accessed 2008).

Dresel, W., Synthetic base oils, in *Lubricants and Lubrications*, Mang, T. and Dresel, W., Eds., Wiley-VCH, Weinheim, 2001.

Henderson, H.E., Chemically modified mineral oils, in *Synthetics, Mineral Oils, and Bio-Based Lubricants Chemistry and Technology*, Rudnick, L.R., Ed., CRC Press, Boca Raton, FL, 2005a.

Henderson, H.E., Gas to liquids, in *Synthetics, Mineral Oils, and Bio-Based Lubricants Chemistry and Technology*, Rudnick, L.R., Ed., CRC Press, Boca Raton, FL, 2005b.

Lakes, S.C., Automotive crankcase oils, in *Synthetic Lubricants and High-Performance Functional Fluids*, 2nd ed., Marcel Dekker Inc., New York, 1999.

Lynch, T.R., *Process Chemistry of Lubricant Base Stocks*, CRC Press, Boca Raton, FL, 2007.

MacAlpine, G.A., Halls, P.M.E., and Asselin, A.E., Wax isomerate having a reduced pour point, CA Patent 2,054,096, 1991.

Mang, T., Base oils, in *Lubricants and Lubrications*, Mang, T. and Dresel, W., Eds., Wiley-VCH, Weinheim, 2001.

McCabe, C., Cui, S., and Cummings, P.T., *Fluid Phase Equilibria*, 183–184, 363–370 (2001).

Pavia, D.L., Lampman, G.M., and Kriz, G.S., *Introduction to Spectroscopy*, Harcourt Brace College Publishers, Orlando, FL, 1996.

Phillipps, R.A., Highly refined mineral oils, in *Synthetic Lubricants and High-Performance Functional Fluids*, 2nd ed., Rudnick, L.R. and Shubkin, R.L., Eds., Marcel Dekker Inc., New York, 1999.

Pillon, L.Z., *Petroleum Science and Technology*, 20(1&2), 101 (2002a).

Pillon, L.Z., *Petroleum Science and Technology*, 20(1&2), 223 (2002b).

Pillon, L.Z., *Petroleum Science and Technology*, 20(3&4), 357 (2002c).

Pillon, L.Z. and Asselin, A.E., Wax isomerate having a reduced pour point, U.S. Patent 5,229,029, 1993.

Pirro, D.M. and Wessol, A.A., *Lubrication Fundamentals*, 2nd ed., Marcel Dekker Inc., New York, 2001.

Prince, R.J., Base oils from petroleum, in *Chemistry and Technology of Lubricants*, 2nd ed., Mortier, R.M. and Orszulik, S.T., Eds., Blackie Academic & Professional, London, U.K., 1997.

Rudnick, L.R. and Shubkin, R.L., Poly(alfa-olefins), in *Synthetic Lubricants and High-Performance Functional Fluids*, 2nd ed., Rudnick, L.R. and Shubkin, R.L., Eds., Marcel Dekker, Inc., New York, 1999.

Sequeira, A. Jr., *Lubricant Base Oil and Wax Processing*, Marcel Dekker, Inc., New York, 1994.

Sharma, B.K. and Stipanovic, A.J., *Industrial and Engineering Chemistry Research*, 42, 1522 (2003).

Sharma, B.K. et al., *Energy and Fuels*, 18, 952 (2004).

Sigma-Aldrich, *Handbook of Fine Chemicals*, Sigma-Aldrich Company, St. Louis, MO, 2007–2008.

Srivastava, S.P. et al., *Petroleum Science and Technology*, 20(7&8), 831 (2002).

Stambaugh, R.L., Viscosity index improvers and thickeners, in *Chemistry and Technology of Lubricants*, 2nd ed., Mortier R.M. and Orszulik S.T., Eds., Blackie Academic & Professional, London, U.K., 1997.

Taylor, R.J., McCormack, A.J., and Nero, V.P., *Pre-Prints Division of Petroleum Chemistry, ACS*, 37(4), 1337, 1992.

Yutai, Q., Huipeng, L., and Shuren, Q., *Petroleum Science and Technology*, 19(3&4), 403 (2001).

4 Oxidation Stabilities of Base Stocks

4.1 OXIDATION BY-PRODUCTS

Most petroleum oils, when exposed to air over a period of time, react with oxygen. The addition of oxygen not only changes the chemistry of hydrocarbon molecules but also increases the boiling point and the melting point. The aliphatic n-octane hydrocarbon, with a C_8H_{18} formula, was reported to have a boiling point of 125°C–127°C and a very low melting point of −57°C. The addition of one oxygen atom can lead to 1-octanol, which has a higher boiling point of 196°C and a higher melting point of −15°C. The addition of two oxygen atoms can lead to octanoic acid, which has a very high boiling point of 237°C and a very high melting point of 15°C–17°C. However, the addition of two oxygen atoms to form hexyl acetate ester leads to a decrease in the boiling point to 168°C–170°C and a drastic decrease in the melting point to −80°C (Sigma-Aldrich, 2007–2008). The aromatic ethyl benzene hydrocarbon, with a C_8H_{10} formula, was reported to have a boiling point of 136°C and a very low melting point of −95°C. The addition of one oxygen atom can lead to 2-phenylethanol, which has a higher boiling point of 219°C–221°C and a higher melting point of −27°C. The addition of two oxygen atoms can lead to benzeneacetic acid, which has a very high boiling point of 265°C and a very high melting point of 76°C–78°C. However, the addition of two oxygen atoms to form methyl benzoate ester leads to a decrease in the boiling point to 198°C–199°C and a drastic decrease in the melting point to −12°C (Sigma-Aldrich, 2007–2008).

The effect of the oxygen on the boiling points and the melting points of the aliphatic and the aromatic hydrocarbons is shown in Table 4.1. The literature reports that in lubricating oils, under low-temperature oxidation conditions, peroxides, alcohols, aldehydes, ketones, and water are formed. Under high-temperature oxidation conditions, acids are formed (Bardasz and Lamb, 2003). The majority of lubricant products use solvent-refined mineral base stocks. The literature reports that the maximum oxidation stability is observed when the base stock contains 10–20 wt% of aromatics and 0.005–0.07 wt% of sulfur. The polynuclear aromatics, together with appropriate sulfur compounds, were reported to retard the rate of oxidation; however, they also contribute to sludge formation (Rasberger, 1997). The early literature reports that the organo-sulfur compounds present in lubricating oil fractions function as antioxidants by decomposing the peroxide species (Deninson, 1944). In conventional refining of crude oils, after distillation and solvent refining, the aromatic and sulfur contents of base stocks were found to vary depending on the crude oil origin and its type.

TABLE 4.1

Effect of Oxygen on the Boiling Points and the Melting Points of Hydrocarbons

Formula	Name	Boiling Point (°C)	Melting Point (°C)
C_8H_{18}	n-Octane	125–127	−57
$C_8H_{18}O$	1-Octanol	196	−15
$C_8H_{16}O_2$	Octanoic acid	237	15–17
$C_8H_{16}O_2$	Hexyl acetate	168–170	−80
C_8H_{10}	Ethyl benzene	136	−95
$C_8H_{10}O$	2-Phenylethanol	219–221	−27
$C_8H_8O_2$	Benzeneacetic acid	265	76–78
$C_8H_8O_2$	Methyl benzoate	198–199	−12

Source: Sigma-Aldrich, *Handbook of Fine Chemicals*, St. Louis, MO, 2007–2008.

TABLE 4.2

Effect of Crude Oil Origin and Type on the Aromatic and the S Contents of Base Stocks

	Crude Oil Origin			
	Gulf Coast	**Pennsylvania**	**California**	
Crude oil type			Naphthenic	Paraffinic
Paraffins (wt%)	58	77	52	72
Naphthenes (wt%)	32	16	33	27
Aromatics (wt%)	10	7	15	1
Sulfur (wt%)	0.2	0.1	0.5	0.2

Source: Lynch, T.R., *Process Chemistry of Lubricant Base Stocks*, CRC Press, Boca Raton, FL, 2007. With permission.

The effect of crude oil origins and types on the aromatic and S contents of solvent-refined base stocks is shown in Table 4.2. The solvent-refined base stock, from Gulf Coast crude oil, was reported to contain 58 wt% of paraffins, 32 wt% of naphthenes, 10 wt% of aromatics, and 0.2 wt% of sulfur. The solvent-refined base stock, from Pennsylvania crude oil, was reported to contain 77 wt% of paraffins, 16 wt% of naphthenes, 7 wt% of aromatics, and 0.1 wt% of sulfur. The solvent-refined base stock, from California naphthenic type crude oil, was reported to contain 52 wt% of paraffins, 33 wt% of naphthenes, 15 wt% of aromatics, and 0.5 wt% of sulfur. Another solvent-refined base stock, from California paraffinic type crude oil, was reported to contain a higher paraffin content of 72 wt%, a lower naphthenic content of 27 wt%, a low aromatic content of 1 wt%, and a lower S content of 0.2 wt%. After distillation, solvent extraction, and solvent dewaxing, the process such as clay treatment or hydrofinishing, also known as hydrofining (HF), are used to remove the molecules that affect the color and the oxidation stability of the base stocks. Different oxidation tests are used to test the oxidation stability of base stocks, and some tests include metal catalysts to cause accelerated aging of the oil and to reduce the testing time (Rasberger, 1997).

The effect of clay treatment and hydrofinishing on the oxidation stability of a lower viscosity mineral base stock, from Western Canadian crude oil, is shown in Table 4.3. As shown by the Indiana oxidation test, after 380 h of oxidation at 110°C and in the presence of copper (Cu) used as catalyst, a lower viscosity clay-treated base stock was reported to have a total acid number (TAN) of 2.5 and it contained 90 mg of sludge. Under the same oxidation conditions, the hydrofinished base stock was found to have a lower TAN of 2.2 and it contained a lower sludge content of 47 mg indicating an increase in the oxidation stability. The TAN value of oil is the weight of potassium hydroxide (KOH) required to neutralize 1 g of oil. Fresh hydrofined mineral base stocks have a low TAN of 0.01 mg KOH/g and an increase in TAN usually indicates an increase in acid type molecules. The ASTM D 943 test, also known as turbine oxidation stability test (TOST), is used to measure the condition of the lubricant after the specific defined test period by measuring several aging parameters, such as TAN, viscosity change, and sludge formation (Braun and Omeis, 2001). Under the ASTM D 943 test conditions, at 95°C and in the presence of water and metals, such as Cu and Fe used as oxidation catalysts, the same lower viscosity clay-treated base stock was reported to have a TAN of 2 after 1325 h. Under the same oxidation conditions, the hydrofinished base stock was reported to have a TAN of 2 after a longer time of 2050 h indicating an increase in the oxidation stability.

The effect of clay treatment and hydrofinishing on the oxidation stability of a higher viscosity mineral base stock, from Western Canadian crude oil, is shown in Table 4.4. As shown by the Indiana oxidation test, after 380 h of oxidation at 110°C and in the presence of Cu, a higher viscosity clay-treated mineral base stock was reported to have a TAN of 3 and it contained 69 mg of sludge. Under the same

TABLE 4.3
Effect of Finishing on the Oxidation Stability of Lower Viscosity Mineral Base Stock

Mineral Base Stock (KV of 33.5 cSt at 38°C)	Oxidation Test Conditions	After Clay Treatment	After Hydrofinishing
Indiana test	110°C/Cu/380 h		
TAN (mg KOH/g)		2.5	2.2
Sludge (mg/10 g)		90	47
ASTM D 943 test	95°C/Cu/Fe/water		
TAN (mg KOH/g)		2	2
TOST life (h)		1325	2050

Source: Lynch, T.R., *Process Chemistry of Lubricant Base Stocks*, CRC Press, Boca Raton, FL, 2007. With permission.

TABLE 4.4

Effect of Finishing on the Oxidation Stability of Higher Viscosity Mineral Base Stock

Mineral Base Stock (KV of 139 cSt at 38°C)	Oxidation Test Conditions	After Clay Treatment	After Hydrofinishing
Indiana test	110°C/Cu/380 h		
TAN (mg KOH/g)		3	2
Sludge (mg/10 g)		69	17
ASTM D 943 test	95°C/Cu/Fe/water		
TAN (mg KOH/g)		2	2
TOST life (h)		1080	1250

Source: Lynch, T.R., *Process Chemistry of Lubricant Base Stocks*, CRC Press, Boca Raton, FL, 2007. With permission.

oxidation conditions, the hydrofinished base stock was found to have a lower TAN of 2 and it contained a lower sludge content of 17 mg indicating an increase in the oxidation stability. Under the ASTM D 943 test conditions, at 95°C and in the presence of water and metals, a higher viscosity clay-treated mineral base stock was reported to have a TAN of 2 after 1080 h. Under the same oxidation conditions, the hydrofinished base stock was reported to have a TAN of 2 after a longer time of 1250 h indicating an increase in the oxidation stability. For the same viscosity mineral base stocks, the use of hydrofinishing was reported to increase the oxidation stability by decreasing the discoloration and the formation of acidic by-products measured as TAN and sludge content (Lynch, 2007). However, with an increase in the viscosity of clay-treated and hydrofinished mineral base stocks, a significant increase in TAN with a decrease in the sludge formation of oxidized oils is observed indicating the effect of different chemistry molecules found in higher boiling distillate feeds. The literature reports that the nitrogen compounds have a negative impact on the oxidation stability of mineral lubricating base oils (Wang and Li, 2000).

The effect of the boiling range of heavy distillates from California crude oil on the chemistry and the content of N-compounds is shown in Table 4.5. Heavy distillate, having a lower boiling range of 205°C–370°C, was reported to contain mainly pyridine derivatives. With an increase in the boiling range of the heavy distillate, 370°C–455°C, the presence of carbazoles, pyrroles, and quinolines was also reported. Heavy distillate, having a high boiling range of 455°C–540°C, was reported to contain mainly benzocarbazoles and benzoquinolines indicating an increase in the total N content and their aromaticity. The literature reports that the vacuum distillates and the deasphalted oils contain many undesirable molecules, which cause darkening, oxidation, and sludging of lubricant products. The early literature reports that aliphatic amines oxidize easily to form nitrogen oxides, while aromatic and heterocyclic amines require more severe conditions (Lake and Hoh, 1963). The vacuum distillate from Dagang oil was reported to have a negative

TABLE 4.5
Effect of Boiling Range of Heavy Distillates on the Chemistry of N-Compounds

California Crude Oil Heavy Distillate	Boiling Range		
	205°C –370°C	370°C–455°C	455°C–540°C
Indoles (wt%)	0.1	0.6	0.8
Carbazoles (wt%)	0.3	3.4	4.1
Benzocarbazoles (wt%)	0	0.5	1.3
Pyrrole derivatives (wt%)	0.4	4.5	6.1
Pyridines (wt%)	0.4	0.7	1.3
Quinolines (wt%)	0.2	1.7	2
Benzoquinolines (wt%)	0.1	0.3	1.6
Pyridine derivatives (wt%)	0.6	2.7	4.9
Pyridones/quinolones (wt%)	0.2	1.2	2
Azaindoles (wt%)	0	0.1	0.4

Source: Reproduced from Lynch, T.R., *Process Chemistry of Lubricant Base Stocks*, CRC Press, Boca Raton, FL, 2007. With permission.

viscosity index (VI) of −5 and a high TAN of 1.7 mg KOH/g. The literature reports that the TAN of lube oil distillates increase with an increase in the viscosity of lube oil distillate (Wang et al., 2001).

The heteroatom content, and the carbon residue of heavy distillate and deasphalted oil from the same crude oil are shown in Table 4.6. The heavy distillate was reported to have a dewaxed VI of 53, an S content of 2.6 wt%, an N content of 0.12 wt%, and a carbon residue of 0.6 wt%. The deasphalting of vacuum residua and solvent refining is used to produce very high viscosity mineral base stocks, such as

TABLE 4.6
Properties and Composition of Heavy Distillate and Deasphalted Oil

Properties	Heavy Distillate	Deasphalted Oil
KV at 99°C (cSt)	14.1	40.3
Dewaxed VI	53	75
Sulfur (wt%)	2.6	2.1
Nitrogen (wt%)	0.12	0.06
Ramsbottom carbon residue (wt%)	0.6	1.6

Source: Lynch, T.R., *Process Chemistry of Lubricant Base Stocks*, CRC Press, Boca Raton, FL, 2007. With permission.

bright stocks. The deasphalted oil, from the same crude oil, was reported to have a higher dewaxed VI of 75, a lower S content of 2.1 wt%, a lower N content of 0.06 wt% but a higher carbon residue of 1.6 wt%. Different extraction solvents are used to increase the VI of lube oil distillates and deasphalted oils. The literature reports that the highly toxic phenol extraction solvent has a moderate selectivity with a high dissolving power and does not guarantee a high yield of raffinate. In comparison to phenol, N-methylpyrrolidone (NMP) solvent does not form azeotropic mixtures with water, is less toxic, and has a higher selectivity with high dissolving power. The yield of raffinate is increased by 6–8 wt%, and the solvent/feedstock ratio is reduced by 15% (Ivanov, Lazarev, and Yaushev, 2000). Due to an increase in requirements for environmental protection, many refineries have switched from phenol to NMP extraction solvent.

The effect of phenol and NMP extraction solvents on aromatic and polar contents of raffinates is shown in Table 4.7. The 150N lube oil distillate feed, from Middle East crude oil, was reported to have a dewaxed oil VI of 62 and a pour point of 9°C. It contained 48.9 wt% of saturates, 40.7 wt% of aromatics, and 10.4 wt% of polars. After the phenol extraction, the VI of raffinate increased to 104, the aromatic content decreased to 24.1 wt%, and the polar content decreased to 1.1 wt%. After the NMP extraction, the VI of raffinate increased to 104, the aromatic content decreased to 23.9 wt% but the polar content decreased only to 1.8 wt%. The content of oxidation by-products, such as N-methylsuccinimide, was reported to gradually increase from 0.1 to 10 wt% in an NMP circulating solvent. This caused an increased corrosion in the pipelines and the equipment, especially in the NMP vaporization condensation sections. It was recommended to deaerate the feedstock in a vacuum and to conduct the process under an inert gas blanket to reduce the oxidation (Ivanov, Lazarev, and Yaushev, 2000).

The effect of HF on the composition and the basic nitrogen (BN) content of phenol, and NMP solvent-extracted and solvent-dewaxed base stocks is shown in Table 4.8. With an increase in the viscosity of hydrofined phenol-extracted base

---

TABLE 4.7

Effect of Phenol and NMP Extraction Solvents on the Polar Contents of Raffinates

Middle East Crude Oil	150N Distillate	Phenol Solvent	NMP Solvent
Solvent Extraction	Feed	Raffinate	Raffinate
Dewaxed oil VI	62	104	104
Pour point (°C)	9	9	9
Saturates (wt%)	48.9	74.8	74.3
Aromatics (wt%)	40.7	24.1	23.9
Polars (wt%)	10.4	1.1	1.8

Source: Lynch, T.R., *Process Chemistry of Lubricant Base Stocks*, CRC Press, Boca Raton, FL, 2007. With permission.

TABLE 4.8

Effect of Extraction Solvent on the Composition and the BN Content of Hydrofined Oils

Extraction Solvent	Phenol	Phenol	NMP	NMP
After Solvent Dewaxing	HF 150N	HF 600N	HF 150N	HF 600N
KV at 40°C (cSt)	29.5	105.9	29.7	111.4
KV at 100°C (cSt)	5.0	11.3	5.1	11.6
VI	94	92	96	89
Saturates (wt%)	82.8	80.5	86.1	80.4
Aromatics (wt%)	17.2	19.5	13.9	19.6
Sulfur (wt%)	0.09	0.12	0.06	0.11
Nitrogen (ppm)	8	30	36	100
BN (ppm)	4	16	33	88

Source: Pillon, L.Z. et al., Lubricating oil for inhibiting rust formation, U.S. Patent 5,397,487, 1995.

stocks from 150 to 600N, their aromatic content increased from 17.2 to 19.5 wt%, their sulfur content increased from 0.09 to 0.12 wt%, and their N content increased from 8 to 30 ppm.

With an increase in the viscosity of hydrofined NMP-extracted base stocks from 150 to 600N, their aromatic content increased from 13.9 to 19.6 wt%, their sulfur content increased from 0.06 to 0.12 wt%, and their N content increased from 36 to 100 ppm. The oxygen content of petroleum products is difficult to measure. The ASTM D 2896 test method is used to determine the BN content of petroleum oils by titration with perchloric acid. The basic constituents of petroleum include not only BN compounds but also other compounds, such as salts of weak acids (soaps), basic salts of polyacidic bases, and salts of heavy metals. With an increase in the viscosity of hydrofined phenol-extracted base stocks, their BN content increased from 4 to 16 ppm. With an increase in the viscosity of hydrofined NMP-extracted base stocks, their BN content increased significantly from 33 to 88 ppm indicating a higher content of some polar contaminants. Another oxidation test, the ASTM D 2440 test, is used to measure the oxidation stability of lubricating oils at a higher temperature of 120°C and in the presence of Cu.

The effect of extraction solvent on the oil/water interfacial tension (IFT) of oxidized different viscosity hydrofined mineral base stocks is shown in Table 4.9. After oxidation, the uninhibited hydrofined mineral base stocks, not containing any antioxidants, were analyzed for an increase in acidity, sludge, and a change in their IFT indicating the presence of surface active molecules that are capable of adsorbing at the oil/water interface. When compared to the same viscosity oxidized HF phenol-extracted base stocks, the oxidized HF NMP-extracted base stocks were found to produce a higher content of volatile acidity, a higher content of soluble acidity, and a higher content of sludge. The IFT of oxidized HF

TABLE 4.9
Effect of Extraction Solvent on the IFT of Oxidized Hydrofined Mineral Base Stocks

Extraction Solvent	Phenol	NMP	Phenol	NMP
After Solvent Dewaxing	HF 150N	HF 150N	HF 600N	HF 600N
Fresh oils, IFT (mN/m)	43	33	43	39
ASTM D 2440 (120°C/24 h)				
Oxidized oils, IFT (mN/m)	5–10	<5	10–20	5–10

Sources: Pillon, L.Z. and Asselin, A.E., Method for improving the demulsibility of base oils, U.S. Patent 5,282,960, 1994; Pillon, L.Z., *Petrol. Sci. Technol.*, 21(9–10), 1469, 2003.

150N NMP-extracted base stock was lower indicating an increase in oxidation by-products with surface active properties. The IFT of oxidized HF 600N NMP-extracted base stock was also lower indicating an increase in oxidation by-products having surface active properties. When oxidized, clay-treated and hydrofined solvent-refined base stocks produce acidic by-products and sludge; however, the use of HF increases their stability. When compared to the same viscosity hydrofined mineral base stocks, the oxidized NMP-extracted base stock was also found to produce a higher content of volatile acidity, a higher content of soluble acidity but showed no significant increase in the content of sludge. Under the same oxidation conditions, higher viscosity mineral base stocks produce a higher content of surface active acidic by-products but less sludge.

4.2 EFFECT OF HYDROPROCESSING

The literature reports that the processing of paraffinic base stocks, including bright stocks, requires steps that include vacuum distillation, deasphalting, solvent extraction, and solvent dewaxing. The literature reports that color is darker for higher viscosity oils and, while it has no effect on lubricant performance, it might indicate the presence of contamination (Shell, 2001). Many modern plants use hydroprocessing and catalytic dewaxing to upgrade the distillate. Earlier plants involved in the processing of naphthenic base stocks used sulfur dioxide extraction, and sulfuric acid and clay treatments. Modern plants use less hazardous hydroprocessing to upgrade the naphthenic distillate (Shell, 2001). The literature also reports that the use of hydrofinishing, as the last processing step, is no longer needed due to an increase in the severity of extraction and the use of hydroconversion where the color and the stability of base oil are improved by treating with hydrogen (Shell, 2001). Based on their saturate content, sulfur content, and VI, different petroleum-derived base stocks and synthetic fluids, which are used in lubricant products, are divided into five American Petroleum Institute (API) categories.

TABLE 4.10

API Base Stock and Synthetic Fluids Categories

API Group	Saturates (wt%)		Sulfur (wt%)	VI
I	<90	and/or	>0.03	80–120
II	>90	and	<0.03	80–120
III	>90	and	<0.03	>120
IV		PAO		
V		Esters		

Source: American Petroleum Institute, *Engine Oil Licensing and Certification System*, 15th ed., API 1509, API, Washington, DC, 2002.

The five API base stock and synthetic fluids categories are shown in Table 4.10. The API group I includes the mineral base stocks, which are conventionally solvent refined and have the VI in the range of 80–120. Another API group II includes the mildly hydrocracked base stocks, which have a higher saturate content, a lower S content, and a VI in the range of 80–120. The API group III includes severely hydrotreated (HT) mineral oils with a VI in excess of 120. The slack wax isomerate base stocks, having a VI above 140, are also classified as API III highly refined mineral oils, which contain practically no aromatics, sulfur, and nitrogen contents. The API group IV includes synthetic polyalfaolefin (PAO) base stocks, while the API group V includes other synthetic fluids, such as esters, polyethers, and polyalkylene glycols.

The hydrocarbon composition and the heteroatom content of different petroleum-derived 150N base stocks, including synthetic PAO fluid, having a similar viscosity are shown in Table 4.11. The use of hydrocracking (HC) leads to an API group II base stock having a low aromatic, and S and N contents. The hydroprocessing of base stocks to obtain VHVI leads to oils having practically no aromatic, and very low S and N contents. The API group III base stocks, with extra high VI (XHVI), have practically no aromatics, and no S and N contents. Petroleum-derived base stocks might contain O-containing molecules but do not report the O content because it is difficult to measure. The synthetic PAO oils have 100% saturate content, and no S and N contents; however, even synthetic oils might contain entrained air, which can affect their oxidation stability.

The severely HT base stocks, produced from vacuum distillate and deasphalted vacuum residue, were reported to form a flocculant precipitate upon prolonged exposure to ultraviolet (UV) light known as daylight stability. A poor daylight stability of HT base stocks was reported to be related to their poor oxidation stability (Bijwaard and Morcus, 1985).

The variation in the daylight stability and the oxidation stability of some severely HT base stocks is shown in Table 4.12. The daylight stability of base stocks was reported to vary from a low of only 2 days to over 15 days. The oxidation of mineral and hydroprocessed base stocks leads to an increase in acidity and formation of sludge. Depending on the severity of oxidation conditions, an increase in viscosity

TABLE 4.11

Hydrocarbon Composition and Heteroatom Content of Different API Base Stocks

API Category	API I	API II	API III	API IV
150N Base Stocks	Solvent Refined	HC	VHVI	Synthetic PAO
Paraffins (wt%)	27.6	33.4	55.5	100
1-Ring naphthenes (wt%)	20.8	30.2	20.4	0
2-Ring naphthenes (wt%)	25.9	17.2	12.1	0
3-Ring naphthenes (wt%)	2.9	9.3	9.1	0
4-Ring naphthenes (wt%)	0.3	5.1	2.1	0
5-Ring naphthenes (wt%)	0	1.1	0	0
Total naphthenes (wt%)	49.9	63.1	43.7	0
Aromatics (wt%)	22.5	3.5	0.8	0
Sulfur (ppm)	5800	300	<10	0
Nitrogen (ppm)	12	4	<1	0

Source: Henderson, H.E., Chemically modified mineral oils, in *Synthetics, Mineral Oils, and Bio-Based Lubricants Chemistry and Technology*, Rudnick, L.R., Ed., CRC Press, Boca Raton, FL, 2005. With permission.

TABLE 4.12

Variation in the Daylight Stability and the Oxidation Stability of Some HT Base Stocks

Fresh HT Base Stocks	Oxidized	Oxidized	Oxidized
Daylight Stability (Days)	Acidity (mg eq/100 g)	Sludge (wt%)	KV Increase (%)
>15	2	0	9
>15	7	0	35
9	3	0.2	17
6	3	0.2	15
6	4	0.5	22
5	3	0.4	20
2	3	0.4	19

Source: Bijwaard H.M.J. and Morcus A., Lubricating base oil compositions, CA Patent 1,185,962, 1985.

might also be observed. The patent literature reports that base stocks with good daylight stability, over 15 days, were found to have no tendency to form sludge when oxidized. Under the same oxidation conditions, HT base stocks, having a daylight stability of less than 9 days, were reported to form sludge (Bijwaard and Morcus, 1985). The term hydroisomerization indicates that the isomerization process predominates over HC. With an increase in the hydroisomerization temperature, an increase in cracking will take place. XHVI and SWI base stocks, with a very high VI over 140 and a high saturate content over 99 wt%, were found to have varying daylight stability (Pillon, 2007).

The properties, compositions, and daylight stabilities of XHVI 5 and SWI 6 base stocks are shown in Table 4.13. XHVI 5 base stock, with a high VI of 144, was reported to contain 99.5 wt% of saturates, 3 ppm of S, a trace of N, and a volatility of 12.9% off. It was found to have an excellent daylight stability of over 70 days. The solvent dewaxed and non-hydrofined SWI 6 base stock, with a higher VI of 146, was reported to contain 99.1 wt% of saturates, practically no S and N contents, and a lower volatility of 6.5% off. It was found to have a daylight stability of only 9 days. The literature reports on the use of nuclear magnetic resonance (NMR) spectroscopy and FTIR to analyze the chemistry of XHVI 5 and SWI 6 base stocks with different daylight stabilities (Pillon, 2002).

The ^{1}H NMR analysis of XHVI 5 and SWI 6 base stocks is shown in Table 4.14. The ^{1}H NMR spectra of XHVI 5 and SWI 6 base stocks were found to have two broad peaks, at 0.8–0.9 ppm and at 1.3 ppm, and a sharp peak at 1.5 ppm. The peaks at 0.8–0.9 ppm were due to saturated hydrogen atoms from the methyl groups ($-CH_3$). The peak at 1.3 ppm was due to saturated hydrogen atoms from methylene ($-CH_2-$) and methine ($-CH-$) groups. The peak at 1.5 ppm was reported to be caused by impurities in water. The ^{1}H NMR analysis of XHVI 5 base stock indicated the presence of a trace of aromatic hydrocarbons at 7.2–7.3 ppm and the presence of polynuclear aromatics. A sharp peak at 8–9 ppm was reported to be characteristic of an unsubstituted coronene molecule, which is a 6-ring polyaromatic structure (Sigma-Aldrich,

TABLE 4.13

Composition and Daylight Stability of XHVI 5 and SWI 6 Base Stocks

	Base Stock	
	XHVI 5	SWI 6
KV at 100°C (cSt)	5.13	5.77
KV at 40°C (cSt)	24.44	28.96
VI	144	146
Dewaxed pour point (°C)	−21	−21
Saturates (wt%)	99.5	99.1
Aromatics (wt%)	0.5	0.9
Sulfur (ppm)	3	<1
Nitrogen (ppm)	2	<1
Volatility (% off)	12.9	6.5
Daylight stability (days)	>70	9

Source: Reprinted from Pillon, L.Z., *Petrol. Sci. Technol.*, 20(1–2), 101, 2002. With permission.

TABLE 4.14

^{1}H NMR Analysis of XHVI 5 and SWI 6 Base Stocks

^{1}H NMR Analysis (ppm)	XHVI 5	SWI 6
0.8–0.9	CH$_3$ (saturate)	CH$_3$ (saturate)
1.3	CH$_2$, CH (saturate)	CH$_2$, CH (saturate)
2.4	CH$_3$ (aromatic)	CH$_3$ (aromatic)
2.6	CH$_2$ (aromatic)	CH$_2$ (aromatic)
6.9		Aromatic
7		Aromatic
7.2–7.3	Aromatic	
8.9	Coronenes	

Source: Pillon, L.Z., *Petrol. Sci. Technol.*, 20(3–4), 357, 2002. With permission.

2007–2008). The ^{1}H NMR analysis of SWI 6 base stock showed small peaks in the 7 ppm region indicating some aromatic content and small peaks at 2.6 and 2.4 ppm, which were assigned to the protons of carbons directly bonded to the aromatic rings. The use of double extraction, where the dimethyl sulfoxide (DMSO) extract is extracted with chloroform (CHCl$_3$), can be used to perform FTIR analysis.

The FTIR analysis of DMSO–CHCl$_3$ extracts from XHVI 5 and SWI 6 base stocks, having different daylight stabilities, is shown in Table 4.15. The FTIR analysis of XHVI 5 base stock, having an excellent daylight stability of over 70 days, confirmed the presence of the coronene molecular structure. The FTIR analysis of the DMSO/CHCl$_3$ extract also showed the presence of other aromatic compounds, ketones, acids, and esters. The FTIR analysis of solvent-dewaxed SWI 6 base stock,

TABLE 4.15

FTIR Analysis of DMSO–CHCl$_3$ Extracts from XHVI 5 and SWI 6 Base Stocks

Base Stocks	XHVI 5	SWI 6
Daylight Stability	**>70 days**	**9 days**
FTIR analysis (DMSO–CHCl$_3$ extract)	Ketones	Ketones
	Acids	Acids
	Esters	Esters
	Aromatics	1–2 ring aromatics
	Coronenes	

Source: Pillon, L.Z., *Petrol. Sci. Technol.*, 19(9–10), 1263, 2001. With permission.

having a poor daylight stability of 9 days, confirmed the absence of coronenes. The FTIR analysis of the DMSO/CHCl₃ extract also indicated the presence of different aromatic compounds, ketones, acids, and esters. An additional MS analysis SWI 6 base stock indicated the presence of 1-ring and 2-ring aromatic molecules (Pillon, 2002). While color and daylight stability of base stocks are not critical when formulating lubricating oils containing additives, the color stability is often tested to identify the possible contamination. Long-term storage of base stocks can lead to discoloration. The color of mineral base stocks can vary, from pale to red and dark, and heavy mineral base stocks are darker in color. The presence of oxygen and sunlight during storage was reported to accelerate the deterioration of mineral base stocks.

The effect of HF on the daylight stability and UV absorbance of solvent-dewaxed SWI 6 base stock, hydroisomerized at a low temperature of 290°C, is shown in Table 4.16. After solvent dewaxing, SWI 6 base stock with a daylight stability of 24 days was found to have an increased UV absorbance. The UV absorbance of the DMSO extract, in the region of 260–350 nm, was found to be varying, from 0.9 to 2.6. After HF, the daylight stability increased to over 50 days, and the UV absorbance of the DMSO extract, in the region of 260–350 nm, decreased to 0.1.

The effect of HF on the daylight stability and the UV absorbance of another solvent-dewaxed SWI 6 base stock, hydroisomerized at a low temperature of 290°C, is shown in Table 4.17. After solvent dewaxing, another SWI 6 base stock with a daylight stability of only 5 days was also found to have an increased UV absorbance. The UV absorbance of the DMSO extract, in the region of 260–350 nm, was found vary, from 0.8 to 2.4. After HF, the daylight stability increased to 25 days, and the UV absorbance of the DMSO extract, in the region of 260–350 nm, decreased to 0.2–0.4.

The effect of HF on the daylight stability and the UV absorbance of another solvent-dewaxed SWI 6 base stock, hydroisomerized at a low temperature of 290°C, is

TABLE 4.16
Effect of HF on the Daylight Stability and the UV Absorbance of SWI 6 Oil

SWI 6 Base Stock	Solvent Dewaxed	After HF
VI	148	148
Pour point (°C)	−21	−21
Daylight stability (days)	24	>50
UV of DMSO extract		
260–285 nm	2.6	0.1
290–299 nm	2.2	0.1
300–329 nm	1.6	0.1
330–350 nm	0.9	0.1

Source: Pillon, L.Z., *Petrol. Sci. Technol.*, 19(9–10), 1263, 2001. With permission.

TABLE 4.17
Effect of HF on the Daylight Stability and the
UV Absorbance of SWI 6 Oil

SWI 6 Base Stock	Solvent Dewaxed	After HF
VI	143	143
Pour point (°C)	−21	−21
Daylight stability (days)	5	25
UV of DMSO extract		
260–285 nm	2.4	0.3
290–299 nm	1.8	0.2
300–329 nm	1.4	0.4
330–350 nm	0.8	0.2

Source: Pillon, L.Z., *Petrol. Sci. Technol.*, 19(9–10), 1263, 2001. With permission.

TABLE 4.18
Effect of HF on the Daylight Stability and the
UV Absorbance of SWI 6 Oil

SWI 6 Base Stock	Solvent Dewaxed	After HF
VI	146	146
Pour point (°C)	−21	−21
Daylight stability (days)	9	38
UV of DMSO extract		
260–285 nm	2.7	0.3
290–299 nm	2.3	0.2
300–329 nm	1.8	0.3
330–350 nm	1.2	0.2

Source: Pillon, L.Z., *Petrol. Sci. Technol.*, 19(9–10), 1263, 2001. With permission.

shown in Table 4.18. After solvent dewaxing, SWI 6 base stock with a daylight stability of only 9 days was found to have an increased UV absorbance. The UV absorbance of the DMSO extract, in the region of 260–350 nm, was found vary, from 1.2 to 2.7. After HF, the daylight stability increased to 38 days and the UV absorbance of the DMSO extract, in the region of 260–350 nm, decreased to 0.2–0.3. The HF will remove the aliphatic trace of contaminants but not of aromatics. The UV absorbance and the daylight stability of hydrofined SWI 6 base stocks were found to vary.

The effects of hydrogen finishing temperature and pressure are highly dependent on the quality of feedstock. An increase in the hydroisomerization temperature was found to decrease the daylight stability of solvent-dewaxed SWI 6 base stocks to only 2–3 days and the use of HF was found to be totally ineffective in improving their daylight stability. The literature reports that hydrorefining can be used to stabilize the slack wax isomerate followed by fractionation to remove light ends and the solvent dewaxing (Sequeira, 1994). Hydrocracked base stocks were also reported to darken and form sediment on exposure to light.

The effect of hydrorefining on the properties and the UV stability of 500N hydrocracked base stock is shown in Table 4.19. The HC 500N base stock, with a conventional VI of 95, was reported to have a UV light stability of only 3 days and a sunlight stability of 12 days. After hydrorefining, a small increase in the VI, from 95 to 97, was reported with a decrease in the S content, a decrease in the TAN, and a decrease in the carbon residue from 0.12 to 0.07 wt%. After hydrorefining of HC 500N oil, the UV light stability increased to 14 days and the sunlight stability increased to over 30 days. The hydrocracked base stocks require stabilization, if used in some specialty products. The HC followed by solvent dewaxing and hydrorefining was reported to stabilize the hydrocracked base stocks. Other methods used to stabilize the hydrocracked base stock, after dewaxing, consist of clay treating and solvent refining. Some refiners use furfural extraction to stabilize the hydrocracked oils against discoloration and sludging (Sequeira, 1994).

The advantages and the disadvantages of the HC process are shown in Table 4.20. HF is a mild hydrofinishing process. The literature reports that an increase in the hydrogen pressure or temperature of an HF process will usually improve neutralization, desulfurization, denitrification, color, and stability of base stocks. An increase in the temperature above a certain maximum might degrade the color, the oxidation stability, and other properties (Sequeira, 1994). The color of base stocks is not critical when formulating lubricating oils containing additives, because color variation can result from the amount and the nature of additives. In finished lubricants, color

TABLE 4.19
Effect of Hydrorefining on the Properties and the UV Stabilities of 500N Hydrocracked Oil

Hydrocracked 500N Oil	Before Hydrorefining	After Hydrorefining
Viscosity at 100°F (SUS)	484	479
KV at 38°C (cSt)	104.8	103.7
VI	95	97
Pour point (°C)	−18	−18
Sulfur (wt%)	0.09	0.008
TAN (mg KOH/g)	0.03	<0.03
Carbon residue (wt%)	0.12	0.07
UV light stability (days)	3	14
Sunlight stability (days)	12	>30

Source: Reproduced from Sequeira, A. Jr., *Lubricant Base Oil and Wax Processing*, Marcel Dekker, Inc., New York, 1994. With permission.

TABLE 4.20

Advantages and Disadvantages of Lube Oil HC Process

HC Process Advantage	HC Process Disadvantage
Use of poor quality crude oils	Catalytically dewaxed bright stocks are hazy
Higher lube oil yields	Tendency to darken on exposure to light
Conversion of residual oils to distillate oils	Tendency to form sludge
Higher VI of lube oils	Additive solubility problems

Source: Reproduced from Sequeira, A. Jr., *Lubricant Base Oil and Wax Processing*, Marcel Dekker, Inc., New York, 1994. With permission.

has little significance except in the case of industrial and medicinal white oils, which need to have a good color stability (Pirro and Wessol, 2001). Organic nitrogen compounds are known to affect the color stability of lube oil base stocks and need to be removed (Gary and Handwerk, 2001). The literature reports on the effect of hydrorefining and hydroisomerization on the properties and the hydrocarbon composition of base stocks (Pillon, 2007).

The effect of hydrorefining and hydroisomerization on the properties and the hydrocarbon composition of base stocks are shown in Table 4.21. The use of hydrorefining decreases the aromatic and the heteroatom contents of mineral base stocks. Hydrorefined technical white oil, having a VI of 106, was found to contain 30.5 wt% of paraffins, which include *n*-paraffins and *iso*-paraffins, 68.8 wt% of 1–6

TABLE 4.21

Effect of Hydrorefining and Hydroisomerization on the Composition of Base Stocks

Processing	Hydrorefined	Hydroisomerized	Synthetic
Base Stocks	White Oil	SWI 6	PAO 6
KV at 40°C (cSt)	32.71	29.38	30.43
KV at 100°C (cSt)	5.55	5.77	5.77
VI	106	143	134
Total saturates (wt%)	>99	>99	100
n- and *iso*-Paraffins (wt%)	30.5	89.9	100
Naphthenes (wt%)	68.8	10.1	0
Aromatics (wt%)	0.7	<0.5	0
Sulfur (ppm)	<1	<1	0
Nitrogen (ppm)	<1	<1	0

Source: Pillon, L.Z. et al., Lubricating oil for inhibiting rust formation, U.S. Patent 5,397,487, 1995.

TABLE 4.22

Effect of Oxidation on the IFT of Hydroprocessed Base Stocks and Synthetic PAO 6

Processing	Hydrorefined	Hydroisomerized	Synthetic
Base Stocks	Technical White Oil	SWI 6	PAO 6
Fresh oils, IFT (mN/m)	39	42	46
ASTM D 2440 (120°C/24 h)			
Oxidized oils, IFT (mN/m)	20–30	10–20	10–20

Sources: Pillon, L.Z. and Asselin, A.E., Method for improving the demulsibility of base oils, U.S. Patent 5,282,960, 1994; Pillon, L.Z., *Petrol. Sci. Technol.*, 21(9–10), 1469, 2003.

ring naphthenes, 0.7 wt% of residual aromatics, and practically no S and N contents. The use of hydroisomerization decreases the aromatic, naphthenic, and heteroatom contents of base stocks. The hydroisomerized SWI 6 base stock, having a very high VI of 143, was found to contain 89.9% of paraffins, only 10.1 wt% of 1–6 ring naphthenes, a trace of aromatics, and practically no S and N contents. Synthetic PAO 6 fluid, having a VI of 134, contains 100 wt% of saturate content, which consists of different chemistry aliphatic hydrocarbons.

The effect of oxidation on the IFT of hydrorefined technical white oil, hydroisomerized SWI 6 base stock, and synthetic PAO 6 fluid is shown in Table 4.22. After oxidation, the IFT of hydrorefined technical white oil was found to decrease from 39 to 20–30 mN/m indicating the formation of surface active oxidation by-products capable of adsorbing at the oil/water interface. After oxidation under the same conditions, the IFT of non-hydrofined SWI 6 base stock was found to decrease, from 42 to 10–20 mN/m, indicating an increase in the formation of surface active oxidation by-products. While the use of hydroprocessing decreases the aromatic and heteroatom contents of petroleum-derived base stocks, the variation in feeds and the severity of processing conditions will affect their color, daylight stability, and oxidation stability. After oxidation, the IFT of synthetic PAO 6 was found to decrease, from 46 to 10–20 mN/m, also indicating an increase in the formation of surface active oxidation by-products. Hydrorefined white oil, non-hydrofined SWI 6 base stock, and synthetic PAO 6 fluid were oxidized under the same conditions of the ASTM D 2440 oxidation test, and the formation of surface active oxidation by-products capable of adsorbing at the oil/water interface was observed.

4.3 PERFORMANCE OF ANTIOXIDANTS IN DIFFERENT BASE STOCKS

The early literature reports that the presence of some naturally occurring compounds, containing sulfur and nitrogen functional groups, was found to improve the oxidation stability of some mineral base stocks. The discovery of sulfurized additives providing oxidation stability was followed by the similar properties identified in phenol, which led to the development of sulfurized phenols. Next certain amines and metal salts of phosphorus and sulfur containing acids were identified as imparting the oxidation stability (Migdal, 2003). Many different chemistry antioxidant classes, oil-soluble, have been developed for use in lubricating oils, which include sulfur compounds, phosphorus compounds, sulfur–phosphorus compounds, hindered phenols, aromatic amines, organo-alkaline earth salts, organo-zinc compounds, organo-copper compounds, and organo-molybdenum compounds. An extensive review of the chemistry and the performance of antioxidants was published (Migdal, 2003). The sulfur content of solvent-refined mineral base stocks is often regarded as an indicator of natural oxidation resistance; however, these natural inhibitors are inferior to synthesized commercial additives (Rasberger, 1997). The commercial additives, known as antioxidants or oxidation inhibitors, are required to improve the oxidation stability of different petroleum-derived base stocks and synthetic fluids.

The chemistry of organic molecules, found to increase the oxidation stability of lube oil base stocks and used as antioxidants, is shown in Table 4.23. The literature reports that the performance of antioxidants improves when used in HT, hydrocracked, and hydroisomerized base stocks. Synthetic PAO base stocks have good thermal stability, but they also require antioxidants to resist the oxidation. The failure of uninhibited synthetic PAO to outperform uninhibited mineral base stocks has been attributed to the presence of natural antioxidants in mineral base stocks. The lack of natural antioxidants in pure synthetic PAO was given as the rationale for the greater responsiveness of PAO to antioxidants than the responsiveness of mineral base stocks (Rudnick and Shubkin, 1999). The early literature reports on the effect of the aromaticity and the S-compounds on the oxidation stability of lube oil base stocks and their response to antioxidants. Their poor oxidation stability was

TABLE 4.23
Chemistry of Commercial Antioxidants

Antioxidant Classes	Examples
Sulfur compounds	Thioether
Phosphorus compounds	Tri-(di-*t*-butylphenyl) phosphite
Sulfur–phosphorus compounds	
Aromatic amine compounds	Dioctyldiphenylamine
Hindered phenolic compounds	2,6-Di-*t*-butyl-*p*-cresol
Organo-alkaline earth salt compounds	
Organo-zinc compounds	
Organo-copper compounds	
Organo-molybdenum compounds	

Source: Migdal, C.A., Antioxidants, in *Lubricant Additives Chemistry and Applications*, Rudnick, L.R., Ed., Marcel Dekker, Inc., New York, 2003. With permission.

TABLE 4.24

Effect of Phenolic Antioxidant on the Oxidation Stability of Different Base Stocks

Processing	Solvent Refined		Commercial	
Base Stocks	For Turbine Oil	For Automotive Oil	Oil #1	Oil #2
KV at 40°C	84.42	85.55	106.9	90.12
KV at 100°C	10.24	10.01	11.6	10.74
VI	102	96	95	103
Pour point (°C)	−4	−12	−15	−15
Paraffins (wt%)	26.9	23.4	14.8	20.9
Naphthenes (wt%)	46.1	53.8	59.4	76.7
Aromatics (wt%)	27.0	22.8	24.8	2.0
Sulfur (ppm)	6800	3500	170	12
Nitrogen (ppm)	42	120	5	11
ASTM D 943 (95°C)				
TAN (mg KOH/g)	2	2	2	2
TOST life (h)	3820	2200	2839	9000

Source: Galiano-Roth, A.S. and Page, N.M., *Lubr. Eng.*, 50(8), 659, 1993. With permission.

reported to be caused by a low S content (Lynch, 2007). The literature also reports on the variation in the oxidation stability of base stocks, having different S contents, and containing the same phenolic type antioxidant (Galiano-Roth and Page, 1993).

The effect of phenolic antioxidant on the oxidation stability of different base stocks is shown in Table 4.24. The solvent-refined base stock, for use in turbine oils, was reported to have a VI of 102, a dewaxed pour point of −4°C, and contained 6800 ppm of S and 42 ppm of N. Under the oxidation conditions of the ASTM D 943 test, the mineral oil, containing the phenolic antioxidant, reached a TAN of 2 after 3820 h. Another solvent-refined base stock, for use in automotive oils, was reported to have a VI of 96, a lower dewaxed pour point of −12°C, a lower S of 3500 ppm and a higher N content of 120 ppm. Under the same oxidation conditions of the ASTM D 943 test, the mineral oil, containing the same phenolic antioxidant, reached a TAN of 2 after only 2200 h indicating a significant decrease in the oxidation stability. The commercial base stock #1 was reported to have a VI of 95, a dewaxed pour point of −15°C, and contained a low S content of 170 ppm and a low N content of only 5 ppm. Under the oxidation conditions of the ASTM D 943 test, the commercial oil #1, containing the same phenolic antioxidant, reached a TAN of 2 after 2839 h. Another commercial base stock #2 was reported to have a VI of 103, a dewaxed pour point of −15°C, and contained a very low S content of 12 ppm and a low N content of 11 ppm. Under the same oxidation conditions of the ASTM D 943 test, the commercial oil #2, containing the same phenolic antioxidant, reached a TAN of 2 after 9000 h indicating a drastic increase in the oxidation stability. Despite low aromatic and heteroatom contents, the variation in the oxidation stability of different hydrocracked base stocks, containing phenolic antioxidant, was reported (Galiano-Roth and Page, 1993).

The effect of the phenolic antioxidant on the oxidation stability of hydrocracked base stocks from HC of LVGO is shown in Table 4.25. The hydrocracked oil #1, with a conventional VI of 98, was reported to have an aromatic content of <5 wt%, 24 ppm of S, and 1 ppm of N. Under the oxidation conditions of the ASTM D 943 test, the hydrocracked oil #1 containing the phenolic antioxidant reached a TAN of 2 after 4558 h. Another hydrocracked oil #2, with a conventional VI of 99, was reported to have an aromatic content of 10.5 wt%, only 2 ppm of S, and 16 ppm of N. Under the oxidation conditions of the ASTM D 943 test, the hydrocracked oil #2 containing, the same phenolic antioxidant, reached a TAN of 2 after 6500 h indicating an increase in the oxidation stability. The hydrocracked oil #3, with a higher VI of 103, was reported to have an aromatic content of 4.6 wt%, 8 ppm of S, and 3 ppm of N. Under the oxidation conditions of the ASTM D 943 test, the hydrocracked oil #3 containing the same phenolic antioxidant reached a TAN of 2 after 6912 h also indicating an increase in the oxidation stability. Another hydrocracked oil #4, with a conventional VI of 99, was reported to have an aromatic content of <5 wt%, 5 ppm of S, and 1 ppm of N. Under the oxidation conditions of the ASTM D 943 test, the hydrocracked oil #4 containing the same phenolic antioxidant reached a TAN of 2 after 6579 h also indicating an increase in the oxidation stability.

The effect of the phenolic antioxidant on the oxidation stability of hydrocracked base stocks from HC of HVGO is shown in Table 4.26. The hydrocracked base stock #1, with a conventional VI of 97, was reported to have an aromatic content of 14.7 wt%, 32 ppm of S, and 10 ppm of N. Under the oxidation conditions of the ASTM D 943 test, the hydrocracked oil #1 containing phenolic antioxidant reached a TAN of 2 after 3050 h. Another hydrocracked base stock #2, also with a conventional VI

TABLE 4.25
Effect of Phenolic Antioxidant on the TOST Life of HC Base Stocks

	HC of LVGO			
	Oil #1	Oil #2	Oil #3	Oil #4
Viscosity at 100°F (SUS)	134	424	378	146
KV at 38°C (cSt)	28.4	91.8	81.8	31.2
VI	98	99	103	99
Pour point (°C)	−18	−15	−15	−18
Aromatics (wt%)	<5	10.5	4.6	<5
Sulfur (ppm)	24	2	8	5
Nitrogen (ppm)	1	16	3	1
ASTM D 943 (95°C)				
TAN (mg KOH/g)	2	2	2	2
TOST life (h)	4558	6500	6912	6579

Source: Galiano-Roth, A.S. and Page, N.M., *Lubr. Eng.*, 50(8), 659, 1993. With permission.

TABLE 4.26

Effect of Phenolic Antioxidant on the TOST Life of Different HC Base Stocks

	HC of HVGO			
	Oil #1	Oil #2	Oil #3	Oil #4
Viscosity at 100°F (SUS)	426	448	437	141
KV at 38°C (cSt)	80.2	87.0	84.6	30.0
VI	97	97	98	98
Pour point (°C)	−15	−15	−15	−18
Aromatics (wt%)	14.7	8.4	8.2	<5
Sulfur (ppm)	32	26	5	9
Nitrogen (ppm)	10	13	11	1
ASTM D 943 (95°C)				
TAN (mg KOH/g)	2	2	2	2
TOST life (h)	3050	6340	6775	3980

Source: Galiano-Roth, A.S. and Page, N.M., *Lubr. Eng.*, 50(8), 659, 1993. With permission.

of 97, was reported to have an aromatic content of 8.4 wt%, 26 ppm of S, and 13 ppm of N. Under the oxidation conditions of the ASTM D 943 test, the hydrocracked oil #2 containing the same phenolic antioxidant reached a TAN of 2 after 6340 h indicating an increase in the oxidation stability. The hydrocracked base stock #3, with a conventional VI of 98, was reported to have an aromatic content of 8.2 wt%, 5 ppm of S, and 11 ppm of N. Under the oxidation conditions of the ASTM D 943 test, the hydrocracked oil #3 containing the same phenolic antioxidant also reached a TAN of 2 after 6775 h indicating an increase in the oxidation stability. The hydrocracked base stock #4, also with a conventional VI of 98, was reported to have an aromatic content of <5 wt%, 9 ppm of S, and 1 ppm of N. However, under the oxidation conditions of the ASTM D 943 test, the hydrocracked oil #4 containing the same phenolic antioxidant reached a TAN of 2 after only 3980 h indicating a decrease in the oxidation stability. The variation in the feed and the severity of HC process was found to lead to a variation in the oxidation stability of base stocks containing the same phenolic type antioxidant.

The effect of antioxidant chemistry on the oxidation stability of different petroleum-derived base stocks and synthetic PAO fluids is shown in Table 4.27. To shorten the time and lower the cost of testing, many different laboratory bench tests were developed and are used to assess the oxidation stability of lubricating oils. The ASTM D 2272 test method is used to compare the oxidation stability of steam turbine oils with the same composition in terms of base stocks and additives. The test oil, water, and copper catalyst are placed in a vessel equipped with a pressure gauge. The vessel is charged with oxygen, placed in an oil bath at 150°C and rotated. The time, in minutes, required to reach a specific drop in gauge pressure is the rotary bomb oxidation

TABLE 4.27
Effect of Antioxidant Chemistry on the Oxidation
Stability of Different Base Stocks

API Category	API I	API II	API III	API IV
Base Stocks	Mineral	HC Oil	SWI	PAO
ASTM D 2272 (150°C)				
RBOT life (min)				
Uninhibited oil	<50	<50	<50	<50
Hindered phenol (0.5 wt%)	100	200	200	200
Diphenylamine (0.5 wt%)	450	1300	1700	1800

Source: Migdal, C.A., Antioxidants, in *Lubricant Additives Chemistry and Applications*, Rudnick, L.R., Ed., Marcel Dekker, Inc., New York, 2003. With permission.

test (RBOT) life of the test sample. Following the ASTM D 2272 test, without an antioxidant, the RBOT life of mineral, hydrocracked, hydroisomerized, and synthetic PAO base stocks was found to be less than 50 min. The use of 0.5 wt% of hindered phenol, C_7–C_9 branched alkyl ester of 3, 5-di-*tert*-butyl-4-hydroxy-dihydrocinnamic acid, was found to increase the RBOT life of mineral base stock to 100 min. The use of 0.5 wt% of diphenylamine, a mixture of butyl and octyl diphenylamines, was found to further increase the RBOT life of mineral base stocks to 450 min. The use of the same 0.5 wt% of hindered phenol further increased the RBOT life of hydrocracked, hydroisomerized, and synthetic PAO base stocks to 200 min. However, the use of the same 0.5 wt% of diphenylamine mixture was found to further increase the RBOT life of hydrocracked base stock to 1300 min, hydroisomerized oil to 1700 min, and synthetic PAO fluid to 1800 min.

The chemistry and the product names of some commercial antioxidants are shown in Table 4.28. Many different chemistry molecules, with antioxidant properties, have been commercialized and used in different lubricant products. At temperatures below 93°C, hindered alkylated phenols and aromatic amines are used to prevent oil oxidation (Pirro and Wessol, 2001). Some antioxidants, beneficial at low temperatures, were reported to cause heavy deposits when they break down at higher temperatures (Rasberger, 1997). The literature also reports on the growing demand for aminic type antioxidants with a free radical scavenging ability in the temperature range from 100°C to 250°C (*Chemical News*, 2003). The use of antioxidants only delays the oxidation process. The oxidation degradation of base stocks leads to acids, which cause corrosion and the formation of insolubles known as sludge (Rasberger, 1997). With an increase in the oxidation time, a gradual increase in the TAN indicating the formation of acidic oxidation by-products and an increase in the sludge content will be observed. While hundreds of different antioxidants are available, their uses in different base stocks will lead to variation in their oxidation stability.

TABLE 4.28
Chemistry and Product Names of Some
Commercial Antioxidants

Antioxidant Chemical Name	Commercial Product
2,6-Di-*t*-butyl-*p*-cresol	Additin 7110
Phenol derivative sterically hindered	Additin 7115
Dioctyldiphenylamine	Additin 7001
Phenyl-alfa-naphthylamine	Additin 7130
1,2-Dihydro-4-trimethylquinoline	Additin 7010
2,6-Di-*t*-butyl-phenol	Hitec 4701
Alkylated diphenylamine	Hitec 4777
Hindered bis-phenol	Irganox L67
2,6-Di-*t*-butyl-*p*-cresol	Naugard BHT
Phenyl-alfa-naphthylamine	Naugard PANA
Dioctyl diphenylamine	Naugalube 438
Styrenated diphenylamine	Naugalube 635
2,6-Di-*t*-butyl-*p*-cresol	Vanlube PCX
Dinonyl diphenylamine	Vanlube DND
Dioctyl diphenylamine	Vanlube 81

Source: Migdal, C.A., Antioxidants, in *Lubricant Additives Chemistry and Applications*, Rudnick, L.R., Ed., Marcel Dekker, Inc., New York, 2003. With permission.

4.4 ANTIOXIDANTS' SYNERGISM

The good oxidation stability of lubricants is necessary to maximize their service life by minimizing the formation of acids and sludge. The literature reports that the undesirable molecules responsible for poor color, poor thermal and oxidation stability, poor water separation properties, and poor electrical insulating properties of mineral base stocks tend to be nitrogen, oxygen, and to lesser extent sulfur containing compounds. The oxidation degradation of base stocks was reported to cause viscosity changes and the loss of electrical resistivity of some lubricating oils (Rasberger, 1997). The patent literature reports on the effect of different heterocyclic N-compounds on the oxidation stability of mineral electrical insulating oils, containing phenolic type antioxidant. The electrical insulating oils were reported to be mineral oils, having a low viscosity of 8.68 cSt at 40°C, which contained 2,6-di-*t*-butyl-*p*-cresol antioxidant and pour point depressants, such as chlorinated wax/naphthalene diluted in mineral oil (Wright, MacDonald, and MacAlpine, 1984). The heterocyclic N-molecule, 7H-imidazo(4,5-d)pyrimidine, is known as purine, while 6-amino-8-azapurine is known as 8-azaadenine (Sigma-Aldrich, 2007–2008).

The effect of 2,6-di-*t*-butyl-*p*-cresol antioxidant and different heterocyclic N-compounds on the oxidation stability of low viscosity mineral oils is shown in

TABLE 4.29

Effect of Butyl-*p*-Cresol and Heterocyclic N-Compounds on Oxidation Stability

	Low Viscosity Mineral Oils		
	Oil #1	Oil #2	Oil #3
2,6-Di-*t*-butyl-*p*-cresol (wt%)	0.06		0.06
Purine (wt%)		0.01	
8-Azaadenine (wt%)			0.05
ASTM D 2272 (150°C)			
RBOT life (min)	109	190	440

Source: Wright, P.G. et al., Selected heteroaromatic nitrogen compounds as antioxidant/metal deactivators/electrical insulators in hydrocarbon compositions, CA Patent 1,172,237, 1984.

Table 4.29. The low viscosity mineral oil #1, containing only 0.06 wt% of 2,6-di-*t*-butyl-*p*-cresol antioxidant, was reported to have the RBOT life of 109 min. The more recent literature reports that high viscosity 900N mineral base stock, containing 0.8 wt% of 2,6-di-*t*-butyl-*p*-cresol, was found to have the RBOT life of 97 min (Wang and Li, 2000). The low viscosity mineral oil #2, containing 0.01 wt% of purine, was found to have the RBOT life of 190 min indicating a significant improvement in the oxidation stability. Another low viscosity mineral oil #3, containing 0.06 wt% of 2,6-di-*t*-butyl-*p*-cresol and 0.05 wt% of 8-azaadenine, was found to have the RBOT life of 440 min indicating a drastic improvement in oxidation stability. While many different oxidation tests are used to determine the oxidation stability of lubricants, the oxidation stability of mineral transformer oils is assessed by following the ASTM D 2440 test method. The oxidation stability of oil is measured, at 120°C and in the presence of Cu catalyst, by the amount of acid products and the sludge formed. After a specific oxidation time, the TAN, the IFT, and the sludge content of oxidized oils are measured.

The effect of purine on the TAN, the IFT, and the sludge content of oxidized low viscosity mineral oils, containing 2,6-di-*t*-butyl-*p*-cresol antioxidant, is shown in Table 4.30. After 24 h of oxidation, the low viscosity mineral oil #1 containing 0.06 wt% of 2,6-di-*t*-butyl-*p*-cresol antioxidant and 0.07 wt% of the pour point depressant, was found to have a TAN of 1.46 mg KOH/g and a low IFT of 10.4 mN/m indicating an increase in the surface active acidic by-products. After 164 h of oxidation, the TAN value was found to further increase to 3.46 mg KOH/g and 0.8 wt% of sludge was measured. After 24 h of oxidation, the low viscosity mineral oil #2, containing 0.06 wt% of 2,6-di-*t*-butyl-*p*-cresol antioxidant, 0.001 wt% of purine, and 0.07 wt% of pour point depressant, was found to have a TAN of 0.01 mg KOH/g and a high IFT of 31.4 mN/m indicating no increase in the acidic oxidation by-products. After 164 h of oxidation, the TAN value increased to only 0.46 mg KOH/g and only 0.05 wt% of sludge was measured.

TABLE 4.30

Effect of Purine on the Oxidation of Mineral Oils Containing 2,6-Di-t-Butyl-p-Cresol

	Low Viscosity Mineral Oil		
	Oil #1	Oil #2	Oil #3
2,6-Di-*t*-butyl-*p*-cresol (wt%)	0.06	0.06	0.06
Purine (wt%)		0.001	0.003
Pour point depressant (wt%)	0.07	0.07	0.07
ASTM D 2440 (120°C/24h)			
TAN (mg KOH/g)	1.46	0.01	0.01
IFT (mN/m)	10.4	31.4	31.4
ASTM D 2440 (120°C/164h)			
TAN (mg KOH/g)	3.46	0.46	0.06
Sludge (wt%)	0.8	0.05	0.01

Source: Wright, P.G. et al., Selected heteroaromatic nitrogen compounds as antioxidant/metal deactivators/electrical insulators in hydrocarbon compositions, CA Patent 1,172,237, 1984.

After 24h of oxidation, another low viscosity mineral oil #3, containing 0.06 wt% of 2,6-di-*t*-butyl-*p*-cresol antioxidant, 0.003 wt% of purine, and 0.07 wt% of pour point depressant, was also found to have a TAN of 0.01 mg KOH/g and a high IFT of 31.4 mN/m indicating absence of the acidic oxidation by-products. After 164h of oxidation, the TAN value increased to only 0.06 mg KOH/g, and a further decrease in the sludge content to 0.01 wt% was reported. An increase in the purine content, from 0.001 to 0.003 wt%, further increased the oxidation stability of mineral oils containing phenolic type antioxidant.

The effect of thiopurine and benzotriazole derivatives on the oxidation stability of low viscosity mineral oils, containing 2,6-di-*t*-butyl-*p*-cresol antioxidant, is shown in Table 4.31.

After 24h of oxidation, the low viscosity mineral oils #1–2, containing 0.06 wt% of 2,6-di-*t*-butyl-*p*-cresol antioxidant, 0.003–0.005 wt% of thiopurine derivative, and 0.07 wt% of pour point depressant, were found to have a TAN of 0.02–0.03 mg KOH/g and an IFT of 27.8–29.6 mN/m indicating a small increase in the acidic oxidation by-products. After 164h of oxidation, their TAN values increased to 0.3–0.6 mg KOH/g and 0.04–0.08 wt% of sludge was formed. After 24h of oxidation, the low viscosity mineral oil #3, containing 0.06 wt% of 2,6-di-*t*-butyl-*p*-cresol antioxidant, 0.005 wt% of benzotriazole, and 0.07 wt% of pour point depressant, was found to have a TAN of 0.01 mg KOH/g and an IFT of 30.2 mN/m indicating the absence of surface active oxidation by-products. However, after 164h of oxidation, the TAN drastically increased to 2.64 mg KOH/g and 0.23 wt% of sludge was reported. After 24h of oxidation, the low viscosity mineral oil #4, containing 0.06 wt% of 2,6-di-*t*-butyl-*p*-cresol antioxidant, 0.015 wt% of benzotriazole, and 0.07 wt% of pour point

TABLE 4.31

Effect of Thiopurine and Benzotriazole on the Performance of 2,6-Di-*t*-Butyl-*p*-Cresol

	Low Viscosity Mineral Oils			
	Oil #1	Oil #2	Oil #3	Oil #4
2,6-Di-*t*-butyl-*p*-cresol (wt%)	0.06	0.06	0.06	0.06
Thiopurine derivative (wt%)	0.003	0.005		
Benzotriazole derivative (wt%)			0.005	0.015
Pour point depressant (wt%)	0.07	0.07	0.07	0.07
ASTM D 2440 (120°C/24 h)				
TAN (mg KOH/g)	0.02	0.03	0.01	0.01
IFT (mN/m)	29.6	27.8	30.2	27.9
ASTM D 2440 (120°C/164 h)				
TAN (mg KOH/g)	0.6	0.27	2.64	1.53
Sludge (wt%)	0.08	0.04	0.23	0.14

Source: Wright, P.G. et al., Selected heteroaromatic nitrogen compounds as anti-oxidant/metal deactivators/electrical insulators in hydrocarbon compositions, CA Patent 1,172,237, 1984.

depressant, was found to have a TAN of 0.01 mg KOH/g and a IFT of 27.9 mN/m. After 164 h of oxidation, the TAN increased to 1.53 mg KOH/g and 0.14 wt% of sludge was reported. The use of thiopurine and benzotriazole derivatives, while effective in improving the performance of 2,6-di-*t*-butyl-*p*-cresol antioxidant, was found to be less effective than the use of the heterocyclic N-compound, such as purine.

The effect of purine and its derivatives on the TAN and the sludge content of oxidized mineral oils, containing 2,6-di-*t*-butyl-*p*-cresol antioxidant, is shown in Table 4.32. After 164 h of oxidation, the low viscosity mineral oil #1, containing 0.06 wt% of 2,6-di-*t*-butyl-*p*-cresol antioxidant and 0.07 wt% of pour point depressant, was reported to have a TAN of 2.1 mg KOH/g and 0.39 mg of sludge. Another oxidized mineral oil #2, containing 0.06 wt% of 2,6-di-*t*-butyl-*p*-cresol antioxidant, 0.005 wt% of purine, and 0.07 wt% of pour point depressant, was found to have a TAN of 0.01 mg KOH/g and a very low sludge content of 0.01 mg. After 164 h of oxidation, the oxidized mineral oils #3–4, containing 0.06 wt% of 2,6-di-*t*-butyl-*p*-cresol antioxidant, 0.005 wt% of different purine derivatives #1–2, and 0.07 wt% of pour point depressant, were found to have an increased TAN of 0.04–0.06 mg KOH/g and a sludge content of 0.01–0.02 mg thus indicating no improvement over the use of purine. The addition of heterocyclic N-compounds, such as purine, was found to improve the performance of phenolic type antioxidant, 2,6-di-*t*-butyl-*p*-cresol, by decreasing the formation of acidic by-products and sludge. In high temperature applications, antioxidants and additional molecules, known as metal deactivators, are used to protect the metal surface.

The chemistry and the product names of some commercial metal deactivators are shown in Table 4.33. Metal deactivators prevent the corrosion of copper and copper

TABLE 4.32

**Effect of Purines on Oxidized Mineral Oils Containing
2,6-Di-*t*-Butyl-*p*-Cresol**

	Low Viscosity Mineral Oil			
	Oil #1	Oil #2	Oil #4	Oil #3
2,6-Di-*t*-butyl-*p*-cresol (wt%)	0.06	0.06	0.06	0.06
Purine (wt%)		0.005		
Purine derivative #1 (wt%)			0.005	
Purine derivative #2 (wt%)				0.005
Pour point depressant (wt%)	0.07	0.07	0.07	0.07
ASTM D 2440 (120°C/164 h)				
TAN (mg KOH/g)	2.1	0.01	0.04	0.06
Sludge (wt%)	0.39	0.01	0.01	0.02

Source: Wright, P.G. et al., Selected heteroaromatic nitrogen compounds as antioxidant/metal deactivators/electrical insulators in hydrocarbon compositions, CA Patent 1,172,237, 1984.

TABLE 4.33

**Chemistry and Product Names of Some
Commercial Metal Deactivators**

Metal Deactivators	Product Name
4,5,6,7-Tetrahydrobenzotriazole	Reomet SBT
1-Dialkylaminomethyl benzotriazole	Reomet 38
Triazole derivative	Irgamet 30
Tolutriazole derivative	Irgamet 39
Water soluble tolutriazole derivative	Irgamet 42
Tolutriazole in oil	Vanlube 887
Tolutriazole in ester	Vanlube 887E
2,5-Dimercapto-1,3,4-thiadiazole	Vanlube 881-P

Source: Migdal, C.A., Antioxidants, in *Lubricant Additives Chemistry and Applications*, Rudnick, L.R., Ed., Marcel Dekker, Inc., New York, 2003. With permission.

alloys, such as bronze and brass. They function by adsorbing onto the metal surface of copper alloys, forming a protective layer that inhibits attack by acids formed during the oxidation of the bulk lubricant. Conversely, the metal deactivators can protect the oil from the catalytic effects of the copper alloy, which could accelerate bulk oil oxidation (Lee and Harris, 2003). There are two types of metal deactivators, namely, chelating agents and film forming agents. The chelating agents form a stable complex with metal ions, reducing their catalytic activity and showing their antioxidant effect. Disalicylidene propylene diamine is an example of the chelating agent. The film forming agents coat the metal surface and prevent the metal ions from entering the oil and minimize corrosion by physically restricting the access of the corrosive acids to the metal surface. Derivatives of benzotriazole are examples of a film forming agent (Migdal, 2003).

The formation of oxidation by-products increases with an increase in the temperature, the amount of oxygen contacting the oil, and the catalytic effects of metals. In most cases, the lubricant products contain a synergistic blend of two different antioxidants, such as hindered phenol and aromatic amine derivatives, to increase their oxidation resistance. Some high temperature lubricant applications require the use of a more complex antioxidant system, which includes one or two antioxidants, and metal deactivator or metal passivator. The metal deactivators used to protect the oil from the catalytic effects of the metals are called copper passivators or metal passivators (Lee and Harris, 2003). While the use of different chemistry oxidation inhibitors cannot entirely prevent oil oxidation, a significant improvement in oxidation resistance can be achieved.

REFERENCES

American Petroleum Institute, *Engine Oil Licensing and Certification System, API 1509*, 15th ed., API, Washington, DC, 2002.

Bardasz, E.A. and Lamb, G.D., Additives for crankcase lubricant applications, in *Lubricant Additives Chemistry and Applications*, Rudnick, L.R., Ed., Marcel Dekker, Inc., New York, 2003.

Bijwaard, H.M.J. and Morcus, A., Lubricating base oil compositions, CA Patent 1,185,962, 1985.

Braun, J. and Omeis, J., Additives, in *Lubricants and Lubrications*, Mang, T. and Dresel, W., Eds., Wiley-VCH GmbH, Weinheim, Germany, 2001.

Chemical News, Petroleum additives plans to increase capacity at Canadian facility, March 2003.

Deninson, G.H., *Industrial and Engineering Chemistry*, 36, 477 (1944).

Galiano-Roth, A.S. and Page, N.M., *Lubrication Engineering*, 50(8), 659 (1993).

Gary, J.H. and Handwerk, G.E., *Petroleum Refining, Technology and Economics*, 4th ed., Marcel Dekker, Inc., New York, 2001.

Henderson, H.E., Chemically modified mineral oils, in *Synthetics, Mineral Oils, and Bio-Based Lubricants Chemistry and Technology*, Rudnick, L.R., Ed., CRC Press, Boca Raton, FL, 2005.

Ivanov, A.V., Lazarev, N.P., and Yaushev, R.G., *Chemistry and Technology of Fuels and Oils*, 36(50), 352 (2000).

Lake, D.B. and Hoh, G.L.K., *Journal of the American Oil Chemist's Society*, 40, 628 (1963).

Lee, F.L. and Harris, J.W., Long-term trends in industrial lubricant additives, in *Lubricant Additives Chemistry and Applications*, Rudnick, L.R., Ed., Marcel Dekker, Inc., New York, 2003.

Lynch, T.R., *Process Chemistry of Lubricant Base Stocks*, CRC Press, Boca Raton, FL, 2007.

Migdal, C.A., Antioxidants, in *Lubricant Additives Chemistry and Applications*, Rudnick, L.R., Ed., Marcel Dekker, Inc., New York, 2003.

Pillon, L.Z., *Petroleum Science and Technology*, 19(9&10), 1263 (2001).

Pillon, L.Z., *Petroleum Science and Technology*, 20(1&2), 101 (2002).

Pillon, L.Z., *Petroleum Science and Technology*, 20(3&4), 357 (2002).

Pillon, L.Z., *Petroleum Science and Technology*, 21(9&10), 1469 (2003).

Pillon, L.Z., *Interfacial Properties of Petroleum Products*, CRC Press, Boca Raton, FL, 2007.

Pillon, L.Z. and Asselin, A.E., Method for improving the demulsibility of base oils, U.S. Patent 5,282,960, 1994.

Pillon, L.Z., Reid, L.E., and Asselin, A.E., Lubricating oil for inhibiting rust formation, U.S. Patent 5,397,487, 1995.

Pirro, D.M. and Wessol, A.A., *Lubrication Fundamentals*, 2nd ed., Marcel Dekker Inc., New York, 2001.

Rasberger, M., Oxidative degradation and stabilisation of mineral oil based lubricants, in *Chemistry and Technology of Lubricants*, 2nd ed., Mortier, R.M. and Orszulik, S.T., Eds, Blackie Academic & Professional, London, 1997.

Rudnick, L.R. and Shubkin, R.L., Poly(alfa-olefins), in *Synthetic Lubricants and High-Performance Functional Fluids*, 2nd ed., Marcel Dekker Inc., New York, 1999.

Sequeira, A. Jr., *Lubricant Base Oil and Wax Processing*, Marcel Dekker, Inc., New York, 1994.

Shell, http://www.shell-lubricants.com (accessed 2001).

Sigma-Aldrich, *Handbook of Fine Chemicals*, Sigma-Aldrich Company, St. Louis, MO, 2007–2008.

Wang, Y. and Li, R., *Petroleum Science and Technology*, 18(7&8), 965 (2000).

Wang, Y. et al., *Petroleum Science and Technology*, 19(7&8), 923 (2001).

Wright, P.G., MacDonald, J.M., and MacAlpine, G.A., Selected heteroaromatic nitrogen compounds as antioxidant/metal deactivators/electrical insulators in hydrocarbon compositions, CA Patent 1,172,237, 1984.

5 Interfacial Properties of Base Stocks

5.1 FOAMING

5.1.1 Viscosity Effect

The literature reports that pure liquids allow entrained air to escape with no delay and no foam is produced (Ross, 1987). The pure organic liquids such as heptane, octane, and toluene do not foam when mixed with air. Crude oils are mixtures of different hydrocarbons and heteroatom-containing molecules and foam when mixed with gas or air (Poindexter, 2002). Gulf of Mexico crude oils with different viscosity containing varying asphaltene content were found to foam. Their foaming tendency was reported to increase with an increase in their viscosity. Arab Heavy crude oil, containing asphaltenes, was found to have a high foaming tendency. Another heavy mineral oil containing no asphaltenes was also reported to foam; however, its foaming tendency was lower. With an increase in the viscosity and an increase in the asphaltene content, the foaming tendency of crude oils decrease. Very high viscosity crude oils, containing asphaltenes, do not foam indicating the temperature–viscosity effect. The ASTM D 892 foam test is used to measure the foaming characteristics of base stocks and lubricating oils at 24°C and 93.5°C. The foaming tendency indicates the volume of foam that is generated after bubbling air through the oil for 5 min at a constant rate and temperature. The foam stability is the volume of foam still left 10 min after the bubbling of air through the oil has stopped.

The foaming resistance of different viscosity synthetic PAO fluids is shown in Table 5.1. At the temperatures of 24°C and 93.5°C, different viscosity synthetic PAO base stocks were found to have a total resistance to foaming. Foaming occurs when air mixes with petroleum oils. Crude oils foam when mixed with air. Lube oil distillates also foam when mixed with air. The early literature reports on the viscosity effect on the foaming of mineral base stocks (Tourret and White, 1952).

In industrial and automotive applications, the lubricating oils are exposed to varying temperatures. The viscosity is affected by the temperature and the literature reports on the effect of the temperature on the foaming of paraffinic and naphthenic lubricating oils. With an increase in the temperature from −10°C to 150°C, their foaming was reported to decrease and no foaming was observed at high temperatures (Hubmann and Lanik, 1988). The early literature reports that the foaming tendency and the foam stability of different lube oil distillates and residual oils were the same when compared at the same viscosity (Kichkin, 1966).

TABLE 5.1

Foaming Resistance of Different Viscosity Synthetic PAO Fluids

	Synthetic Fluids		
	PAO 4	PAO 6	PAO 8
ASTM D 892 foam test			
Seq. I at 24°C (mL)	0	0	0
Seq. II at 93.5°C (mL)	0	0	0

Source: Pillon, L.Z., *Interfacial Properties of Petroleum Products*, CRC Press, Boca Raton, FL, 2007. With permission.

TABLE 5.2

Variation in the Viscosity of Different Lube Oil Distillates and Residual Oils

Feedstock	Treatment	Temperature (°C)	Viscosity (cSt)
Distillate	Solvent	50	6.5
Distillate	Acid	50	8.5
Distillate-spindle oil	Acid	50	12
Distillate-spindle oil	Acid	50	20
Distillate-machine oil	Acid	50	50
Distillate	Solvent	100	11
Residual oil	Solvent	100	16
Residual oil	Solvent	100	34
Residual oil	Acid	100	22

Source: Kichkin, G.I., *Chem. Technol. Fuels Oils*, 4, 49, 1966. With permission.

The variation in the viscosity of different lube oil distillates and residual oils at different temperatures is shown in Table 5.2. The ASTM D 2896 test method is used to determine the basic nitrogen (BN) content of petroleum oils by titration with perchloric acid. The basic constituents of petroleum include BN compounds and other compounds such as salts of weak acids (soaps), basic salts of polyacidic bases, and salts of heavy metals. The patent literature reports that the foaming tendency of lube oil distillates, containing polar BN content, varies depending on their viscosity and the temperature subsequent to the ASTM D 892 test (Butler and Henderson, 1990).

The effect of the temperature on the foaming tendency of different viscosity lube oil distillates is shown in Table 5.3. At the temperature of 24°C, a lower viscosity 60N lube oil distillate was reported to have a foaming tendency of 120–160 mL, which decreased to 30 mL at a higher temperature of 93.5°C.

TABLE 5.3
Effect of Temperature on the Foaming Tendency of Different Viscosity Distillates

	Lube Oil Distillates	
	60N	150N
BN content	128 ppm	11 ppm
ASTM D 892 foam test		
Foaming tendency		
Seq. I at 24°C (mL)	160	535
Seq. II at 93.5°C (mL)	30	20
Seq. III at 24°C (mL)	120	470

Source: Butler, K.D. and Henderson, H.E., Adsorbent processing to reduce base stock foaming, U.S. Patent 4,600,502, 1990. With permission.

At the same temperature of 24°C, a higher viscosity 150N lube oil distillate was reported to have a higher foaming tendency of 470–535 mL, which decreased to 20 mL at a higher temperature of 93.5°C. With an increase in the viscosity of lube oil distillates, their BN content was reported to decrease, while their foaming tendency at 24°C increased thus confirming the viscosity effect on foaming. The patent literature reports on the effect of temperature on the foaming tendency of different viscosity solvent-refined mineral base stocks, ranging from a low viscosity 100N oil to a high viscosity bright stock (Butler and Henderson, 1990).

The effect of temperature on the foaming tendency of different viscosity solvent-refined base stocks and a high viscosity bright stock is shown in Table 5.4. At the temperature of 24°C, a low viscosity solvent-refined 100N base stock was reported

TABLE 5.4
Effect of Temperature on the Foaming Tendency of Different Viscosity Base Stocks

	Solvent-Refined Oils		
	100N	150N	Bright Stock
ASTM D 892			
Foaming tendency			
Seq. I at 24°C (mL)	385	535	30
Seq. II at 93.5°C (mL)	25	20	220
Seq. III at 24°C (mL)	370	470	120

Source: Butler, K.D. and Henderson, H.E., Adsorbent processing to reduce base stock foaming, U.S. Patent 1,266,007, 1990. With permission.

to have a foaming tendency of 370–385 mL, which decreased to 25 mL at a higher temperature of 93.5°C. At the same temperature, a higher viscosity solvent-refined 150N base stock was reported to have a higher foaming tendency of 470–535 mL, which also decreased to 20 mL at a higher temperature of 93.5°C. However, at the temperature of 24°C, a very high viscosity bright stock was reported to have a lower foaming tendency of 30–120 mL, which increased to 220 mL at a higher temperature of 93.5°C. Low viscosity base stocks have a low foaming tendency at high temperatures because their viscosity is very low and the foams are too unstable to be measured. Very high viscosity bright stocks have a low foaming tendency at low temperatures because their viscosity is too high and they are too viscous to foam. The patent literature also reports on the use of different adsorbents to decrease their surface-active content and their tendency to foam. The mineral base stocks were mixed with an equal volume of 1:1 heptane/toluene (v/v) to reduce their viscosity and, at the temperature of 21°C; the diluted oils were stirred with selected adsorbents for 3 h, followed by filtration. After the treatment, only 1 wt% of base stocks was reported to adsorb onto adsorbents (Butler and Henderson, 1990).

The effect of different adsorbents on the BN content and the foaming tendency of diluted mineral base stocks are shown in Table 5.5. The use of calcium oxide adsorbent having basic properties was found not effective in reducing the BN content; however, a decrease in the foaming tendency of 150N mineral base stock was observed. The use of alumina adsorbent was reported effective in reducing the BN content of 150N mineral base stock, from 28 to 9 ppm, and a practically total resistance to foaming was observed. The use of basic type alumina adsorbent was found to decrease the foaming tendency of mineral base stocks without removing all BN molecules indicating a selective adsorption of the acidic type surface-active molecules. The use of activated calcium oxide in higher viscosity 600N mineral base stock did not change the BN content; however, a drastic decrease in the foaming tendency was observed. The use of fluorosil adsorbent, which is an activated magnesium silicate,

TABLE 5.5

Effect of Adsorbents on the BN Content and Foaming Tendency of Mineral Base Stocks

Diluted Mineral Base Stocks	BN (ppm)	Foam at 24°C Seq. I (mL)	Foam at 93.5°C Seq. II (mL)	Foam at 24°C Seq. III (mL)
Diluted 150N feed	28	355	80	435
Calcium oxide	26	40	20	80
Activated calcium oxide	27	80	20	80
Alumina	9	0	10	0
Diluted 600N feed	42	460	100	600
Calcium oxide	42	5	25	10
Fluorosil (MgO_3Si)	0	0	20	0

Source: Butler, K.D. and Henderson, H.E., Adsorbent processing to reduce base stock foaming, U.S. Patent 4,600,502, 1990. With permission.

TABLE 5.6
Effect of HF on the Composition and Foaming Tendency of 150N Base Stock

Solvent-Refined 150N Oil	Before HF	After HF
KV at 40°C (cSt)	29.64	29.6
KV at 100°C (cSt)	4.99	4.99
VI	89	90
Saturates (wt%)	78.0	79.1
Aromatics (wt%)	22.0	20.9
Sulfur (wt%)	0.33	0.25
Nitrogen (ppm)	72	57
ST (mN/m)	33.5	33.5
ASTM D 892 foam test		
Seq. I at 24°C (mL)	310/0	305/0
Seq. II at 93.5°C (mL)	35/0	30/0

Source: Reproduced from Pillon, L.Z., *Interfacial Properties of Petroleum Products*, CRC Press, Boca Raton, FL, 2007. With permission.

removed the BN content and a practically total resistance to foaming was reported. The use of adsorbents was found more effective in decreasing the foaming tendency of higher viscosity 600N base stock indicating the presence of different chemistry or different molecular weight surface-active contaminants.

The effect of hydrofining (HF) on the composition and the foaming tendency of a solvent-refined 150N base stock are shown in Table 5.6. Solvent-dewaxed 150N mineral base stock, having a VI of 89, was found to contain 22 wt% of aromatics, 0.33 wt% of S, 72 ppm of N, and have a surface tension (ST) of 33.5 mN/m measured at room temperature (RT). Following the ASTM D 892 foam test, it was found to have a high foaming tendency of 310 mL at 24°C and a low foaming tendency of 35 mL at 93.5°C. For low viscosity base stocks, their foam stability is less than 10 min. After HF, a small increase in the VI to 90, with a decrease in the aromatic content to 20.9 wt% was observed indicating that some hydrogen saturation of aromatics was taking place. Hydrofined 150N base stock was also found to have a lower sulfur content of 0.25 wt% and a lower nitrogen content of 57 ppm; however, no change in the ST and no significant decrease in the foaming tendency were observed. The effect of different temperatures on the foaming tendency of low viscosity 100N and 150N base stocks indicated that solvent-refined base stocks foam in a specific viscosity range. A lower viscosity 100N base stock was found to have a maximum foaming tendency at −10°C while a higher viscosity 150N base stock was found to have a maximum foaming tendency at a higher temperature of 0°C where their viscosity was found to be almost the same.

The effect of HF on the composition and the foaming characteristics of a solvent-refined 600N base stock is shown in Table 5.7. A solvent-dewaxed 600N mineral

TABLE 5.7

Effect of HF on the Composition and Foaming Characteristics of 600N Oil

Solvent-Refined 600N Oil	Before HF	After HF
KV at 40°C (cSt)	112.33	108.36
KV at 100°C (cSt)	11.82	11.68
VI	93	95
Saturates (wt%)	77.7	78.9
Aromatics (wt%)	22.3	21.1
Sulfur (wt%)	0.3	0.18
Nitrogen (ppm)	41	36
ST (mN/m)	34.3	34.3
ASTM D 892 foam test		
Seq. I at 24°C (mL)	560/280	560/260
Seq. II at 93.5°C (mL)	60/0	65/0

Source: Reproduced from Pillon, L.Z., *Interfacial Properties of Petroleum Products*, CRC Press, Boca Raton, FL, 2007. With permission.

base stock having a VI of 93 was found to contain 22.3 wt% of aromatics, 0.3 wt% of S, 41 ppm of N, and have a higher ST of 34.5 mN/m measured at RT. Following the ASTM D 892 foam test, it was found to have a high foaming tendency of 560 mL and a high foam stability of 280 mL measured at 24°C. High viscosity mineral base stocks form stable foams that last for more than 10 min. However, with an increase in the temperature and a decrease in the viscosity, a lower foaming tendency of 60 mL and no foam stability at 93.5°C was measured. After HF, an increase in the VI to 95, with a decrease in the aromatic content was also observed indicating that some hydrogen saturation of aromatics was taking place. The hydrofined 600N base stock was also found to possess a lower sulfur content of 0.18 wt% and a lower nitrogen content of 36 ppm; however, no change in the ST and no significant change in the foaming tendency and the foam stability were observed. The effect of different temperatures on the foaming tendency of high viscosity 600N and 1400N base stocks confirmed that the solvent-refined base stocks foam in a specific viscosity range. The hydrofined 600N base stock was found to have a maximum foaming tendency at 24°C while a higher viscosity 1400N base stock was found to have a maximum foaming tendency at a higher temperature of 30°C–40°C where their viscosity was found to be similar.

Different processing options used to produce lube oil base stocks are shown in Table 5.8. To remove unstable and low VI components from mineral base stocks, conventional refining uses different extraction solvents. The commonly used extraction solvents are phenol and *N*-methylpyrrolidone (NMP). After solvent dewaxing, HF is used. While the NMP-extracted base stocks have higher BN content indicating an increase in the polar basic type contaminants, no significant effect of the

TABLE 5.8
**Different Processing Options Used to Produce Lube
Oil Base Stocks**

Conventional	Nonconventional	Nonconventional
VGO-feed	VGO-feed	HT slack wax-feed
Solvent extraction	HT or HC	Hydroisomerization
Solvent dewaxing	Solvent dewaxing	Solvent dewaxing
HF	HF	HF optional

Source: Reproduced from Pillon, L.Z., *Interfacial Properties of Petroleum
    Products*, CRC Press, Boca Raton, FL, 2007. With permission.

extraction solvent on the foaming characteristics of base stocks was observed. After
HF, mineral base stocks continue to foam and their foaming characteristics vary
depending on their viscosity. The literature reports on different processing options
used to produce nonconventional base stocks, which include the use of vacuum
gas oil (VGO) as feed followed by hydrotreatment (HT) upgrade, hydrodewaxing,
and hydrofinishing (Pirro and Wessol, 2001). Depending on the feed, the VGO HT
upgrade might involve more severe hydrocracking (HC) conditions. The literature
reports that the conventional VI base stocks, obtained by the HC processes, are similar to the solvent-refined base stocks with the exception of lower aromatic, sulfur,
and nitrogen contents. Their color is lighter and their performance in formulated
products containing additives is improved (Sequeira, 1994). The feed and the severity of hydroprocessing can affect the VI, the composition and a trace polar content of
lube oil base stocks leading to variation in their daylight stability.

The effect of HF on the properties and the foaming tendency of HC 100N and
SWI 6 base stocks are shown in Table 5.9. After HF, HC 100N base stock having a
VI of 103, was found to contain 7.7 wt% of aromatics, 82 ppm of S, and 2 ppm of N.
Following the ASTM D 892 foam test, a low viscosity HC 100N base stock was
found to have a very low foaming tendency of 30 mL at 24°C and no foaming tendency at a higher temperature of 93.5°C. After solvent dewaxing, SWI 6 base stock
having a high VI of 143 was found to contain only 0.9 wt% of aromatics, practically
no S and N, and have an ST of 32.8 mN/m measured at RT. Following the ASTM D
892 foam test, a higher viscosity SWI 6 base stock was found to have a high foaming
tendency of 105 mL and a low foaming tendency of 10 mL at a higher temperature of
93.5°C. After HF, a small decrease in the aromatic content to 0.7 wt% with no change
in ST was observed. Hydrofined SWI 6 base stock was found to have an increased
foaming tendency of 130 mL at 24°C indicating an increase in the surface-active
molecules capable in adsorbing at the oil/air interface and stabilizing the foam. The
effect of different temperatures on the foaming tendency of low viscosity HC 100N
and SWI 6 base stocks also indicated that they foam in a specific viscosity range.
A low viscosity HC 100N base stock was found to have a high foaming tendency
above 400 mL at −10°C where the base stock viscosity increased to about 300 cSt.
A higher viscosity SWI 6 base stock was found to have a high foaming tendency,

TABLE 5.9

Effect of HF on the Foaming Tendency of HC 100N and SWI 6 Base Stocks

Hydroprocessing	HC 100N	SWI 6	HF SWI 6
KV at 40°C (cSt)	18.29	28.94	28.84
KV at 100°C (cSt)	3.87	5.72	5.71
VI	103	143	143
Dewaxed pour point (°C)	—	−21	−21
Saturates (wt%)	92.6	99.1	99.3
Aromatics (wt%)	7.7	0.9	0.7
Sulfur (ppm)	82	<1	<1
Nitrogen (ppm)	2	<1	<1
ST (mN/m)	—	32.8	32.8
ASTM D 892 foam test			
Seq. I at 24°C (mL)	30/0	105/0	130/0
Seq. II at 93.5°C (mL)	0/0	10/0	10/0

Source: Pillon, L.Z., *Interfacial Properties of Petroleum Products,* CRC Press, Boca Raton, FL, 2007. With permission.

above 400 mL at a higher temperature of −5°C where the base stock viscosity also increased to about 300 cSt. The use of hydroprocessing improves the quality of mineral base stocks by decreasing their aromatic and their heteroatom contents and, while these base stocks also foam, their foaming tendency is lower and their foams are less stable when compared with similar viscosity mineral base stocks.

5.1.2 FOAM INHIBITION

Lubricants can foam as a result of fluid agitation and a variation in the temperature that can lead to cavitations, air binding in oil pumps, or overflow of tanks and sumps (Exxon Mobil Chemical, 2001). The literature also reports that foams are made by the agitation of petroleum oil with air and, under high temperature conditions, foaming of lubricants accelerate the oxidation process (Denis, Briant, and Hipeaux, 2000). Lubricating oils need to have resistance to foaming.

To prevent foaming of lubricating oils, the addition of antifoaming agents is required. According to the literature, they are two modes of action by which antifoaming agents prevent foam formation. After entering the oil/air interface, an antifoaming agent will either form mixed layers with stabilizing surface-active molecules or will spread over the surface (Rome, 2001).

The conditions required to enter (E) an interface can be defined in terms of the ST of the oil ST(oil), the ST of antifoaming agents ST(a/f), and the interfacial tension (IFT) between the oil and the antifoaming agent. The early literature reports on highly purified mineral oil having an ST of 31 mN/m at 20°C (Bondi, 1951). The

ST of different organic molecules was reported to increase with an increase in their molecular weight and their aromaticity (Pillon, 2007). For an antifoaming agent to enter the oil/air interface, the value of E needs to be >0, and, for an antifoaming agent to spread on the oil surface, the value of S needs to be >0 (Rome, 2001).

$$E = ST(oil) - ST(a/f) + IFT$$

$$S = ST(oil) - ST(a/f) - IFT$$

The effect of the viscosity and the aromatic content on the ST of hydrofined mineral base stocks is shown in Table 5.10.

Mineral lube oil base stocks contain a mixture of different hydrocarbons having a different molecular weight. Low viscosity 100N mineral base stocks containing 15.2 wt% of aromatics were found to have an ST of 33.2 mN/m measured at 24°C. Higher viscosity 150N base stocks containing 17.2–20.1 wt% of aromatic content were found to have a higher ST of 33.7–33.8 mN/m measured at 24°C. High viscosity of 600N mineral base stocks having a similar aromatic content of 19.5–19.6 wt% were found to have a higher ST of 34.7–34.8 mN/m at 24°C, indicating the effect of higher molecular weight molecules. A very high viscosity 2500N bright stock containing a high aromatic content of 34–35 wt% was found to have a significantly higher ST of 35–36 mN/m confirming the effect of higher molecular weight molecules. The ST of mineral base stocks increases with an increase in their viscosity and their aromatic content.

The most common type antifoaming agents used in petroleum applications are silicone polymers. Dimethyl silicones have the lowest ST values and these are largely independent of viscosity (Quinn, Traver, and Murthy, 1999). The polydimethylsiloxane used as silicone antifoaming agents are nonpolar and nonionic and have been commercially available since 1943. Silicone antifoaming agents are used in applications that require a high surface activity and a great spreading power. Water has

---

TABLE 5.10
Effect of Viscosity and Aromatic Content on the ST of Solvent-Refined Base Stocks

Solvent Refining	Aromatics (wt%)	ST at 24°C (mN/m)
100N base stock	15.2	33.2
150N base stock	17.2	33.7
150N base stock	20.1	33.8
600N base stock	19.5	34.7
600N base stock	19.6	34.8
2500N bright stock	34–35	35–36

Source: Reproduced from Pillon, L.Z., *Interfacial Properties of Petroleum Products*, CRC Press, Boca Raton, FL, 2007. With permission.

a high ST of 72.8 mN/m at 20°C and early literature reports on a high spreading coefficient of 10–12 mN/m of 5–35 cSt silicone antifoaming agents on the surface of water (Bondi, 1951). The literature reports that the combination of flexible siloxane backbone and the presence of short chain organic methyl groups makes silicones surface active toward the oil/air surface (Pape, 1981). The degree of polymerization (*n*) of silicones, which can vary from 0 to 2500 is used to control their molecular weight (Pape, 1983).

$$(CH_3)_3Si-O-[Si(CH_3)_2-O-]_n\,Si(CH_3)_3$$

Polydimethylsiloxane (silicone fluids)

Different viscosity silicone fluids are available, they are identified by referring to their viscosity in cSt at RT, and Si 12500 is usually used in lubricant products. The surface activity of Si 12500 on the oil surface of different 150N mineral base stocks is shown in Table 5.11. At 24°C, in 150N mineral base stock #1 having an ST of 33.7 mN/m, Si 12500 having an ST of 24 mN/m, was found to have a low IFT of 1.5 mN/m. A high efficiency to enter the oil/air interface (*E*) of 11.2 mN/m and a high efficiency to spread on the oil surface (*S*) of 8.2 mN/m was calculated. At the same temperature of 24°C in another 150N mineral base stock #2 having a similar ST of 33.7 mN/m, Si 12500 was also found to have a low IFT of 1.8 mN/m. A high efficiency to enter the oil/air interface (*E*) of 11.5 mN/m and a high efficiency to spread on the oil surface (*S*) of 7.9 mN/m was calculated. With an increase in the viscosity of silicones from 300 to 60,000 cSt, only a small increase in their ST from 23.9 to 24.1 mN/m at 24°C was measured (Pillon, 2007). The patent literature reports that only 1–2 ppm of different viscosity silicones are required to prevent foaming in 150N mineral base stocks (Pillon and Asselin, 1998).

The effect of different viscosity silicones on the foaming tendency of 150N mineral base stocks is shown in Table 5.12. 150N mineral base stocks were found to have

---

TABLE 5.11

Surface Activity of Si 12500 on the Oil Surface of 150N Mineral Base Stocks

Oil/Air Interface at 24°C	Mineral 150N Oil #1	Mineral 150N Oil #2
Neat mineral 150N oil ST (mN/m)	33.7	33.7
Neat Si 12500 ST (mN/m)	24.0	24.0
Oil/Si 12500 IFT (mN/m)	1.5	1.8
Oil/Si 12500 entering (*E*) (mN/m)	11.2	11.5
Oil/Si 12500 spreading (*S*) (mN/m)	8.2	7.9

Source: Reproduced from Pillon, L.Z., *Interfacial Properties of Petroleum Products*, CRC Press, Boca Raton, FL, 2007. With permission.

TABLE 5.12

Effect of Different Viscosity Silicones on the Foaming Tendency of 150N Base Stock

Mineral 150N Base Stock	Silicones (ppm)	Foaming at 24°C Seq. I (mL)	Foaming at 93.5°C Seq. II (mL)	Foaming at 24°C Seq. III (mL)
Neat 150N	0	280/0	20/0	240/0
Si 350	1	35/0	30/0	290/0
	2	5/0	25/0	10/0
	3	10/0	35/0	10/0
Si 12500	1	45/0	10/0	45/0
	2	0/0	10/0	5/0
	3	0/0	10/0	5/0
Si 60000	1	65/0	10/0	15/0
	2	5/0	10/0	15/0
	3	25/0	15/0	25/0

Source: Pillon, L.Z. and Asselin, A.E., Antifoaming agents for lubricating oils, U.S. Patent 5,766,513, 1998.

a high foaming tendency of 240–280 mL at 24°C and a low foaming tendency of 20 mL at a higher temperature of 93.5°C. At 24°C, the use of 1 ppm of low viscosity silicone, Si 350, was found to decrease the Seq. I foaming tendency to 35 mL but an increase in the Seq. III foaming tendency to 290 mL was observed. The use of 2–3 ppm of Si 350 was found effective in decreasing the Seq. I foaming tendency to 5–10 mL and the Seq. III foaming tendency to 10 mL. However, at 93.5°C the use of 1–3 ppm of low viscosity Si 350 was found to increase the Seq. II foaming tendency from 20 to 30–35 mL. The use of 1 ppm of a higher viscosity Si 12500 was found effective in decreasing the Seq. I and Seq. III foaming tendency to 45 mL. The use of 2–3 ppm of Si 12500 was found effective in decreasing the Seq. I and Seq. III foaming tendency to below 5 mL and no increase in the Seq. II foaming was observed. The use of 1 ppm of the high viscosity Si 60000 was found effective in decreasing the Seq. I and Seq. III foaming tendency to 15–65 mL. The use of 2–3 ppm of Si 60000 was found effective in decreasing the Seq. I and Seq. III foaming tendency to 5–25 mL and no increase in the Seq. II foaming was observed. Higher viscosity silicones were more effective at 93.5°C.

The effect of hydroprocessing on VI and the hydrocarbon composition of different 100N base stocks are shown in Table 5.13. Solvent-refined 100N mineral base stocks having a VI of 95–105, were reported to have 15–65 wt% of paraffins, 45–72 wt% of naphthenes, and 4–30 wt% of aromatic contents. Hydroprocessing leads to modified mineral 100N mineral base stocks having a higher VI of 109–129, which also have a higher paraffinic content and lower naphthenic and aromatic contents. A further increase in the VI to 130–145 leads to a further increase in the paraffinic content to 70–85 wt%, a decrease in the naphthenic content to 10–25 wt%, and a decrease in the aromatic content to 1–10 wt%.

TABLE 5.13

Effect of Hydroprocessing on VI and Hydrocarbon Composition of Base Stocks

100N Base Stocks	Solvent Refined	Hydroprocessing	Hydroprocessing
Processing	Conventional	Nonconventional	Nonconventional
VI	95–105	109–129	130–145
Dewaxed pour point (°C)	−7 to −18	−7 to −18	−7 to −18
Paraffins (wt%)	15–65	25–75	70–85
Naphthenes (wt%)	45–72	20–60	10–25
Aromatics (wt%)	4–30	6–15	1–10
Sulfur (wt%)	0–0.75	0–0.20	0–0.1

Source: Sequeira, A. Jr., *Lubricant Base Oil and Wax Processing*, Marcel Dekker, Inc., New York, 1994. With permission.

TABLE 5.14

Effect of Hydroprocessing on Aromatic Content and ST of Base Stocks

Base Stocks	KV at 40°C (cSt)	Aromatics (wt%)	ST at 24°C (mN/m)
150N	29.48	17.2	33.7
150N	29.60	20.1	33.8
SWI 6	29.38	<0.5	32.7
SWI 6	31.35	<0.5	32.8
PAO 6	30.43	0	32.4

Source: Reproduced from Pillon, L.Z., *Interfacial Properties of Petroleum Products*, CRC Press, Boca Raton, FL, 2007. With permission.

The effect of hydroprocessing on the aromatic content and the ST of base stocks having a similar viscosity is shown in Table 5.14. The use of HC and hydroisomerization-refining technologies produce base stocks having a low or practically no aromatic and heteroatom content that leads to a lower ST. With a significant decrease in the aromatic content to below 0.5 wt%, the ST of SWI 6 base stocks was found to decrease to 32.7–32.8 mN/m, which is lower by about 1 mN/m when compared to a similar viscosity 150N mineral base stock. With a change in the saturate hydrocarbon chemistry, a further decrease in the ST of synthetic PAO 6 base stock having a similar viscosity to 32.4 mN/m is observed.

The surface activity of different viscosity silicones on the oil surface of SWI 6 base stocks is shown in Table 5.15. At 24°C, in SWI 6 base stock having an ST of 32.8 mN/m, a low viscosity Si 350 having an ST of 23.9 mN/m was found to have a

TABLE 5.15

Surface Activity of Different Viscosity Silicones on the Oil Surface of SWI 6 Oil

Oil/Air Interface at 24°C	Si 350	Si 12500	Si 12500	Si 60000
Neat SWI 6 ST (mN/m)	32.8	32.8	32.8	32.8
Neat silicone ST (mN/m)	23.9	24.0	24.0	24.1
Oil/silicone IFT (mN/m)	0.8	1.8	2.2	2.2
Oil/silicone (E) (mN/m)	9.7	10.6	11.0	10.9
Oil/silicone (S) (mN/m)	8.1	7.0	6.6	6.5

Source: Reproduced from Pillon, L.Z., *Interfacial Properties of Petroleum Products*, CRC Press, Boca Raton, FL, 2007. With permission.

low IFT of 0.8 mN/m. A high efficiency to enter the oil/air interface (E) of 9.7 mN/m and a high efficiency to spread on the oil surface (S) of 8.1 mN/m was calculated. At the same temperature of 24°C, a higher viscosity Si 12500 having an ST of 24 mN/m was found to have a higher IFT of 1.8 mN/m. A high efficiency to enter the oil/air interface (E) of 10.6 mN/m but a lower efficiency to spread on the oil surface (S) of 7 mN/m was calculated. A higher viscosity Si 12500 was also found to have a higher IFT of 2.2 mN/m in another SWI 6 base stock. A high efficiency to enter the oil/air interface (E) of 11 mN/m and a lower efficiency to spread on the oil surface (S) of 6.6 mN/m was calculated. At 24°C, a high viscosity Si 60000 having an ST of 24.1 mN/m was also found to have a higher IFT of 2.2 mN/m. A high efficiency to enter the oil/air interface (E) of 10.9 mN/m but a lower efficiency to spread on the oil surface (S) of 6.5 mN/m was calculated. Some variation in the surface activity of silicones is observed.

The effect of Si 12500 on the foaming tendency of different batches of SWI 6 base stocks is shown in Table 5.16. At the treat rate of only 1 ppm, Si 12500 was

TABLE 5.16

Performance of Si 12500 in Different Batches of SWI 6 Base Stocks

Hydroisomerized Base Stocks	Silicone (ppm)	Foam at 24°C Seq. I (mL)	Foam at 93.5°C Seq. II (mL)	Foam at 24°C Seq. III (mL)
SWI 6 #1	0	140/0	20/0	145/0
Si 12500	1	15/0	0/0	10/0
Si 12500	3	0/0	0/0	0/0
SWI 6 #2	0	175/0	25/0	170/0
Si 12500	1	15/0	5/0	5/0
Si 12500	3	5/0	0/0	0/0

Source: Pillon, L.Z. and Asselin, A.E., Antifoaming agents for lubricating oils, U.S. Patent 5,766,513, 1998.

found effective in decreasing the Seq. I and Seq. III foaming tendency of SWI 6 base stock #1 from 140–145 mL to 10–15 mL measured at 24°C. The use of 3 ppm of Si 12500 was sufficient to develop a total resistance to foaming. At the treat rate of only 1 ppm, Si 12500 was also found effective in decreasing the Seq. I and Seq. III foaming tendency of SWI 6 base stock #2 from 170–175 mL to 5–15 mL measured at 24°C. The use of 3 ppm of Si 12500 was also sufficient to develop a total resistance to foaming. While 2 ppm of Si 12500 was sufficient to prevent foaming of mineral 150N base stocks, the use of 3 ppm was required to prevent foaming of SWI 6 base stocks, having a lower ST by about 1 mN/m. A decrease in the ST of base stocks leads to a decrease in the surface activity of silicone antifoaming agents and higher treat rates might be required.

The term "silicone" covers the family of organosiloxane polymers containing Si–O–Si bonds of the same nature as in silicates but with organic radicals fixed on the silicon. The siloxane Si–O–Si bond is longer, more polar, and has a higher energy than a C–C bond. The literature reports that the methyl radicals in polydimethylsiloxane could be substituted by many other organic groups such as hydrogen, alkyl, allyl, trifluoropropyl, glycol ether, epoxy, alkoxy, carboxy, and amino (Rome, 2001). Fluorosilicone fluids were commercialized and are used in applications in which the metal surface and the lubricant are exposed to hydrocarbon solvents (Quinn, Traver, and Murthy, 1999). The literature also reports that replacing some of the methyl groups with fluorine-containing alkyl groups required a hydrocarbon bridge to be placed between the CF_3-group and the siloxane backbone to achieve adequate thermal stability (Owen, 1980). The inclusion of $-(CH_2)_2-$ group was reported to affect the properties of the molecule to the extent that the trifluoropropylmethyl siloxanes have a higher ST and a different solubility (Pape, 1981).

$$(CH_3)_3Si-O-[Si(CH_3)(CH_2CH_2CF_3)-O-]_n Si(CH_3)_3$$

Trifluoropropylmethyl siloxane (fluorosilicone fluids)

For the same viscosity polymer of 300 mm²/s, an increase in ST from 20.4 to 24 mN/m for trifluoropropylmethyl siloxane was reported (Rome, 2001). The surface activity of FSi 300 having a higher ST of 24 mN/m on the oil surface of 150N mineral and SWI 6 base stocks is shown in Table 5.17. The use of different chemistry silicone, such as FSi 300, having a higher ST was found to significantly increase the IFT to 5.8 mN/m in mineral 150N base stock. A high efficiency to enter the oil/air interface (E) of 13.3 mN/m but a very low efficiency to spread on the oil surface (S) of 1.7 mN/m was calculated. The use of FSi 300 was found to further increase the IFT to 6.8 mN/m in SWI 6 base stock. A high efficiency to enter the oil/air interface (E) of 13.4 mN/m but a negative spreading coefficient (S) of −0.2 mN/m was calculated. Replacing some of the methyl groups with fluorine-containing alkyl groups increased the ST of fluorosilicones and also increased the IFT leading to a decrease in their efficiency to spread on the oil surface of mineral 150N and SWI 6 base stocks.

The effect of FSi 300 on the foaming tendency of different 150N mineral base stocks is shown in Table 5.18. The use of only 1–2 ppm of Si 12500 was found effective in preventing the foaming of different mineral 150N base stocks.

TABLE 5.17

Surface Activity of FSi 300 on the Oil Surface of Mineral 150N and SWI 6 Base Stocks

Oil/Air Interface at 24°C	Mineral 150N Base Stock	SWI 6 Base Stock
Neat base stock ST (mN/m)	33.7	32.8
Neat FSi 300 ST (mN/m)	26.2	26.2
Oil/FSi 300 IFT (mN/m)	5.8	6.8
Oil/FSi 300 entering (E) (mN/m)	13.3	13.4
Oil/FSi 300 spreading (S) (mN/m)	1.7	−0.2

Source: Pillon, L.Z. and Asselin, A.E., Antifoaming agents for lubricating oils, U.S. Patent 5,766,513, 1998.

TABLE 5.18

Performance of FSi 300 in Different Mineral 150N Base Stocks

Solvent-Refined Base Stocks	FSi 300 (ppm)	Foam at 24°C Seq. I (mL)	Foam at 93.5°C Seq. II (mL)	Foam at 24°C Seq. III (mL)
HF 150N (Phenol)	0	345/0	30/0	—
FSi 300	3	340/0	25/0	—
HF 150N (NMP)	0	280/0	20/0	240/0
FSi 300	3	300/0	0/0	265/0

Source: Pillon, L.Z. and Asselin, A.E., Antifoaming agents for lubricating oils, U.S. Patent 5,766,513, 1998.

The use of 3 ppm of low viscosity FSi 300 was found not effective in the prevention of foaming and only a small decrease in foaming tendency was observed. To decrease the Seq. I foaming tendency of 150N base stocks below 50 mL the use of 10 ppm of FSi 300 was required. However, the use of only 3–5 ppm of FSi 300 was needed to prevent foaming in higher viscosity mineral 600N base stocks, having a higher ST of 34.7 mN/m, indicating an increase in its surface activity. The literature reports that under certain conditions of temperature and concentration, the solubility of certain components in a medium might decrease leading to an increase in their surface activity (Ross, 1987).

The effect of FSi 300 on the foaming tendency of SWI 6 base stock is shown in Table 5.19. At the treat rate of 3 ppm, FSi 300 was found effective in decreasing the Seq. I foaming tendency of SWI 6 base stock from 175 to 35 mL, and, at a higher treat rate of 5 ppm, FSi 300 was found effective in decreasing the Seq. I foaming tendency of SWI 6 base stock to 20 mL and no foaming tendency at 93.5°C was observed. Despite a higher ST and a negative spreading coefficient of −0.2 mN/m, FSi 300 was found effective in preventing the foaming of SWI 6 base stock. The high saturate content and the poor solvency properties of SWI 6

TABLE 5.19
Performance of FSi 300 in SWI 6 Base Stock

Hydroisomerized Base Stock	FSi 300 (ppm)	Foam at 24°C Seq. I (mL)	Foam at 93.5°C Seq. II (mL)
Neat SWI 6	0	175/0	15/0
FSi 300	3	35/0	0/0
FSi 300	5	20/0	0/0

Source: Pillon, L.Z. and Asselin, A.E., Antifoaming agents for lubricating oils, U.S. Patent 5,766,513, 1998.

TABLE 5.20
Effect of Higher Temperature on Surface Activity of Different Viscosity Silicones

SWI 6 Oil/Air Interface Silicones	Spreading (S) at 24°C (mN/m)	Spreading (S) at 93.5°C (mN/m)
Si 350	8.9	0.9
Si 12500	6.6	1.6
Si 60000	7.9	7.1
Si 100000	7.7	6.1

Source: Pillon, L.Z. et al., Method for reducing foaming of lubricating oils, U.S. Patent 6,090,758, 2000.

base stock might lead to an increase in its adsorption at the oil/air interface. In some applications, lubricating oils foam when used at higher temperatures. The effect of high temperatures on the foam inhibition requires that the measurement of the ST of oil, the ST of antifoaming agents, and the oil/antifoaming IFT are taken at these high temperatures. One of the instruments commercially available, Processor Tensiometer K12 allows measurements to be taken at different temperatures ranging from −10°C to 100°C.

The effect of higher temperature on the surface activity of different viscosity silicones on the oil surface of SWI 6 base stock is shown in Table 5.20. With an increase in the temperature from 24°C to 93.5°C, the spreading coefficient of low viscosity Si 350 drastically decreased from 8.9 to only 0.9 mN/m, on the oil surface of SWI 6 base stock. With an increase in the temperature, the spreading coefficient of higher viscosity Si 12500 decreased from 6.6 to 1.6 mN/m. With an increase in the temperature, the spreading coefficient of high viscosity Si 60000 decreased from 7.9 to only 7.1 mN/m, and the spreading coefficient of a very high viscosity Si 100000 decreased from 7.7 to only 6.1 mN/m. With an increase in the temperature, there is a drastic decrease in the surface activity of low viscosity silicones, whereas only a small decrease in the surface activity of very high viscosity silicones is observed. When mixed with air the solvent-refined base stocks foam. While the foaming tendency of petroleum-derived base stocks is mostly affected by their viscosity, the presence of surface-active contaminants at the oil/air interface might lead to an increase in the foam stability. In industrial applications, the lubricating oils are exposed to varying temperatures. At high temperatures, the ST of base stocks decreases, which can lead to a decrease in the surface activity of antifoaming agents. Engine oils require foam inhibition at low temperatures and at very high temperatures.

5.2   AIR ENTRAINMENT

Unlike foaming, air entrainment is a dispersion of small air bubbles, which rise slowly to the oil surface and can lead to hazy or cloudy appearance. According to the literature, pure liquids allow entrained air to escape with no delay. Certain solutes are able to form a thin layer known as lamella at the liquid/air interface and adversely affect the air separation properties of liquids (Ross, 1987). The literature reports that the air present in petroleum oil known as air entrainment might be dispersed as very fine bubbles or dissolved (Denis, Briant, and Hipeaux, 2000). The presence of entrained air can cause the disruption of the lubricant film that can lead to excessive wear and compressibility problems in hydraulic systems (Exxon Mobil Chemical, 2001). The literature reports on the effect of viscosity and the temperature on foaming and the air entrainment of mineral and synthetic oils (Hubman and Lanik, 2006). The use of NMP as an extraction solvent leads to an increase in the N and the BN content of hydrofined base stocks. The recent literature reports on the effect of the viscosity and the extraction solvent on the air entrainment of different hydrofined mineral base stocks (Pillon, 2007). The air release time of lube oil base stocks is measured following the ASTM D 3427 method which is the same as the IP-313 test. The air release time of oil indicates its ability to separate entrained air and is defined as the time needed to reduce air entrainment to 0.2% of its original volume at a specific temperature, usually at 50°C.

The effect of the viscosity and the extraction solvent on the air release time of different hydrofined mineral base stocks is shown in Table 5.21. After HF, phenol-extracted and solvent-dewaxed 150N base stocks were found to have an air release time of 1.6 min measured at 50°C. After HF, NMP-extracted and solvent-dewaxed 150N base stocks were found to have an air release time of 2.1 min also measured at 50°C. The higher viscosity hydrofined 600N phenol-extracted and solvent-dewaxed base stocks were found to have a drastically longer air release time of 6–7 min at 50°C. The higher viscosity hydrofined 600N NMP-extracted and solvent-dewaxed base stocks were found to have a further increased air release time of 8–9 min at 50°C. After HF, NMP-extracted base stocks were found to have higher BN contents and longer air release times. The air release time of base stocks increases with an increase in their viscosity and with an increase in their BN contents. The literature reports that the use of hydroprocessing leads to an increase in the VIs of naphthenic oils, a lower aromatic content, and a lower volatility; however, a poor stability against ultraviolet (UV) light and a rapid color degradation when exposed to sunlight was also reported (Han et al., 2005).

The total acid number (TAN) and the polar content of hydrotreated (HT) naphthenic oil are shown in Table 5.22. The literature reports that the HT naphthenic

TABLE 5.21

Effect of Viscosity and Extraction Solvent on the Air Release Time of Different Oils

Extraction Solvent	Phenol	Phenol	NMP	NMP
Hydrofined Base Stocks	**150N**	**600N**	**150N**	**600N**
KV at 40°C (cSt)	29.5	105.9	29.7	111.4
KV at 100°C (cSt)	5	11.3	5.1	11.6
VI	94	92	96	89
Saturates (wt%)	82.8	80.5	86.1	80.4
Aromatics (wt%)	17.2	19.5	13.9	19.6
Sulfur (wt%)	0.09	0.12	0.06	0.11
Nitrogen (ppm)	8	30	36	100
BN (ppm)	4	16	33	88
ASTM D 892				
Seq. I at 24°C (mL)	345/0	630/165	280/0	620/250
Seq. II at 93.5°C (mL)	30/0	40/0	20/0	50/0
ASTM D 3427 (50°C)				
Air release time (min)	1.6	6–7	2.1	8–9

Sources: Pillon, L.Z. et al., Lubricating oil for inhibiting rust formation, U.S. Patent 5,397,487, 1995; Pillon, L.Z. and Asselin, A.E., Antifoaming agents for lubricating oils, U.S. Patent 5,766,513, 1998.

TABLE 5.22

TAN and Polar Content of HT Naphthenic Oil

HT Naphthenic Base Oil	Properties	Test Method
Boiling range (°C)	380–450	
KV at 40°C (cSt)	130.6	ASTM D 4052
KV at 100°C (cSt)	9.67	
Pour point (°C)	−27	ASTM D 97
TAN (mg KOH/g)	0.01	ASTM D 974
UV absorbance at 260 nm	0.07	ASTM D 2008
Silica gel separation		ASTM D 2140
Saturates (wt%)	95.6	
Aromatics (wt%)	4.29	
Polars (wt%)	0.11	

Source: Reprinted from From Han, S. et al., *Energy Fuels*, 19, 625, 2005. With permission.

oil was found to have a TAN of 0.01 mg KOH/g indicating no increase in acidic oxidation by-products; however, an increase in the UV absorbance was reported. To identify the polar content of HT naphthenic base stock, the oil was passed over silica gel, and petroleum ether and benzene were used to wash off saturates and aromatics. The polar material retained on silica was removed using ethanol and analyzed (Han et al., 2005). The analysis of the polar content indicated a mixture of complex heteroatom-containing compounds, mainly O-containing molecules, which did not form sediment due to the good solvency properties of naphthenic oils. Some oxidation by-products are surface active and can adsorb at the oil/air interface leading to an increase in the foam stability and air entrainment.

According to the early literature, the long chain aliphatic amine oxides are surface active and are used as foam stabilizers in liquid dishwashing formulations. In the presence of acids, amine oxides become protonated to form salts which further increases their surface activity and foam stabilizing properties (Lake and Hoh, 1963).

The BN content and the air release time of different SWI 6 base stocks are shown in Table 5.23. After hydroisomerization and solvent dewaxing, SWI 6 base stock #1 having a kinematic viscosity of 29.4 cSt at 40°C and no BN content was found to have a short air release time of only 1.1 min measured at 50°C. After hydroisomerization and solvent dewaxing, another SWI 6 base stock #2 having an increased kinematic viscosity of 31.4 cSt at 40°C and no BN content was also found to have a short air release time of 1.2 min measured at 50°C. The presence of surface-active molecules on the oil surface is very difficult to measure. The early literature reports that the surface-active molecules decrease the ST of petroleum oils, which can be

TABLE 5.23
BN Content and Air Release Time of Different SWI 6 Base Stocks

Hydroisomerization	SWI 6 #1	SWI 6 #2
KV at 40°C (cSt)	29.4	31.4
KV at 100°C (cSt)	5.8	6.0
VI	143	138
Saturates (wt%)	>99.5	>99.5
Aromatics (wt%)	<0.5	<0.5
Sulfur (ppm)	1	17
Nitrogen (ppm)	1	2
BN (ppm)	0	0
ASTM D 892		
Seq. I at 24°C (mL)	145/0	180/0
Seq. II at 93.5°C (mL)	10/0	25/0
ASTM D 3427 (50°C)		
Air release time (min)	1.1	1.2

Source: Pillon, L.Z. and Asselin, A.E., Antifoaming agents for lubricating oils, U.S. Patent 5,766,513, 1998.

TABLE 5.24

Theoretical Effect of Polydimethylsiloxane and Fatty Acids on the ST of Mineral Oil

Surface-Active Molecules	Concentration in Mineral Oil (ppm)	ST Depression (Calculated) (mN/m)
PDMS 1000 cSt	0.5	0.09
Polydimethylsiloxane	3	0.20
	10	0.72
Span 20	55	0.09
Fatty acid mixture	150	0.18
	400	0.65

Source: Ross, S., Foaminess and capilarity in apolar solutions, in *Interfacial Phenomena in Apolar Media*, Eicke, H.F. and Parfitt, G.D., Eds., Marcel Dekker Inc., New York, 1987.

measured when used at very high concentrations (Bondi, 1951). Silicone antifoaming agents are used at very low concentrations below 3 ppm and their content is too low to affect the ST of oil. The literature reports on the theoretical calculations of the effect of surface-active molecules, such as polydimethylsiloxane and fatty acids, on the ST of mineral oil (Ross, 1987).

The theoretical effect of polydimethylsiloxane and fatty acids on the ST of mineral oil is shown in Table 5.24. Based on theoretical calculations, when used at a treat rate of 3 ppm, the highly surface-active polydimethylsiloxane would depress the ST of the mineral oil by only 0.2 mN/m, which would be difficult to measure. Span 20 contains different chemistry fatty acids and is a mixture of about 50% of lauric acid, myristic acid, palmitic acid, and linolenic acid (Sigma-Aldrich, 2007–2008). Based on theoretical calculations, at a higher treat rate of 150 ppm, the addition of Span 20 would depress the ST of mineral oil also by about 0.2 mN/m indicating a lower surface activity toward oil/air interface when compared to polydimethylsiloxane, used as an antifoaming agent.

The effect of different viscosity silicone antifoaming agents on the air release time of mineral 150N base stock is shown in Table 5.25. With an increase in the treat rate of low viscosity Si 350 from 1 to 3 ppm, the foaming tendency of 150N mineral base stocks decreased indicating its adsorption at the oil/air interface and the air release time was found to increase from 2.1 to 5.2–6.3 min measured at 50°C. With an increase in the treat rate of Si 12500 from 1 to 3 ppm, the foaming tendency of 150N mineral base stocks also decreased indicating its adsorption at the oil/air interface and the air release time was also found to increase from 2.1 to 4.2–5.5 min measured at 50°C. With an increase in the treat rate of high viscosity Si 60000 from 1 to 3 ppm, the foaming tendency of 150N mineral base stocks also decreased indicating its adsorption at the oil/air interface and the air release time was also found to increase from 2.1 to 3.2–5.5 min measured at 50°C. Silicone antifoaming agents cause air entrainment.

TABLE 5.25
Effect of Different Viscosity Silicones on the Air Release Time of Mineral 150N Oil

150N Mineral Base Stock	Air Release Time at 50°C (min)	Air Release Time at 50°C (min)	Air Release Time at 50°C (min)
Silicones (ppm)	Si 350	Si 12500	Si 60000
0	2.1	2.1	2.1
1	5.2	4.2	3.2
2	6.2	5.0	4.2
3	6.3	5.5	5.5

Source: Pillon, L.Z. and Asselin, A.E., Antifoaming agents for lubricating oils, U.S. Patent 5,766,513, 1998.

TABLE 5.26
Effect of FSi 300 on the Air Release Time of Different Mineral 150N Oils

Extraction Solvent	Phenol	NMP
Air Release Time at 50°C (min)	150N Base Stock	150N Base Stock
FSi 300 (ppm)		
0	1.6	2.1
3	3.7	4.3
5	4.8	4.8
10	5.1	5.1

Source: Pillon, L.Z. and Asselin, A.E., Antifoaming agents for lubricating oils, U.S. Patent 5,766,513, 1998.

The effect of FSi 300 on the air release time of different mineral 150N base stocks is shown in Table 5.26. FSi 300 has a higher ST and is less surface active in mineral base stocks. A high treat rate of 5–10ppm is required to decrease the foaming tendency of 150N mineral base stocks. With an increase in the treat rate of low viscosity FSi 300 from 1 to 10ppm, the foaming tendency of phenol-extracted and solvent-dewaxed 150N base stock gradually decreased indicating its adsorption at the oil/air interface and the air release time was found to increase from 1.6 to 5.1 min measured at 50°C. With an increase in the treat rate of low viscosity FSi 300 from 1 to 10ppm, the foaming tendency of NMP-extracted and solvent-dewaxed 150N base stock also gradually decreased indicating its adsorption at the oil/air interface, and the air release time was also found to increase from 2.1 to 5.1 min measured at 50°C. The use of a less surface-active FSi 300 also leads to the air entrainment of mineral base stocks.

TABLE 5.27

Effect of Si 12500 and FSi 300 on the Foaming and Air Release Time of SWI 6 Oil

SWI 6 Base Stock Different Silicones	Foam at 24°C Seq. I (mL)	Foam at 93.5°C Seq. II (mL)	Air Release Time at 50°C (min)
Neat SWI 6	175/0	25/0	1.2
Si 12500 (3 ppm)	0/0	0/0	2.8
FSi 300 (3 ppm)	35/0	0/0	1.6

Source: Pillon, L.Z. et al., Method for reducing foaming of lubricating oils, U.S. Patent 6,090,758, 2000.

The effect of Si 12500 and FSi 300 on the foaming tendency and the air release time of SWI 6 base stock is shown in Table 5.27. At the treat rate of 3 ppm, Si 12500 is effective in preventing the foaming of SWI 6 base stock and an increase in the air release time from 1.2 to 2.8 min measured at 50°C is observed. At the same treat rate of 3 ppm, FSi 300 is less effective in preventing the foaming of SWI 6 base stock and an increase in the air release time from 1.2 to only 1.6 min measured at 50°C is observed. The use of a less surface-active antifoaming agent is less effective in the prevention of foaming and only a small increase in air release time is observed. While silicones are the most surface-active and effective antifoaming agents, their use in lubricants needs to be strictly controlled to minimize an increase in the air entrainment. Some lubricants, such as turbine oils, are required to have good air separation properties and only non-silicone type antifoaming agents are used to meet the air release specification. There are no additives to prevent air entrainment and antifoaming agents and their treat rates need to be carefully selected. Many other additives used in some lubricant products are highly surface-active and can adsorb at the oil/air interface leading to an increase in the air entrainment. Their use should also be controlled to prevent oxidation degradation.

5.3　STABILITY OF OIL/WATER INTERFACE

In many applications, petroleum products might become contaminated with water and lube oil base stocks are required to have good demulsibility properties by which is meant the ability to separate the water. The early literature reports that the oil/water IFT is very sensitive to even traces of hydrophilic impurities and will greatly vary for the same technical materials (Bondi, 1951). Acid-alkali refining, called wet refining, is a chemical process in which lube oils are contacted with sulfuric acid followed by neutralization with aqueous or alcoholic alkali (Sequeira, 1994). The lube oil distillates contain paraffins, naphthenes, aromatics, and heteroatom-containing molecules. Olefins are normally not present but they can be present as a result of cracking. The literature reports that the paraffins and naphthenes are relatively nonreactive but aromatics and olefins are readily attacked by sulfuric acid.

TABLE 5.28

Effect of Different Treatments on the IFT of Lube Oil Distillate

Lube Oil Distillate	Oil/Distilled Water	Oil/Aqueous KCl
	IFT (mN/m)	
Lube oil distillate	22.9	23.9
After treatments		
NaOH	19.9	23.2
3% H_2SO_4/3% clay	30.7	31.6
5% H_2SO_4/15% clay	34.7	39.4

Source: Bondi, A., *Physical Chemistry of Lubricating Oils*, Reinhold Publishing Corporation, New York, 1951.

The sulfur compounds are mainly thiophenes and converted to thiophene-sulfonic acid and removed as acid sludge. The nitrogen compounds consist of amines, amides, and minor amounts of amino acids, which are converted to salts and removed with the acid sludge. The oxygen compounds are primary acids with minor amounts of aldehydes and alcohols that are oxidized to acids and removed (Sequeira, 1994).

The effect of different treatments on the IFT of lube oil distillate is shown in Table 5.28. The early chemical-refining process was based on the sulfuric acid and the clay treatment of lube oil distillate. The action of sulfuric acid on lube oil distillates was reported to be complex and varied depending on the acid concentration, the reaction temperature, and the residence time (Sequeira, 1994). Clays are natural compounds of silica and alumina, also containing oxides of sodium, potassium, magnesium, calcium, and other alkaline earth metals. The literature describes the Attapulgite clay as magnesium–aluminum silicate while porocel is described as hydrated aluminum oxide known as bauxite. Attapulgite clay was reported effective in decolorizing and neutralizing any petroleum oil. Porocel clay was reported effective in decolorization, reducing organic acidity, effective deodorizer, and also adsorbing aromatic type molecules and polar compounds containing sulfur and nitrogen (Sequeira, 1994).

After NaOH treatment, the lube oil distillate was reported to have an IFT of about 20 mN/m in the presence of distilled water and an increased IFT of about 24 mN/m in the presence of aqueous KCl solution. After the 3% H_2SO_4/3% clay treatment, the IFT of lube oil distillate drastically increased to 30.7 mN/m indicating a decrease in surface-active molecules capable of adsorbing at the oil/water interface. After the 5% H_2SO_4/15% clay treatment, the IFT of lube oil distillate further increased to 34.7 mN/m indicating a further decrease in surface-active molecules capable of adsorbing at the oil/water interface. The use of aqueous KCl solution additionally increased the IFT to 39.4 mN/m indicating a further decrease in the surface-active content adsorbing at the oil/water interface.

The effect of petroleum resins and naphthenic acids on the IFT of white oil is shown in Table 5.29. The early literature reports on white oil having a high IFT of

TABLE 5.29
Effect of Petroleum Resins and Naphthenic Acids on the IFT of White Oil

White Oil/Water Interface at 20°C	Oil/Distilled Water	Oil/Aqueous KCl
	IFT (mN/m)	
Neat white oil	52	52.2
Petroleum resins (1.4%)	19.8	7.8
Petroleum resins (15%)	36.6	29.4
Naphthenic acids (1.4%)	24.5	11.8
Naphthenic acids (6%)	30.4	12.7
Petroleum resins (0.5%)/ naphthenic acids (1%)	22.9	9.9

Source: Bondi, A., *Physical Chemistry of Lubricating Oils*, Reinhold Publishing Corporation, New York, 1951.

52 mN/m, when tested in the presence of distilled water, and having a high IFT of 52.2 mN/m, when tested in the presence of water containing KCl. In the presence of distilled water, the addition of only 1.4% of petroleum resins decreased the IFT to 19.8 mN/m while the addition of 15% of petroleum resins actually increased the IFT to 36.6 mN/m. In the presence of aqueous KCl solution, the addition of 1.4% of petroleum resins decreased the IFT to 7.8 mN/m, while the addition of 15% of petroleum resins increased the IFT to 29.4 mN/m. In the presence of distilled water, the addition of only 1.4% of naphthenic acid decreased the IFT to 24.5 mN/m, while the addition of 6% of naphthenic acid actually increased the IFT to 30.4 mN/m. In the presence of aqueous KCl solution, the addition of 1.4% of naphthenic acid decreased the IFT to 11.8 mN/m, while the addition of 6% of naphthenic acid increased the IFT to only 12.7 mN/m. The addition of only 0.5% of petroleum resin and only 1% of naphthenic acid decreased the IFT to 22.9 mN/m, in the presence of distilled water, and 9.9 mN/m in the presence of aqueous KCl solution indicating many different interactions taking place at the oil/water interface.

As a final step, the majority of refineries use solvent refining and HF to remove impurities introduced during the various processing steps. After HF, mineral base stocks were reported to have improved color, color stability, and water separation properties (Sequeira, 1994). The literature reports that when two pure immiscible liquids are stirred, one will disperse in the other in the form of droplets. When the stirring stops the droplets will coalesce into larger drops, and reform the two separate phases under the effect of gravity. In the presence of an emulsifier, the droplets of water dispersed in oil are smaller, and the coalescence on resting is very slow (Denis, Briant, and Hipeaux, 2000). The ASTM D 1401 method for the water separability of petroleum oils and synthetic fluids requires the mixing of oil with distilled water at 54°C or 82°C, depending on the oil viscosity. The demulsibility of oil is the ability to separate from water after being mixed with water and agitated so that an emulsion is

TABLE 5.30

Effect of Viscosity on the IFT and the Demulsibility of Hydrofined Base Stocks

Extraction Solvent	Phenol	Phenol	Phenol
Mineral Base Stocks	**HF 150N**	**HF 600N**	**HF 1400N**
KV at 40°C (cSt)	29.48	105.91	301.71
KV at 100°C (cSt)	5.03	11.32	22.01
VI	94	92	89
Aromatics (wt%)	16.1	19.5	28.2
Sulfur (wt%)	0.09	0.12	0.19
Nitrogen (ppm)	8	30	141
BN (ppm)	4	16	51
IFT (mN/m) (after 30 min)	43.1	42.8	40.8
ASTM D 1401 (54°C)			
O/W/E (mL)	40/40/0	41/28/12	44/36/2
Time (min)	2	30	45

Source: Pillon, L.Z. and Asselin, A.E., Method for improving the demulsibility of base oils, U.S. Patent 5,282,960, 1994.

formed. The time needed to separate the water-in-oil emulsion or the volume of oil/water/emulsion (O/W/E) left after 30 min on standing is recorded.

The effect of viscosity on the IFT and the demulsibility properties of hydrofined base stocks are shown in Table 5.30. The IFT of lube oil base stocks is usually measured after 30 min of the equilibrium time. HF 150N phenol-extracted and solvent-dewaxed base stock was found to have an IFT of 43.1 mN/m and required only 2 min at 54°C to separate the emulsion indicating excellent water separation properties. The higher viscosity HF 600N phenol-extracted and the solvent-dewaxed base stock was found to have an IFT of 42.8 mN/m and, after 30 min standing time did not separate the emulsion at 54°C. The very high viscosity HF 1400N phenol-extracted and solvent-dewaxed base stock was found to have a lower IFT of 40.8 mN/m and required 45 min to separate the emulsion at 54°C. The IFT and the demulsibility properties of different viscosity hydrofined phenol-extracted and solvent-dewaxed base stocks were found to vary. The ASTM D 1401 method is performed at 82°C in a case of high viscosity oils, which reduces their viscosity and improves their water separation properties.

The IFT and the demulsibility properties of different batches of the hydrofined NMP-extracted and solvent-dewaxed base stocks are shown in Table 5.31. The HF 150N NMP-extracted and solvent-dewaxed base stock, batch #1, was found to have a lower IFT of 33.4 mN/m and, after 60 min on standing a stable emulsion was still present. Another batch of the HF 150N NMP-extracted and solvent-dewaxed base stock, batch #2, was found to require 10 min to separate the emulsion indicating good demulsibility properties. The higher viscosity HF 600N NMP-extracted and solvent-dewaxed base stock was found to have a higher IFT of 39.4 mN/m, and, after only 5 min on standing,

TABLE 5.31

IFT and Demulsibility of the Different Batches of Hydrofined Base Stocks

Extraction Solvent	NMP	NMP	NMP
Mineral Base Stocks	**HF 150N #1**	**HF 150N #2**	**HF 600N**
KV at 40°C (cSt)	29.71	29.6	111.4
KV at 100°C (cSt)	5.07	4.99	11.6
VI	96	90	89
Aromatics (wt%)	14.0	20.1	19.6
Sulfur (wt%)	0.06	0.12	0.11
Nitrogen (ppm)	36	66	100
BN (ppm)	33	53	88
IFT (mN/m) (after 30 min)	33.4	—	39.4
ASTM D 1401 (54°C)			
O/W/E (mL)	41/25/14	40/40/0	39/40/1
Time (min)	60	10	5

Source: Reproduced from Pillon, L.Z., *Interfacial Properties of Petroleum Products*, CRC Press, Boca Raton, FL, 2007. With permission.

the emulsion was separated indicating excellent demulsibility properties. The IFT and the demulsibility properties of the hydrofined NMP-extracted and solvent-dewaxed base stocks were found to vary from batch to batch. The literature reports that the storage of mineral base stocks caused degradation in their water separation properties. The ASTM D 2440 oxidation test can be used to measure the effect of oxidation on the demulsibility properties of uninhibited hydrofined mineral base stocks at the temperature of 120°C and in the presence of copper, used as a catalyst.

The effect of oxidation on the IFT and the demulsibility of hydrofined mineral base stocks is shown in Table 5.32. After the oxidation of hydrofined 150N phenol-extracted base stock the IFT decreased from 43 to 10–20 mN/m, and the emulsion separation time increased from 2 to over 30 min, indicating an increase in the surface-active contaminants capable of adsorbing at the oil/water interface and having emulsifying properties. After the oxidation of hydrofined 600N phenol-extracted base stock, the IFT also decreased from 43 to 20–30 mN/m, but no further degradation in the demulsibility was observed indicating the saturation at the oil/water interface. After the oxidation of hydrofined 150N NMP-extracted base stock, the IFT decreased from 34 to 5–10 mN/m, but no further degradation in the demulsibility was observed indicating the saturation at the oil/water interface. After oxidation of hydrofined 600N NMP-extracted base stock, the IFT also decreased from 39 to 5–10 mN/m, and the emulsion separation time increased from 5 to over 30 min, indicating an increase in the surface-active contaminants having emulsifying properties.

The early literature reports that the HF followed by the clay treatment was effective in improving the demulsibility of mineral base stocks (Watanabe, Fukushima,

TABLE 5.32

Effect of Oxidation on the IFT and the Demulsibility of Hydrofined Mineral Base Stocks

HF Mineral Base Stocks	150N	600N	150N	600N
Extraction Solvent	Phenol	Phenol	NMP	NMP
Fresh oils				
IFT (mN/m)	43	43	33	39
ASTM D 1401 demulsibility (54°C)				
Emulsion separation time (min)	2	45	60	5
ASTM D 2440 oxidation (120°C/7.5 h)				
IFT (mN/m)	10–20	20–30	5–10	5–10
ASTM D 1401 demulsibility (54°C)				
Emulsion separation time (min)	>30	>45	>60	>30

Sources: Pillon, L.Z. and Asselin, A.E., Method for improving the demulsibility of base oils, U.S. Patent 5,282,960, 1994; Pillon, L.Z., *Petrol. Sci. Technol.*, 21(9–10), 1469, 2003.

and Nose, 1976). Porocel clay has been known for years to be a good refining agent for mineral oils used as turbine and transformer oils. Attapulgite clay was reported less effective in removing odorous compounds and metals and does not adsorb aromatics (Sequeira, 1994). While some refiners use clay treatment to finish lube oil base stocks, spent clay disposal problems, and other operating restrictions have caused the clay treating processes to be replaced by hydrofinishing. To meet the quality requirements of lubricating oils, the emulsion needs to be separated in less than 30 min. Additives known as demulsifiers are used to improve the water separation properties of hydrofined mineral base stocks.

The effect of demulsifier on the IFT and the demulsibility of different viscosity hydrofined mineral base stocks is shown in Table 5.33. The literature reports that

TABLE 5.33

Effect of Demulsifier on IFT and Demulsibility of Hydrofined Mineral Base Stocks

Solvent-Refined Base Stocks	Demulsifier (wt%)	IFT (mN/m)	Emulsion Separation Time at 54°C (min)
HF 150N (demulsifier)	0.004	26–27	4
HF 150N (demulsifier)	0.006	19–25	5
HF 600N (demulsifier)	0.004	39–41	15
HF 600N (demulsifier)	0.006	32–34	10

Source: Pillon, L.Z., *Interfacial Properties of Petroleum Products*, CRC Press, Boca Raton, FL, 2007. With permission.

in the chemical demulsification, the first challenge for a demulsifier is to reach the oil/water interface. The second challenge is the displacement of the surface-active molecules adsorbed at the oil/water interface by a more surface-active demulsifier. Molecules used as demulsifiers need to concentrate at the oil/water interface and form weak intermolecular interactions. In some cases, by strongly adsorbing at the oil/water interface, demulsifiers can build up the layer around droplets of water and delay their flocculation (Rome, 2001). Some additives, such as demulsifiers, can have detrimental side effects if the dosage is excessive.

The HF 150N base stock containing 0.004 wt% of demulsifier was found to have an IFT of 26–27 mN/m and required 4 min to separate the emulsion. The same HF 150N base stock containing 0.006 wt% of the same demulsifier was found to have a lower IFT of 19–25 mN/m indicating its increased adsorption at the oil/water interface, and required 5 min to separate the emulsion. An overtreatment needs to be avoided to prevent an increase in the emulsion stability. The HF 600N base stock containing 0.004 wt% of demulsifier was found to have an IFT of 39–41 mN/m and required 15 min to separate the emulsion. The same HF 600N base stock, containing 0.006 wt% of the same demulsifier, was found to have a lower IFT of 32–34 mN/m, indicating its increased adsorption at the oil/water interface and required only 10 min to separate the emulsion.

The IFT and the demulsibility properties of the hydroprocessed base stocks and the synthetic PAO 6 fluid having a similar viscosity are shown in Table 5.34. After hydrorefining the technical white oil was found to have an IFT of 39.0 mN/m and required 3 min to separate the emulsion. After solvent dewaxing, non-hydrofined SWI 6 base stock was found to have a lower IFT of 24.4 mN/m and required 4 min to

TABLE 5.34
IFT and Demulsibility of Hydroprocessed Base Stocks and Synthetic PAO 6

	Processing Base Stocks			
	Hydrorefined White Oil	Solvent-Dewaxed SWI 6	Hydrofined SWI 6	Synthetic PAO 6
KV at 40°C (cSt)	32.7	28.9	28.8	30.4
KV at 100°C (cSt)	5.6	5.7	5.7	5.8
VI	106	143	143	134
Aromatics (wt%)	0.9	1.4	<0.5	0
Sulfur (ppm)	1	6	0	0
Nitrogen (ppm)	1	0	0	0
IFT (mN/m)	39.0	24.4	45.6	41.9
ASTM D 1401 (54°C)				
O/W/E (mL)	40/40/0	40/38/2	40/40/0	40/40/0
Time (min)	3	4	2	1

Source: Pillon, L.Z. and Asselin, A.E., Method for improving the demulsibility of base oils, U.S. Patent 5,282,960, 1994.

TABLE 5.35

Effect of Demulsifier on the Demulsibility of HF SWI 6 Base Stock and PAO 6 Oil

Highly Saturated Base Stocks	Demulsifier (wt%)	O/W/E Volume (mL)	Separation Time (min)
HF SWI 6	0	40/40/0	2
Demulsifier	0.005	40/40/0	2
Synthetic PAO 6	0	40/40/0	1
Demulsifier	0.005	40/40/0	5

Source: Reproduced from Pillon, L.Z., *Interfacial Properties of Petroleum Products*, CRC Press, Boca Raton, FL, 2007. With permission.

separate the emulsion. After HF, the IFT of SWI 6 base stock drastically increased to 45.6 mN/m and the time required to separate the emulsion decreased to 2 min indicating the removal of surface-active molecules adsorbing at the oil/water interface and having emulsifying properties. The FTIR analysis of HF SWI 6 base stock indicated a decrease in the carboxylate acid content.

The synthetic PAO 6 oil was found to have an IFT of 41.9 mN/m and required only 1 min to separate the emulsion indicating the absence of surface-active molecules.

The effect of a demulsifier on the demulsibility of the HF SWI 6 base stock and the synthetic PAO 6 oil is shown in Table 5.35. The HF SWI 6 was found to require 2 min to separate the emulsion indicating very good water separation properties. At a treat rate of 0.005 wt%, the addition of a demulsifier was found to have no significant effect on the stability of the oil/water interface. The synthetic PAO 6 oil having a similar viscosity was found to require only 1 min to separate the emulsion. At the same treat rate of 0.005 wt%, the addition of the same demulsifier was found to increase the time required to separate the emulsion from 1 to 5 min indicating its emulsifying properties in PAO fluid. To prevent overtreatment the lubricant products usually use only 0.004 wt% of demulsifiers. The literature reports that the demulsifiers have complex chemistry ranging from polymers to nonionic emulsifiers. In many cases, mixtures of two or more demulsifiers are needed to destabilize the oil/water interface (Exxon Mobil Chemical, 2001). The use of demulsifiers, while usually effective in destabilizing the oil/water interface, can lead to an increase in the stability of emulsions when their treat rate is too high or strong interactions with the surface-active molecules, adsorbed at the oil/water interface take place.

5.4 METAL SURFACE PROTECTION

5.4.1 RUST INHIBITION

The combined action of water and oxygen on the iron surface leads to the formation of ferrous oxides known as rust. Solvent-refined mineral base stocks and synthetic fluids require additives, known as rust inhibitors, to protect the metal surface from rusting. The literature reports that rust inhibitors typically contain a hydrocarbon "tail" attached to a polar group such as carboxylic acids and their salts, amines, phosphates, polyhydric alcohols, and metal sulfonates. The polar functionality attaches to iron and steel surfaces and the hydrocarbon "tails" extend to out into oil to provide both solubility and a protective layer (Exxon Mobil Chemical, 2001). The early literature reports that low concentrations of magnesium palmitate, zinc naphthenate, aluminum, and barium petroleum sulfonate have been effective in protecting the metal surface against distilled water and salt-containing water (Bondi, 1951). The early literature also reports that the rust inhibitors were effective in mineral oil, when used at the treat rate of 0.2 wt%, but their efficiency to protect the metal surface varied depending on their chemistry and the temperature (Bondi, 1951).

The effect of the chemistry and the temperature on the performance of rust inhibitors in mineral oil at the treat rate of 0.2 wt% is shown in Table 5.36. At a low temperature of maximum 30°C, ester and amine derivatives of fatty acids were reported effective in protecting the metal surface. At increased temperature of maximum 60°C, fatty acids, such as oleic acid, were reported more effective. At a higher temperature of maximum 74°C, higher molecular weight fatty acids are required. At a high temperature of maximum 88°C, barium, magnesium, and zinc salts of fatty

TABLE 5.36

Effect of Chemistry and Temperature on the Performance of Rust Inhibitors

Mineral Oil (Rust Inhibitors 0.2 wt%)	Chemistry Type	Performance Maximum Temp. (°C)
Mono palmitin	C_{16}-ester	32
Mono octadecyl amine	C_{18}-amine	32
Lanolin	Fatty esters/acids	60
Undecylic acid	C_{12}-acid	60
Myristic acid	C_{14}-acid	74
Oleic acid (C18:1)	C_{18}-acid	74
Ricinoleic acid	C_{18}-Hydroxyoleic acid	74
Sorbitan mono oleate	Cyclic fatty ester	88
Tetrahydroabietic acid	4-Ring cyclic acid	88
Stearic acid (C18:0)	Fatty acid	88
Barium petroleum sulfonate	Salt	88
Magnesium naphthenate	Salt	88
Magnesium stearate	Salt	88
Zinc stearate	Salt	88
Zinc laurate	Salt	88

Source: Bondi, A., *Physical Chemistry of Lubricating Oils*, Reinhold Publishing Corporation, New York, 1951. With permission.

acids are needed (Bondi, 1951). Some rust inhibitors were found effective only at low temperatures. The rust-preventive properties of lubricating oils are usually tested following the ASTM D 665 method for rust preventing characteristics of inhibited mineral oils in the presence of water. A mixture of 300 mL of the oil is stirred with 30 mL of distilled water (method A) or 30 mL of synthetic sea water (method B) at 60°C with a cylindrical steel test rod completely immersed therein. After 24 h, the test rod is observed for signs of rusting. Usually the evaluation pass/fail is used and the oil passes the test only if no rusting is observed, not even one rust spot. More recent literature reports on many different alkyl derivatives of succinic acid, used as rust inhibitors, which were found effective in protecting the metal surface against distilled water and salt-containing water (Pillon, 2007).

$$HOOC-CH_2-CHR-COOH$$

Alkylsuccinic acid (ASA)

The effect of the ASTM D 665 rust test severity on the rust-preventive properties of white oil and 150N mineral base stocks containing rust inhibitor are shown in Table 5.37. Following the ASTM D 665 test method, the uninhibited white oil is not capable of preventing rust formation, even in the presence of distilled water, and very severe rusting covering over 75% of the metal surface is observed. However, the inhibited white oil containing 0.015 wt% of dodecyl succinic acid (DSA), can prevent rust formation in the presence of distilled water but not in the presence of synthetic water. Following the ASTM D 665 test method, the uninhibited 150N mineral oil is also not capable of preventing rust formation, even in the presence of distilled water, and very severe rusting of the metal surface is observed. However, the inhibited 150N mineral base stocks containing 0.04 wt% of succinic acid derivative can prevent rust formation in the presence of distilled water and even in the presence of synthetic water containing salts. In many cases, an increased treat rate of rust

TABLE 5.37
Effect of ASTM D 665 Rust Test Severity on Rust Inhibitor Performance

Petroleum Derived Base Stocks Rust Inhibition	ASTM D 665A Distilled Water	ASTM D 665B Synthetic Water
Uninhibited white oil	Fail	Fail
DSA (0.015 wt%)	Pass	Fail
Uninhibited 150N mineral base stock	Fail	Fail
Succinic acid derivative (0.04 wt%)	Pass	Pass

Source: Reproduced from Pillon, L.Z., *Interfacial Properties of Petroleum Products*, CRC Press, Boca Raton, FL, 2007. With permission.

inhibitors is required to pass the more severe ASTM D 665B rust test. The patent literature reports on the partially esterified alkylsuccinic acid (EASA), which is a mixture of about 75 wt% of unreacted tetrapropenylsuccinic acid and about 25 wt% of a partially esterified tetrapropenylsuccinic acid as an effective rust inhibitor in salt-containing water (Pillon and Asselin, 1993).

$$C_{12}H_{23}-CH\ (COOH)-CH_2-COOH$$

Tetrapropenylsuccinic acid (~75 wt%)

$$C_{12}H_{23}-CH\ (COOH)-CH_2-COO(CH_2)_3OH$$

Partially esterified tetrapropenylsuccinic acid (~25 wt%)

The effect of rust inhibitor partially EASA on the rust-preventive properties and the IFT of hydrorefined white oil are shown in Table 5.38. The severity of rusting can be described as light, moderate, or severe. Light rusting indicates no more than six rust spots. Moderate rusting indicates more than six rust spots but confined to less than 5% of the surface of the test rod. Severe rusting indicates more than 5% of the surface is covered with rust (Pillon, 2007). In order to pass the test, not even one rust spot can be present on the surface of the test rod. With an increase in the treat rate of the partially EASA from 0 to 0.09 wt%, a visual evaluation of the iron pin showed a gradual decrease in the rusting severity from severe to moderate and light. Inhibited white oil containing 0.1 wt% of EASA, passed the severe ASTM D 665 B rust test. Water is present during the ASTM D 665 rust test and the amount of the rust inhibitor adsorbed at the oil/water interface will decrease the amount available for adsorption on the metal surface. With an increase in the treat rate of the partially EASA

TABLE 5.38

Effect of EASA on the Rust-Preventive Properties and the IFT of White Oil

White Oil EASA (wt%)	ASTM D 665B Rusting	ASTM D 665B Rating	Oil/Distilled Water IFT (mN/m)
0	Severe	Fail	45.1
0.03	Severe	Fail	19.0
0.04	Severe	Fail	16.0
0.05	Severe	Fail	13.5
0.07	Moderate	Fail	11.9
0.08	Moderate	Fail	11.4
0.09	Light	Fail	10.9
0.10	None	Pass	9.3

Source: Pillon, L.Z. and Asselin, A.E., Lubricating oil having improved rust inhibition and demulsibility, U.S. Patent 5,227,082, 1993.

TABLE 5.39
Performance of EASA Rust Inhibitor in Different Viscosity Mineral Base Stocks

Mineral Base Stocks	ASTM D 665 B	ASTM D 665 B	ASTM D 665 B	ASTM D 665 B
EASA Rust Inhibitor	0.03 wt%	0.04 wt%	0.05 wt%	0.15 wt%
150N (phenol)	Fail	Pass		
600N (phenol)	Fail	Fail	Pass	
1400N (phenol)	Fail	Fail	Fail	Fail
150N (NMP) #1	Pass			
150N (NMP) #2	Fail	Pass		
600N (NMP)	Pass			

Source: Pillon, L.Z. and Asselin, A.E., Lubricating oil having improved rust inhibition and demulsibility, U.S. Patent 5,227,082, 1993.

from 0 to 0.1 wt%, the IFT of white oil gradually decreased from 45.1 to 9.3 mN/m, indicating its adsorption at the oil/water interface of white oil.

The performance of the partially EASA in different viscosity mineral base stocks is shown in Table 5.39. The 150N phenol-extracted base stock was found to require a min 0.04 wt% of the partially EASA to pass the more severe ASTM D 665 B rust test. The higher viscosity 600N phenol-extracted base stock was found to require a min 0.05 wt% of the partially EASA to prevent rust formation. The performance of partially EASA was found to decrease with an increase in the viscosity of phenol-extracted base stocks and, even at such a high treat rate of 0.15 wt%, it was found not effective in preventing rust formation in very high viscosity 1400N base stock. The 150N NMP-extracted base stocks were found to require treat rates of 0.03–0.04 wt% of the same EASA rust inhibitor to prevent rust formation. The higher viscosity 600N NMP-extracted base stock was found to require treat rate of 0.03 wt% of the same EASA rust inhibitor to pass the more severe ASTM D 665 B rust test. In many cases, NMP-extracted base stocks containing higher N contents were found to have improved rust-preventive properties. The FTIR analysis indicated the presence of amines and pyridines that can form salts in the presence of acids, including acidic rust inhibitors, and increase their surface activity in protecting the metal surface (Pillon, 2007).

The performance of the rust inhibitor, EASA in SWI 6 base stock and synthetic PAO 6 fluid is shown in Table 5.40. With an increase in the treat rate from 0.05 to 0.15 wt%, the partially EASA was found not effective in SWI 6 base stock, having over 99 wt% saturate content, and a very severe rusting of the metal pin was observed. With an increase in the treat rate from 0.05 to 0.15 wt%, the partially EASA was also found not effective in synthetic PAO 6 base stock having 100% saturate content and a very severe rusting of the metal pin was observed. The chemistry of rust inhibitors can vary from derivatives of ASA to different sulfonates (Pillon, 2007). Petroleum sulfonates are by-products of the production of white oils by treatment with oleum. The resulting acid tar contains long chain alkylarylsulfonic acids

TABLE 5.40

Performance of Rust Inhibitor EASA in SWI 6 and Synthetic PAO 6 Oils

Base Stocks	SWI 6	PAO 6
EASA (wt%)	ASTM D 665B	ASTM D 665B
0	Fail	Fail
0.05	Fail	Fail
0.10	Fail	Fail
0.15	Fail	Fail

Source: Pillon, L.Z. and Asselin, A.E., Lubricating oil having improved rust inhibition and demulsibility, U.S. Patent 5,227,082, 1993.

that can be neutralized with lyes. To assure more constant quality of sulfonates, more expensive synthetic alkylbenzene sulfonates are available (Braun and Omeis, 2001).

The effect of ASA derivatives and different sulfonates on the rust-preventive properties and the IFT of SWI 6 base stock are shown in Table 5.41. The SWI 6

TABLE 5.41

Effect of ASA Derivatives and Sulfonates on Rust Prevention and IFT of SWI 6 Oil

Rust Inhibitors in SWI 6	Treat Rate (wt%)	ASTM D 665 B
Neat SWI 6 base stock	0	Fail
ASA	0.05	Fail
	0.10	Fail
	0.15	Fail
DSA	0.05	Fail
	0.15	Fail
	0.25	Fail
Sodium sulfonate	0.15	Fail
	0.30	Fail
	0.50	Fail
	0.70	Fail
Calcium sulfonate	0.15	Fail
	0.30	Fail
	0.50	Fail
	0.70	Fail

Source: Pillon, L.Z. et al., Lubricating oil having an average ring number of less than 1.5 per mole containing succinic anhydride amine rust inhibitor, U.S. Patent 5,225,094, 1993.

base stock containing 0.05–0.15 wt% of ASA did not prevent rust formation and a severe rusting of the metal pin was observed. At a high treat rate of 0.15 wt%, a visible precipitation of ASA from SWI 6 base stock was also observed indicating its poor solubility. The SWI 6 base stock containing 0.05–0.25 wt% of dodecyl ASA (DSA) also did not prevent rust formation and a severe rusting of the metal pin was observed. The SWI 6 base stock containing 0.15–0.7 wt% of Na sulfonate also did not prevent rust formation and a severe rusting of the metal pin was observed. The SWI 6 base stock containing 0.15–0.7 wt% of Ca sulfonate also did not prevent rust formation and a severe rusting of the metal pin was observed.

Highly polar additives, such as rust inhibitors, are known to have limited solubility in nonpolar highly saturated base stocks. To prevent rust formation, rust inhibitors need to be in the monomer form to adsorb on the metal surface. The literature reports that the equilibrium at the interface might take a fraction of a second or even a few days because the adsorption process involves the diffusion of the solute molecule from the bulk phase to the surface first (Ross, 1987). The standard Processor Tensiometer K12 can repeat the measurements until the equilibrium at the oil/water is established.

The effect of different chemistry rust inhibitors on the IFT of SWI 6 base stock is shown in Table 5.42. Many hydrofined mineral base stocks required 1 h or even 2 h equilibrium time to reach a constant IFT. The nonpolar SWI 6 base stock containing over 99% saturate content and practically no S and N contents, was found to quickly reach the equilibrium IFT in less than 5 min. After 2 h of equilibrium time, the SWI 6 base stock was found to have a very high IFT of 54.8 mN/m measured at RT. Rust inhibitors are highly polar molecules which are used to adsorb at the oil/metal interface and protect the metal surface. In SWI 6 base stock when mixed with water, highly surface-active rust inhibitors were found to quickly reach the oil/water interface and a drastic decrease in the IFT is observed. The IFT of SWI 6 base stock

TABLE 5.42

Effect of Different Chemistry Rust Inhibitors on the IFT of SWI 6 Base Stock

Equilibrium Time at RT	After 5 min	After 30 min	After 60 min
Rust Inhibitors (0.15 wt%)		IFT (mN/m)	
Neat SWI 6 base stock	54.7	55.4	54.8
EASA	7.5	7.2	7.1
ASA	8.3	8.3	8.1
DSA	7.8	7.1	7.4
Sodium sulfonate	5.9	6.1	6.1
Calcium sulfonate	3.8	4.9	4.7

Source: Pillon, L.Z. et al., Lubricating oil having an average ring number of less than 1.5 per mole containing succinic anhydride amine rust inhibitor, U.S. Patent 5,225,094, 1993.

containing 0.15 wt% of EASA decreased to 7.1 mN/m indicating its adsorption at the oil/water interface. The IFT of SWI 6 base stock containing 0.15 wt% of ASA decreased to 8.3 mN/m and the IFT of SWI 6 base stock containing 0.15 wt% of DSA decreased to 7.1 mN/m also indicating their adsorption at the oil/water interface. The IFT of SWI 6 base stock containing 0.15 wt% of Na sulfonate decreased to 6.1 mN/m and the IFT of SWI 6 base stock containing 0.15 wt% of Ca sulfonate, further decreased to 4.9 mN/m also indicating their adsorption at the oil/water interface. The poor solubility of rust inhibitors in highly saturated nonpolar base stocks, such as SWI 6, might lead to precipitation or micelle formation. In the presence of water, their adsorption at the oil/water interface is taking place and a drastic decrease in the IFT is observed.

5.4.2 CORROSION INHIBITION

In industrial equipment, rusting is the most common type of corrosion which can weaken the metal surface and lead to an enhanced wear. The presence of acidic components in crude oils, petroleum feedstocks, and lube oil base stocks leads to corrosion. The literature reports on the Bohai crude oil from China having a high TAN of 3.34 and causing a severe corrosion. The diesel distillate was found to have TAN of 1.1 mg KOH/g and the acid content of 1.47 wt%. The lube oil distillate having a higher boiling range was found to have a higher TAN of 2.7 mg KOH/g and a higher acid content of 3.53 wt% (Qi et al., 2004). The TAN value of a petroleum product is the weight of potassium hydroxide (KOH) required to neutralize 1 g of oil. The usual components of TAN are not only naphthenic acids but also organic soaps, soaps of heavy metals, organic nitrates, and oxidation by-products. The acids present in Bohai crude oil distillates were isolated by aqueous alcoholic sodium hydroxide and converted into methyl esters with BF_3 as catalyst to be analyzed. The acids found in the diesel distillate were 1–3 ring naphthenic acids and carboxylic acids containing S and N heteroatoms. The acids found in higher boiling lube oil distillate were 1–6 ring naphthenic acids and carboxylic acids also containing S and N heteroatoms and having a higher molecular weight (Qi et al., 2004).

The effect of boiling range on the chemistry of acids found in different distillates from the Bohai crude oil is shown in Table 5.43. Clay refining can be used to remove acids from lube oils if the TAN is below 0.1 mg KOH/g. At a higher TAN, a high volume of clay is required and the waste clay disposal problem makes the use of clay not practical (Wang et al., 2001). The literature reports that lubricating oils are in contact with air and often at elevated temperatures. The temperature can range from ambient for transformer oils to over 300°C in the sump of a car engine. Under high temperature conditions and in the presence of metals, the oil will oxidize leading to an increase in the viscosity and the acidic oxidation by-products causing corrosion (El-Ashry et al., 2006). Many lubricant products contain metal deactivators that form a protective film on metals, such as copper and bronze, and prevent metal ions from entering the oil and extending the oxidation process. They also minimize corrosion by physically restricting the access of the corrosive acids to the metal surface (Raab and Hamid, 2003). The early literature reports that, for the same viscosity oils, with a decrease in their oil/water IFT, a delay in the wetting of the metal surface is

TABLE 5.43

Effect of Boiling Range on the Chemistry of Acids in Different Distillates

Petroleum Fractions	Diesel Distillate	Lube Oil Distillate
Boiling Range	200°C–370°C	370°C–550°C
TAN (mg KOH/g)	1.1	2.7
Acid content (wt%)	1.47	3.53
Acid chemistry		
Naphthenic acids	1–3 rings	1–6 rings
Carboxylic acids	$C_{16}H_{27}O_2SN$	$C_{34}H_{51}O_2SN$
Molecular weight	253	489

Source: Reprinted from Qi, B. et al., *Petrol. Sci. Technol.*, 22(3–4), 463, 2004. With permission.

TABLE 5.44

Effect of IFT on Corrosion Severity in Uninhibited Base Stocks

	Uninhibited Base Stocks		
	White Oil	Mineral Oil #1	Mineral Oil #2
Oil/aqueous H_2SO_4			
Initial IFT (mN/m)	46	35	35
IFT (after 1 h) (mN/m)	45	28	26
IFT (after 2 h) (mN/m)	45	26	25
Metal/aqueous H_2SO_4			
Interface at 25°C (after 18 h)	Severe corrosion	Corrosion	Corrosion

Source: Hughes, R.I., *Corros. Sci.*, 9, 535, 1969. With permission.

observed. A lower IFT leads to a smaller size dispersed water droplets and their rate of approaching the metal surface is lower (Hughes, 1969).

The effect of the IFT on the corrosion severity on the metal surface immersed in different uninhibited base stocks is shown in Table 5.44. White oil containing practically 100% saturate content, has a high IFT of 46 mN/m which, after 2 h of equilibrium time decreased to only 45 mN/m. Mineral oils containing aromatic and polar heteroatom molecules were reported to have a lower IFT of 35 mN/m which gradually decreased to 25–26 mN/m, after the equilibrium time of 2 h, indicating that it takes time for surface-active molecules to adsorb at the oil/water interface. After 18 h at 25°C, the metal surface immersed in uninhibited white oil and uninhibited mineral base stocks were reported to show signs of corrosion. The presence of H_2SO_4 in water was reported to produce etched and pitted scars on the metal surface and the most extensive corrosion was observed in white oil, having the highest

TABLE 5.45

Effect of Acid Chemistry on Rust and Corrosion Prevention in White Oil

White Oil at 22°C	Acid Chemistry	Distilled Water	Aqueous H_2SO_4
Additive (0.1 wt%)	Formula	Rusting (0–8)	Corrosion (0–8)
ASA	$HOOCCH_2CHRCOOH$	0	3
Stearic acid (C_{18})	$CH_3(CH_2)_{16}COOH$	0	6–7
Lauric acid (C_{12})	$CH_3(CH_2)_{10}COOH$	1	5
Caproic acid (C_8)	$CH_3(CH_2)_6COOH$	3	6

Source: Hughes, R.I., *Corros. Sci.*, 9, 535, 1969. With permission.

oil-aqueous H_2SO_4 IFT (Hughes, 1969). Corrosion is a more severe form of rusting caused by the presence of acid compounds in water. The early literature reports on ASA effective in the prevention of rusting, when tested in the presence of distilled water, but not effective in preventing corrosion, when tested in the presence of aqueous H_2SO_4. The visual standard ratings varied from "0," indicating no rusting and no corrosion, to "8" indicating that the entire metal surface is covered with rust or corrosion (Hughes, 1969).

The effect of the acid chemistry on the rust and corrosion-preventive properties of white oil is shown in Table 5.45. At 22°C, white oil containing a typical rust inhibitor, such as ASA, was reported effective in preventing rust formation when tested in the presence of distilled water; however, it is not effective in preventing corrosion when tested and in the presence of aqueous H_2SO_4. At 22°C, white oil containing a stearic acid, having a long C_{18} hydrophobic chain, was also reported effective in preventing rust formation when tested in the presence of distilled water; however, it is not effective in preventing corrosion when tested in the presence of aqueous H_2SO_4. At the same low temperature, white oil containing the same treat rate of 0.1 wt% of lauric acid, having a shorter C_{12} hydrophobic chain, was found not effective in the prevention of rusting and corrosion. White oil containing another monocarboxylic acid, such as caproic acid having a short C_8 hydrophobic chain, was found also not effective in the prevention of rusting and corrosion and actually more rust and corrosion was found on the metal surface. The chemistry of effective corrosion inhibitors vary from amides and imides, which are reaction products of saturated and unsaturated acids with alkylamines and alkanolamines to polyamines (PAs). Imidazoline derivatives were also reported effective as anticorrosive additives in mineral lubricating oils (Kelarev et al., 1998). The patent literature reports that succinic anhydride amine (SAA) was found effective in preventing the rust formation in different viscosity mineral base stocks, including very high viscosity 1400N oils (Pillon, Asselin, and MacAlpine, 1993).

The effect of SAA on the rust-preventive properties of different viscosity mineral base stocks is shown in Table 5.46. The 150N and 600N phenol-extracted base stocks were found to require 0.1 wt% of the SAA, to prevent the rust formation. The very high viscosity 1400N phenol-extracted base stock required a higher treat rate

TABLE 5.46

Effect of SAA on Rust Prevention in Different Viscosity Mineral Base Stocks

Mineral Oils	ASTM D 665B	ASTM D 665B	ASTM D 665B	ASTM D 665B
SAA (wt%)	0.04	0.05	0.1	0.15
150N (phenol)	Fail	Fail	Pass	
600N (phenol)	Fail	Fail	Pass	
1400N (phenol)	Fail	Fail	Fail	Pass

Source: Pillon, L.Z. et al., Lubricating oil for inhibiting rust formation, U.S. Patent 5,397,487, 1995.

TABLE 5.47

Effect of SAA on Rust Prevention in Highly Saturated Base Stocks

Saturated Oils	ASTM D 665B	ASTM D 665B	ASTM D 665B	ASTM D 665B
SAA (wt%)	0.05	0.06	0.1	0.15
White oil	Fail	Fail	Fail	Pass
SWI 6	Fail	Pass		
PAO 6	Fail	Pass		

Source: Pillon, L.Z. et al., Lubricating oil having an average ring number of less than 1.5 per mole containing succinic anhydride amine rust inhibitor, U.S. Patent 5,225,094, 1993.

0.15 wt% of the SAA, to prevent rust formation. In comparison with the typical acidic type rust inhibitor, such as partially esterified alkylsuccinic derivative (EASA), the use of basic corrosion inhibitor, such as SAA, is more effective in preventing rust formation in the very high viscosity 1400N mineral base stocks. The patent literature also reports that SAA was found effective in preventing rust formation in nonpolar highly saturated base stocks (Pillon, Asselin, and MacAlpine, 1993).

The effect of basic corrosion inhibitor, SAA on the rust-preventive properties of different highly saturated base stocks is shown in Table 5.47. The typical acidic rust inhibitors were developed for conventionally processed mineral base stocks which contain significant amounts of aromatic and polar compounds and have good solvency properties. The hydrorefined white oil having over 99% saturate content and containing 0.1 wt% of SAA, was found effective in preventing rust formation. SWI 6 oil having over 99% saturate content and containing 0.06 wt% of SAA was also found effective in preventing rust formation. The synthetic PAO 6 oil having 100% saturate content and containing 0.06 wt% of SAA, was also found effective in preventing rust formation. However, the use of another basic type molecule, such as PA, in SWI 6 base stock was found not effective in preventing the rust formation.

The effect of SAA and PA on the IFT, the demulsibility, and the rust-preventive properties of SWI 6 base stock are shown in Table 5.48. Rust and corrosion inhibitors are surface-active molecules that adsorb at the oil/water and the oil/metal interface.

TABLE 5.48

Effect of Different Basic Inhibitors on IFT, Demulsibility, and Rusting in SWI 6

Interfacial Properties	Neat SWI 6	SAA	PA
BN Content	0	Higher	Lower
Basic inhibitors (wt%)	0	0.15	0.15
IFT (mN/m) (after 5 min)	54.7	2.6	4.4
IFT (mN/m) (after 30 min)	55.4	2.5	3.6
IFT (mN/m) (after 60 min)	54.8	2.2	3.7
Basic inhibitors (wt%)	0	0.09	0.09
ASTM D 1401 (54°C)			
Emulsion separation time (min)	4	20	10
ASTM D 665B (60°C/24 h)	Fail	Pass	Fail

Source: Pillon, L.Z. et al., Lubricating oil having an average ring number of less than 1.5 per mole containing succinic anhydride amine rust inhibitor, U.S. Patent 5,225,094, 1993.

The IFT of SWI 6 base stock containing 0.15 wt% of SAA, drastically decreased from 54.8 to 2.2 mN/m, indicating its adsorption at the oil/water interface. Following the ASTM D 1401 demulsibility test, the addition of 0.09 wt% of SAA increased the emulsion separation time from 4 to 20 min at 54°C, indicating its emulsifying properties. Under the ASTM D 665 B rust test conditions, SWI 6 base stock containing 0.09 wt% of SAA was found to have good rust-preventive properties and passed the rust test. The IFT of SWI 6 base stock containing 0.15 wt% of PA also decreased to low value of 3.7 mN/m indicating its adsorption at the oil/water interface. Following the ASTM D 1401 demulsibility test, the addition of 0.09 wt% of PA increased the emulsion separation time to only 10 min at 54°C indicating less emulsifying properties. Under the ASTM D 665B rust test conditions, SWI 6 base stock containing 0.09 wt% of PA was found to have poor rust-preventive properties and failed the rust test. The chemistry and the treat rate of rust and corrosion inhibitors need to be carefully selected to prevent degradation in demulsibility properties of lubricant products.

The types and the chemistry of different corrosion inhibitors are shown in Table 5.49. Rust and corrosion inhibitors are used in nearly every lubricant to protect the metal surface of any machinery or the metalworking tool from the attack of moisture, oxygen, and aggressive acidic products. The acidic products may be formed by the thermal decomposition or the oxidation degradation of lubricating oils or caused by the specific application (Braun and Omeis, 2001).

Rust inhibitors are used to protect ferrous metals from rusting. Corrosion inhibitors are used to protect nonferrous metals from corrosion. The literature reports on the chemistry of some corrosion inhibitors which are effective in protecting cast iron and steel without causing any corrosion of copper, lead, or various alloys. The use

TABLE 5.49
Types and Chemistry of Corrosion Inhibitors

Corrosion Inhibitor Type	Corrosion Inhibitor Type
C_5–C_{20} fatty acid/amine chemistry	Sulfonic acid/fatty acid/amine chemistry
Caprylic acid with aliphatic amine	Derivatives of octylamine
Trimellitic anhydride with aliphatic amine	Derivatives of ethylhexylaminopropylamine
Itaconic acid with aliphatic amine	Derivatives of methylhexahydropyrimidine
Gluconic acid with aliphatic amine	Derivatives of butyl-phenylenediamine
Stearic acid with aliphatic amine	Derivatives of benzylamine
Oleic acid with aliphatic amine	Derivatives of octyl-sulfostearate
Fatty acids with aromatic amines	Derivatives of sulfonated fatty acids
Polyamides of fatty acids	Derivatives of alkylarylsulfonic acid
Mixtures of fatty acids with amines	Mixtures of amine sulfonates and fatty esters

Source: Milovanov, V.D. and Chkheidze, O.Ya., *Chem. Technol. Fuels Oils*, 13(4), 292, 1977.

of sulfonic acid/fatty acid/amine type was reported effective as a rust inhibitor and a strong detergent where sulfonic acid was reported to be petroleum derived or synthetic (Milovanov and Chkheidze, 1977). Vapor phase corrosion inhibitors for closed systems are compounds with high affinity to metal surfaces and relatively high vapor pressure to guarantee their availability on parts that are not steadily in direct contact with the corrosion inhibited lubricant (Braun and Omeis, 2001). Rust and corrosion are two phenomena that increase wear and might lead to the breakdown of the equipment.

5.4.3 Wear Prevention

The literature reports on common terms related to lubrication and wear. The presence of water and acidic compounds leads to metal corrosion. Corrosive wear is caused by chemical reaction. Abrasive wear is a wear between two surfaces in relative motion due to particles or surface roughness. Adhesive wear is caused by metal-to-metal contact and characterized by local welding and tearing of the surface (Lubrizol, 2001). Fretting is described as the wear resulting from small amplitude motion between two surfaces. Scuffing is an abnormal engine wear due to localized welding and fracture. Pitting is described as surface cavities that may be related to fatigue, overload, or corrosion. Fatigue is described as cracking, flaking, or spalling of a surface due to stresses beyond the endurance limit of the material (Lubrizol, 2001). Friction, which is resistance to motion of one object over another, depends on the smoothness of the contacting surfaces and the force with which they are pressed together (Lubrizol, 2001).

Elastohydrodynamic lubrication is characterized by high unit loads and high speeds in rolling elements where the mating parts deform elastically due to the incompressibility of the lubricant film under very high pressure (Lubrizol, 2001). Boundary lubrication is described as lubrication between two rubbing surfaces without the development of a full fluid lubrication film. Boundary lubrication occurs under high loads and requires the use of antiwear or extreme pressure (EP) additive to prevent metal-to-metal contact. Scuffing can be prevented by using antiwear, EP, and friction modifier additives (Lubrizol, 2001). Antiwear agents or their reaction products form thin, tenacious films on highly loaded parts to prevent metal-to-metal contact. EP agents prevent sliding metal surfaces from seizing under EP conditions (Lubrizol, 2001).

The chemistry and antiwear properties of different esters are shown in Table 5.50. The literature reports on the chemistry and the adsorption properties of esters used as antiwear additives. The dithiophosphates were found to have highest adsorption properties followed by xanthates and dithiocarbamates (Mustafaev, Kulieva, and Kulibekova, 2001). Initially antiwear agents were lead salts of long chain carboxylic acids and were often used in combination with sulfur-containing compounds. Oil-soluble sulfur–phosphorous and chlorinated compounds were also used as antiwear agents (Bardasz and Lamb, 2003). The literature reports that a series of *N,N*-bis(alkylthiomethyl)-*N*-alkylamines were synthesized and were found to have antiwear properties superior to well-known dibenzyl disulfide (Movsumzade et al., 2002).

The chemistry of organomolybdenum compounds having antiwear and EP properties is shown in Table 5.51. The literature reports on the oil-soluble

TABLE 5.50
Chemistry and Antiwear Properties of Different Esters

Ester Type	Chemistry of Antiwear Additive	Wear (mm)
Dithiophosphate (DTP)	*S*-(Acetoxymethyl)-*O,O*-diisopropyl	0.39
DTP	*S*-(Benzoyloxymethyl)-*O,O*-diisopropyl	0.40
DTP	*S*-(Methoxycarbonylmethyl)-*O,O*-diisopropyl	0.53
DTP	*S*-Benzoyl-*O,O*-diisopropyl	0.66
Xanthate (X)	*S*-(Benzoyloxymethyl)-*O*-isopropyl	0.7
X	*S*-(Methoxycarbonylmethyl)-*O*-isopropyl	0.75
Dithiocarbamate (DTC)	*S*-(Methoxycarbonylmethyl)-*N,N*-diethyl	0.82

Source: Mustafaev, N.P. et al., *Chem. Technol. Fuels Oils*, 37(2), 126, 2001.

TABLE 5.51
Chemistry of Organomolybdenum Compounds Having Antiwear and EP Properties

Organomolybdenum Type	Organomolybdenum Chemistry
S-free and P-free (molybdate)	Organic amide molybdenum complex
S-containing and P-free (MoDTC)	Molybdenum dithiocarbamate
S-containing and P-containing (MoDTP)	Molybdenum dithiophosphate

Source: Feng, X. et al., *Lubr. Sci.*, 18, 1, 2006.

TABLE 5.52

Effect of ZnDDP on the Antiwear Properties of Organomolybdenum Compounds

Mo-Compound (wt%)	ZnDDP (wt%)	Load 588N/ Wear (mm)
Molybdate (1.0)	0	0.45
Molybdate (1.0)	0.5	0.40
Molybdate (0.5)	1.0	0.43
MoDTC (1.0)	0	0.76
MoDTC (1.0)	1.0	0.39
MoDTC (0.5)	0.5	0.37
MoDDP (1.0)	0	0.39
MoDDP (1.0)	1.0	0.39
MoDDP (0.5)	0.5	0.4

Source: Feng, X. et al., *Lubr. Sci.*, 18, 1, 2006.

organomolybdenum compounds, as antiwear and EP additives. They are classified into three groups: sulfur- and phosphorous-free compounds, sulfur-containing and phosphorous-free compounds, and sulfur- and phosphorous-containing compounds (Feng et al., 2006). The antiwear properties of different organomolybdenum compounds were tested in 150N paraffinic base stock using the four-ball tester under loads of 392, 588, and 784N for 30 min. After the addition of zinc dialkyl dithiophosphates (ZnDDP), an improvement in their performance was observed. The literature reports on antiwear synergism between different organomolybdenum compounds and ZnDDP molecules (Feng et al., 2006).

The effect of ZnDDP molecule on the antiwear properties of different organomolybdenum compounds is shown in Table 5.52. The literature reports that despite the tendency toward a decrease in the P content of motor oils, for environmental reasons, most oils contain ZnDDPs or zinc diaryl dithiophosphates. They are used because they are multifunctional and have antioxidant, anticorrosion, antiwear, and antiscuffing properties (Evstafev et al., 2001). The literature also reports that some highly alkaline and metallic type additives can lead to an increase in the TAN and the total base number (TBN) of oils (Pirro and Wessol, 2001). The TBN of oil indicates the quantity of acid needed to neutralize the basic type molecules.

The effect of Zn dithiophosphates (ZDTP) on pH, the TBN, the oxidation stability, and the corrosive properties of oil is shown in Table 5.53. The ZDTP additive concentration in oil was decreased from 2.5 to 0.7 wt%, to obtain different level of active ingredient. With an increase in the Zn content from 2.4% to 11.8%, and P content from 2.5% to 10.5% of ZDTP additives, the pH increased while the TBN content increased indicating no presence of acid type molecules. After oxidation for 3 h at 230°C, the oil viscosity and the corrosive properties of oil were measured. With an increase in Zn and P contents of ZDTP molecules, a decrease in the oxidative viscosity, from 68% to 49%, was reported indicating the antioxidant properties.

TABLE 5.53
Effect of ZDTP on TBN, Oxidation Stability, and the Corrosive Properties of Oil

Oil (Zn-Compounds)	ZDTP #1	ZDTP #2	ZDTP #3
Treat Rate (wt%)	**2.5**	**1.4**	**0.7**
Zn (%)	2.4	5.5	11.8
P (%)	2.5	4.8	10.5
pH	3.8	6.1	5.8
TBN (mg/g)	0	1.2	7.7
Oxidation at 230°C/3 h			
Viscosity increase (%)	68	65	49
Corrosion of Pb (%)	8.4	7.2	7

Source: Evstafev, V.P. et al., *Chem. Technol. Fuels Oils*, 37(6), 427, 2001.

TABLE 5.54
Effect of Different Composition Zn-Compounds on Corrosion Inhibition

Zn-Compounds	Zn	P	S	ASTM D 130
Composition	**wt%**	**wt%**	**wt%**	**Corrosion Test**
Mineral oil (1.5%)	5.18	6.49	22.0	1a
Mineral oil (1.5%)	0.27	5.12	10.3	1a
Mineral oil (1.5%)	9.82	—	12.3	1b
Mineral oil (1.5%)	5.14	4.92	6.1	1b
Mineral oil (1.5%)	3.44	—	4.0	1b
Mineral oil (1.5%)	7.85	6.73	12.7	2b
Mineral oil (1.5%)	8.15	6.20	13.7	3b

Source: Dinoiu, V. et al., *Lubr. Sci.*, 18(3), 165, 2006.

With an increase in pH and TBN contents of oil containing ZDTP molecules, the corrosion of Pb decreased from 8.4% to 7%, indicating the corrosion inhibitor properties. The recent literature reports on the composition of different chemistry Zn-compounds and their effect on corrosion inhibition. Their performance was tested in mineral base stocks containing 1.5% of each additive (Dinoiu, Florescu, and Stanescu, 2006).

The effect of different composition Zn-compounds, 1.5% in mineral base stocks, on corrosion inhibition is shown in Table 5.54. The ASTM D 130 test for detection of copper corrosion from petroleum products by the copper strip tarnish test was used to assess the corrosivity of mineral base stocks, containing different chemistry Zn-compounds, and oxidized for 6h at 315°C (Dinoiu, Florescu, and

TABLE 5.55
Effect of Different Composition Zn-Compounds on Wear Prevention

Zn (%)	P (%)	S (%)	Wear Scar (mm)	Weld Load (N)
0.27	5.12	10.29	0.62	1400
5.18	6.49	22.0	0.50	1800
9.82	—	12.3	0.78	1800
3.44	—	4.04	1.41	1500
5.14	4.92	6.14	0.42	1800
7.85	6.73	12.7	0.58	1600
8.15	6.20	13.67	0.98	1700

Source: Dinoiu, V. et al., *Lubr. Sci.*, 18(3), 165, 2006.

Stanescu, 2006). According to the ASTM D 130 test, the freshly polished strip showing slight tarnish is rated 1a–b, moderate tarnish is rated 2a–e, dark tarnish is rated 3a–b, and corrosion is rated 4a–c indicating a gradual change in color from light orange (1a) to jet black (4c) (ASTM D 130, 2007). The Zn-compounds, containing P and a high S content were reported as the most effective in preventing corrosion. The Zn-compounds, containing Zn, P, and S content, were also reported as the most effective in preventing wear while Zn-compounds, containing P and S and without P, were effective in terms of antiwear EP performance (Dinoiu, Florescu, and Stanescu, 2006).

The effect of different composition Zn-compounds, 1.5% in mineral base stocks on wear scar and weld load is shown in Table 5.55. The antiwear and EP additives were reported to function by thermally decomposing to yield compounds that react with the metal surface and form a thin layer (Bardasz and Lamb, 2003). The recent literature reports on the film forming characteristics of ester octadecanoates and their thio and thiopyrophosphoro derivatives on the iron surface. The molecules were reported to be linked to the metal surface through hydroxyl, oxy, thia, oxyphosphoro, and thiophosphoro moieties forming chelated and coordinated complexes (Chibber et al., 2006). Some lubricants also contain friction modifiers. Friction, described as resistance to motion of one object over another, depends on the smoothness of the contacting surfaces and the force with which they are pressed together (Lubrizol, 2001). The literature reports on the oil-soluble organomolybdenum compounds widely used in gear oils and engine lubricants to reduce wear and friction (Feng et al., 2006). The literature also reports on antiwear properties and friction reduction of organic borates. They are classified into three groups: nitrogen-containing, sulfur-containing, and nitrogen- and sulfur-containing compounds (Huang et al., 2006). With the variation in the properties of petroleum-derived base stocks and the chemistry of additives used to protect the metal surface, the variation in the performance of lubricants in terms of rust inhibition, corrosion inhibition, and wear prevention will be observed.

REFERENCES

ASTM, D 130 standard test method for detection of copper corrosion from petroleum products by the copper strip tarnish test, ASTM International, 2007.

Bardasz, E.A. and Lamb, G.D., Additives for crankcase lubricant applications, in *Lubricant Additives Chemistry and Applications*, Rudnick, L.R., Ed., Marcel Dekker, Inc., New York, 2003.

Bondi, A., *Physical Chemistry of Lubricating Oils,* Reinhold Publishing Corporation, New York, 1951.

Braun, J. and Omeis, J., *Additives, in Lubricants and Lubrications*, Mang, T. and Dresel, W., Eds., Wiley-VCH GmbH, Weinheim, 2001.

Butler, K.D. and Henderson, H.E., Adsorbent processing to reduce base stock foaming, U.S. Patent 4,600,502, 1990.

Chibber, V.K. et al., *Lubrication Science*, 18, 63 (2006).

Denis, J., Briant, J., and Hipeaux, J.C., *Lubricant Properties Analysis and Testing*, Editions Technip, Paris, 2000.

Dinoiu, V., Florescu, D., and Stanescu, C., *Lubrication Science*, 18(3), 165 (2006).

El-Ashry, El-S.H. et al., *Lubrication Science*, 18(2), 109 (2006).

Evstafev, V.P. et al., *Chemistry and Technology of Fuels and Oils,* 37(6), 427 (2001).

Exxon Mobil Chemical, http://www.exxon.mobil.co (accessed 2001).

Feng, X. et al., *Lubrication Science*, 18, 1 (2006).

Han, S. et al., *Energy and Fuels*, 19, 625 (2005).

Huang, W. et al., *Lubrication Science*, 18, 77 (2006).

Hubmann A. and Lanik A., *Tribologie und Schmierungestechnik* (in German), 35(3), 138 (1988).

Hubman, A. and Lanik, A., *Journal of Synthetic Lubrication*, 2(2), 121 (2006).

Hughes, R.I., *Corrosion Science*, 9, 535, 1969.

Kelarev, V.I. et al., *Petroleum Chemistry*, 38(5), 348 (1998).

Kichkin, G.I., *Chemistry and Technology of Fuels and Oils*, 4, 49 (1966).

Lake, D.B. and Hoh, G.L.K., *Journal of the American Oil Chemists' Society*, 40, 628 (1963).

Lubrizol, http://www.lubrizol.com (accessed 2001).

Milovanov, V.D. and Chkheidze, O.Ya., *Chemistry and Technology of Fuels and Oils*, 13(4), 292 (1977).

Movsumzade, M.M. et al., *Petroleum Chemistry*, 42(5), 363 (2002).

Mustafaev, N.P., Kulieva, M.A., and Kulibekova, T.N., *Chemistry and Technology of Fuels and Oils*, 37(2), 126 (2001).

Owen, M.J., *Industrial and Engineering Chemistry Product Research and Development*, 19, 97 (1980).

Pape, P.G., Silicones: Unique chemicals for petroleum processing, SPE 10089, *56th Annual Fall Technical Conference and Exhibition of SPE*, San Antonio, 1981.

Pape, P.G., *Journal of Petroleum Technology,* 35, 1197 (1983).

Pillon, L.Z., *Petroleum Science and Technology*, 21(9&10), 1469 (2003).

Pillon, L.Z., *Interfacial Properties of Petroleum Products*, CRC Press, Boca Raton, FL, 2007.

Pillon, L.Z. and Asselin, A.E., Lubricating oil having improved rust inhibition and demulsibility, U.S. Patent 5,227,082, 1993.

Pillon, L.Z. and Asselin, A.E., Method for improving the demulsibility of base oils, U.S. Patent 5,282,960, 1994.

Pillon, L.Z. and Asselin A.E., Antifoaming agents for lubricating oils, U.S. Patent 5,766,513, 1998.

Pillon, L.Z., Asselin, A.E., and MacAlpine, G.A., Lubricating oil having an average ring number of less than 1.5 per mole containing succinic anhydride amine rust inhibitor, U.S. Patent 5,225,094, 1993.

Pillon, L.Z., Reid, L.E., and Asselin, A.E., Lubricating oil for inhibiting rust formation, U.S. Patent 5,397,487, 1995.

Pillon, L.Z., Asselin, A.E., and Vernon, P.D.F., Method for reducing foaming of lubricating oils, U.S. Patent 6,090,758, 2000.

Pirro, D.M. and Wessol, A.A., *Lubrication Fundamentals*, 2nd ed., Marcel Dekker Inc., New York, 2001.

Poindexter, M.K. et al., *Energy and Fuels*, 16, 700 (2002).

Qi, B. et al., *Petroleum Science and Technology*, 22(3&4), 463 (2004).

Quinn, C., Traver, F., and Murthy, K., *Silicones, in Synthetic Lubricants and High-Performance Functional Fluids*, 2nd ed., Rudnick, L.R. and Shubkin, R.L., Eds., Marcel Dekker Inc., New York, 1999.

Raab, M.J. and Hamid, S., *Additives for Food-Grade Lubricant Applications, in Lubricant Additives Chemistry and Applications*, Rudnick, L.R., Ed., Marcel Dekker, Inc., New York, 2003.

Rome, F.C., *Hydrocarbon Asia*, May/June, 42 (2001).

Ross, S., Foaminess and capilarity in apolar solutions, in *Interfacial Phenomena in Apolar Media*, Eicke, H.F. and Parfitt, G.D., Eds., Marcel Dekker Inc., New York, 1987.

Sequeira, A. Jr., *Lubricant Base Oil and Wax Processing*, Marcel Dekker, Inc., New York, 1994.

Sigma-Aldrich, *Handbook of Fine Chemicals*, Sigma-Aldrich Company, St. Louis, MO, 2007–2008.

Tourret, R. and White, N., *Aircraft Engineering*, 24, 122 (1952).

Wang, Y. et al., *Petroleum Science and Technology*, 19(7&8), 923 (2001).

Watanabe, J., Fukushima, T., and Nose, Y., *Proceedings of the JSLE–ASLE International Lubrication Conference*, 641 (1976).

6 Lubricant Formulation

6.1 INDUSTRIAL AND AUTOMOTIVE LUBRICANTS

Lubrication is described as the control of wear and friction between moving surfaces in contact by the introduction of a lubricant that can be a fluid, solid, or plastic substance. Polymer films are used when special surface protection is required in addition to lubricity (Mang, 2001). Friction exists in all moving parts and can be beneficial or destructive. In a moving vehicle, if friction is applied to make it stop, it is constructive. In a machine, friction between moving parts causes wear and it is destructive. In this case, friction needs to be minimized to prolong the life of the machine (Shell, 2002). Applying a thin film of lubricating oil can prevent the rubbing of the metal parts against each other. The continuous presence of an oil film between the moving parts is referred to as lubrication (Shell, 2002). Tribology is described as the science of the interactions between surfaces moving relative to each other (Lubrizol, 2001).

Originally, a lubricant was a nonvolatile vacuum-distilled base stock, which was available in different viscosities. The light oil was used during winter and the heavy oil during summer. The unstabilized oil had to be changed frequently, but it was suitable for all applications. As a result of the lubricant development, there are many different lubricant products for different applications (Kajdas, 1997). Petroleum-derived lubricants include lubricating oils, oil-in-water emulsions, and greases. Usually, a lubricant is made of a petroleum-derived base stock and additives, which are in relatively small quantities. Lubricants are generally divided into two groups known as industrial and automotive products. The names of industrial oils, such as circulating oils, turbine oils, compressor oils, hydraulic oils, gear oils, and metalworking fluids, are derived from their applications (Mang, 2001). Greases are semisolid or solid lubricants, and by far their most important application was reported to be the lubrication of rolling element bearings (Singh, 2008).

The typical industrial lubricants and their applications are shown in Table 6.1. The presence of water in a lubricant is generally undesirable. Lubricating oils, such as pneumatic tool oils, have an excellent resistance to water washing, good rust- and corrosion-protective properties, and high load-bearing characteristics, and form an oil film with moisture as well as an oil-protective film (Apar Industries, 2003). Compound oils are emulsions where the continuous phase is a mineral oil, and the disperse phase is vegetable or animal oil, which can form waxy layers on the metal walls of lubricating systems. These emulsions can lubricate in the presence of water and are used to lubricate textile machinery spindles (Denis, Briant, and Hipeaux, 2000). Water-in-oil emulsions can be used for the lubrication of air compressors. In some applications, the base stock must meet other performance requirements, such as the cooling effect. Rolling oil emulsions are used in high speed rolling

TABLE 6.1

Typical Industrial Lubricants and Their Applications

Industrial Oils	Industrial Specialty Oils	Industrial Greases
Circulating oils	Cylinder oils	EP type
Turbine oils	Thermic fluids	Multipurpose
Hydraulic oils	Sugar mill oils	High temperature
Gear oils	Open gear compounds	Graphite type
Spindle oils	Metalworking fluids	
Machinery oils	Process oils	
Pneumatic tool oils	Agricultural spray oils	
Waylube oils	Shuttering fluids	
Compressor oils		
Vacuum pump oils		

Source: Apar Industries, http://www.aparind.com (accessed 2003).

mills to facilitate the rolling process and are expected to provide cooling as well as lubrication. In the case of metalworking fluids, where strong cooling properties are needed water-in-oil emulsions are often used. The water serves as coolant and the oil as a lubricant. Cutting metal generates a lot of heat, which is why the most widely used cutting lubricants are oil-in-water emulsions (Denis, Briant, and Hipeaux, 2000).

The different functions and properties of hydraulic fluids are shown in Table 6.2. Hydraulic oils perform many other functions in addition to lubrication, such as transmitting pressure and energy, and have different properties. Some hydraulic oils are fire resistant. The important characteristics of hydraulic oils also vary depending on the severity of the service. Antiwear special purpose hydraulic oils are recommended for use in screw compressors, vacuum pumps, mining machinery, machine tools, and high performance electrohydraulic and numerically controlled systems (Apar Industries, 2003). Since viscosity varies with temperature, it is necessary to consider the actual operating temperature of the oil in the machine. When selecting the proper oil for a given application, the viscosity is a primary consideration. Industrial lubricants comprise of a wide range of products including liquid mineral oils, water-based fluids, and greases. The American Gear Manufacturers Association (AGMA) defines nine viscosity ranges for industrial gear oils. Metalworking lubricants are not classified under the viscosity system, since properties other than viscosity, cooling, and lubrication are more important. Greases are classified by the National Lubricating Grease Institute (NLG) USA in accordance with the cone penetration method. Industrial lubricating oil is the most important type of industrial lubricants, and the majority of industrial oils are classified according to the ISO 3448 viscosity grades. Each grade is represented by the whole number, which is obtained by rounding off the midpoint kinematic viscosity at 40°C (Kajdas, 1997).

TABLE 6.2

Different Functions and Properties of Hydraulic Fluids

Hydraulic Fluid Functions	Hydraulic Fluid Properties
Lubricant	Proper viscosity
	Low temperature fluidity
	Thermal and oxidative stability
	Hydrolytic stability/water tolerance
	Cleanliness and filterability
	Demulsibility
	Corrosion control
	Antiwear characteristics
Power transfer	Low compressibility
	Fast air release
	Low foaming tendency
	Low volatility
Heat transfer	Good thermal capacity and conductivity
Sealing medium	Proper viscosity, VI, and shear stability
Pump efficiency	Proper viscosity and high VI
Special function	Fire resistance
	Friction modification
	Radiation resistance
Environmental impact	Low toxicity and biodegradability

Source: Reproduced from Placek, D.G., Hydraulics, in *Synthetics, Mineral Oils, and Bio-Based Lubricants Chemistry and Technology*, Rudnick, L.R., Ed., CRC Press, Boca Raton, FL, 2005. With permission.

The viscosity system for industrial fluid lubricants is shown in Table 6.3. The literature reports that high speed and lightly loaded plain bearings need a low viscosity plain mineral oil. Higher loadings and lower speeds require higher viscosity oils. Fast running grinding spindles with plain bearings require ISO VG 5 or ISO VG 7 mineral oils. The lubrication of bearings for the machine tools requires ISO VG 46 or ISO VG 68 mineral oils. Bearings operating under high loads need ISO VG 68 or ISO VG 100 lubricating oils (Kajdas, 1997). The general machinery oils, classified according to ISO viscosity grades, encompass mineral oil products from light oils for spindle lubrication to heavy black oils for the lubrication of wire ropes. Open gear compounds, which are heavy adhesive-type black bituminous lubricants, have good oil adhesiveness to reduce the oil consumption, good load carrying capability, and protect against rusting during the idle period. They are recommended for open gears, wire rope, and chain drive lubrication. They are also recommended for the lubrication of kiln tyres, operating at high temperatures, in cement mills and chemical plants (Apar Industries, 2003). The most important industrial lubricating oils, by volumes consumed by the industry, are circulating oils including turbine oils, hydraulic oils, and industrial gear oils (Lee and Harris, 2003).

TABLE 6.3

Viscosity System for Industrial Fluid Lubricants

Viscosity System Grade Identification	Midpoint Viscosity at 40°C (cSt)	Viscosity Limit Min at 40°C (cSt)	Viscosity Limit Max at 40°C (cSt)
ISO VG 2	2.2	1.98	2.42
ISO VG 3	3.2	2.88	3.52
ISO VG 5	4.6	4.14	5.06
ISO VG 7	6.8	6.12	7.48
ISO VG 10	10	9	11
ISO VG 15	15	13.5	16.5
ISO VG 22	22	19.8	24.2
ISO VG 32	32	28.8	35.2
ISO VG 46	46	41.4	50.6
ISO VG 68	68	61.2	74.8
ISO VG 100	100	90	110
ISO VG 150	150	135	165
ISO VG 220	220	198	242
ISO VG 320	320	288	352
ISO VG 460	460	414	506
ISO VG 680	680	612	748
ISO VG 1000	1000	900	1100
ISO VG 1500	1500	1350	1650
ISO VG 2200	2200	1980	2420
ISO VG 3200	3200	2880	3520

Source: Reprinted from ASTM D 2422 standard classification of industrial fluid lubricants by viscosity system, ASTM International, 2007. With permission.

The typical performance requirements of industrial lubricants are shown in Table 6.4. The term "circulation system" refers to a system in which oil is delivered from a central reservoir to all bearings, gears, and other elements requiring lubrication. All the oil drains back to a central sump and is reused (Pirro and Wessol, 2001). The performance requirements of circulating oils are to resist oxidation and sludge formation, provide bearing lubrication, remove heat through circulation, serve as hydraulic fluid for governor and other control equipment, lubricate reducing gears, protect against corrosion and rust, allow rapid separation of water, and resist foaming (Lee and Harris, 2003). The turbine oils are required to have a good oxidation stability and excellent interfacial properties, such as resistance to foaming, ability to quickly release entrained air, good demulsibility, and rust preventive properties. Antiwear special purpose hydraulic oils have excellent lubrication and antiwear properties. Industrial gear oils are recommended for all the types of heavy duty enclosed gear drives with circulation or splash lubrication systems operating under heavy or shock-loading conditions up to an operating temperature of 95°C, or, in some cases, of 110°C (Apar Industries, 2003). The extreme pressure (EP) additive is used to protect gears against wear when subjected to shock loading.

TABLE 6.4

Typical Performance Requirements of Industrial Lubricants

Industrial Lubricant	Circulating Oils Turbine Oils	Hydraulic Oils	Industrial Gear Oils	Compressor Oils	Industrial Grease
Oxidation stability	Yes	Yes	Yes	Yes	Yes
Foaming resistance	Yes	Yes	Yes	Yes	
Air separation	Yes	Yes	Yes		
Water separation	Yes	Yes	Yes	Yes	
Rust prevention	Yes	Yes	Yes	Yes	Yes
Antiwear properties		Yes	Yes	Yes	Yes
EP properties			Yes		Yes

Source: Reproduced from Lee, F.L. and Harris, J.W., Long-term trends in industrial lubricant additives, in *Lubricant Additives Chemistry and Applications*, Rudnick, L.R., Ed., Marcel Dekker, Inc., New York, 2003. With permission.

TABLE 6.5

Typical Automotive Lubricants and Their Applications

Automotive Oils	Automotive Specialty Oils	Automotive Greases
2T oils	Gear oils	Wheel bearing
4T oils	Transmission fluids	Chasis
Engine oils	Hyfluids	
Diesel engine oils	Brake fluids	
	Shock absorber oils	
	Radiator coolant	

Source: Apar Industries, http://www.aparind.com (accessed 2003).

The typical automotive lubricants and their applications are shown in Table 6.5. Engine oils fall into three categories, namely, passenger car engine oils, heavy-duty engine oils and small engines (both two and four strokes). The literature reports that there are even different automotive lubricants for the vehicles of different manufacturers (Mills, Lindsay, and Atkinson, 1997). The engine oil market is the largest share of the automotive segment, which represents 50% of the worldwide demand for lubricants. The viscosity of engine oils affects the cold-cranking speed of an automobile engine, which is the ability of the engine to start at low temperature. The Society of Automotive Engineers (SAE) classifies oils for use in automotive engines by viscosities determined at both high and low shear rates; at low temperatures (−5°C to −40°C); and at low shear rates and high temperature (100°C), and at high shear rate and high temperature (150°C). The Cold Cranking Stimulator (CCS) test measures the apparent viscosity of the oil at low temperature and high shear rate.

TABLE 6.6

SAE J300 Engine Oil Viscosity Classification (2004)

SAE Viscosity Grade	Cranking Viscosity (cP) Max at Temp. (°C)	Pumping Viscosity (cP) Max at Temp. (°C)	Kinematic Viscosity (cSt) Min at 100°C	Kinematic Viscosity (cSt) Max at 100°C
0W	6,200 at −35	60,000 at −40	3.8	—
5W	6,600 at −30	60,000 at −35	3.8	—
10W	7,000 at −25	60,000 at −30	4.1	—
15W	7,000 at −20	60,000 at −25	5.6	—
20W	9,500 at −15	60,000 at −20	5.6	—
25W	13,000 at −10	60,000 at −15	9.3	—
20	—	—	5.6	<9.3
30	—	—	9.3	<12.5
40	—	—	12.5	<16.3
50	—	—	16.3	<21.9
60	—	—	21.9	<26.1

Sources: Society of Automotive Engineers, SAE J300 Viscosity Grades for Engine Oils. With permission; Lynch, T.R., *Process Chemistry of Lubricant Base Stocks*, CRC Press, Boca Raton, FL, 2007. With permission.

High temperature–high shear viscosities are measured at 150°C under shear stresses similar to those experienced in engine bearings.

The SAE Standard J300 engine oil viscosity classification is shown in Table 6.6. The literature reports that the viscosity grades with the suffix "W" are intended for use at low ambient temperatures. The viscosity grades, without the suffix "W," are used in applications where low ambient temperatures are not encountered (Pirro and Wessol, 2001). The low temperature, high shear and low shear viscosity measurements were found to correlate with engine starting and engine oil pumpability at low temperatures. Engine oils can be formulated to meet the low temperature requirements of the 5W grade and the 100°C limits for the 30 grade. Such oil, called SAE 5W30 grade, is designated as a multigrade or multiviscosity and can be used for year round service with some exceptions (Pirro and Wessol, 2001). The SAE system is widely used by engine manufacturers to determine suitable oil viscosities for their engines. Specific SAE viscosity grades have been created for automotive gear oils.

The typical performance requirements of automotive lubricants are shown in Table 6.7. With a decrease in the temperature, the viscosity of base stocks increases until at certain temperature, no flow is observed. With an increase in the temperature and the presence of air, acidic oxidation by-products and sludges are formed. Engine oils must have good low temperature fluidity, low volatility, and oxidation resistance. Engine oils must also have good surface active properties, such as resistance to foaming, provide rust and corrosion protection, and wear control. The literature also reports that there is a need to develop gear oils, which have improved thermal stability, oxidation

---

TABLE 6.7
Typical Performance Requirements of Automotive Lubricants

Gasoline and Diesel Engine Oils	Gear Lubricants
Foam resistance	Foam resistance
Oxidation stability	Oxidation stability
Sludge and deposit control	Corrosion resistance
Piston cleanliness	Load-carrying capability
Rust and corrosion resistance	
Wear control	
Fuel economy	

Source: Pirro, D.M. and Wessol, A.A., *Lubrication Fundamentals*, 2nd ed., Marcel Dekker Inc., New York, 2001.

---

resistance, high temperature-EP performance, foam resistance, demulsibility properties, surface fatigue protection, and lower friction (Lee and Harris, 2003). The marine engine lubricants are system oil, cylinder oil, and trunk piston engine oil, and there is no simple system for classifying the marine engine lubricants because they are used in many different service applications on engines burning a wide range of fuels. The literature reports that the marine lubricants are developed through a series of laboratory tests culminating in shipboard trials (Carter, 1997).

The performance requirements of marine diesel engine lubricants are shown in Table 6.8. The system oils used to be simple oxidation-inhibited and rust-inhibited mineral oils, but the use of high sulfur residual oils has required the use of alkaline-type oils to provide adequate corrosion protection. The marine diesel lubricants are required to provide a strong oil film between the cylinder liner and the piston rings, hold the partially burned fuel residues in suspension, and neutralize acids formed by the combustion of the sulfur-containing fuel. The literature reports that the alkalinity of the trunk piston engine oils varies depending on the sulfur content of the fuel

---

TABLE 6.8
Performance Requirements of Marine Diesel Engine Lubricants

System Oil	Cylinder Oil	Trunk Piston Engine Oil
High thermal stability	Neutralize acids	Good demulsibility
Good oxidation stability	Good antiwear properties	Prevent corrosion
Low emulsibility	Prevent scuffing	Control piston deposits
Prevent rust and corrosion	Remove deposits	Good filterability

Source: Carter, B.H., Marine lubricants, in *Chemistry and Technology of Lubricants*, 2nd ed., Mortier, R.M. and Orszulik, S.T., Eds., Blackie Academic & Professional, London, U.K., 1997. With permission.

---

that is used to assure a good corrosion protection (Carter, 1997). The rust and the corrosion reduce the efficiency of the engine and its lifetime (Russo et al., 1996). The basic functions of automotive lubricants are friction reduction, heat removal, and the contaminant suspension. With an increase in speed, load, and temperature, the lubricants require a better resistance to oxidation, an improved wear protection, a better engine cleanliness, and an increased resistance to rust and corrosion.

6.2 BASE STOCK SELECTION

The base stock selection is based on meeting the low temperature fluidity, the volatility, and the oxidation stability requirements of the lubricant products. If the required performance is achieved with the mineral base stocks, they will be used on the basis of minimum cost (Prince, 1997).

When formulating lubricating oils, the viscosity and the viscosity index (VI) are the most important properties of base stocks. The viscosity of the lube oil base stocks is determined by the boiling range of their components, while the VI is related to their chemistry. The viscosity of the oil is essential to ensure a hydrodynamic lubrication. It must be high enough to provide proper lubrication films but not so high that friction losses in the oil will be excessive (Kajdas, 1997). When particularly low pour points are required, the mineral lubricants use raffinates obtained from naphthenic crude oils. In the past, the naphthenic base stocks were widely used because they were cheaper than the paraffinic base stocks (Mang, 2001). The naphthenic oils have poor VI characteristics, however, at high pressures they show a greater increase in the viscosity than paraffinic oils (Kajdas, 1997). At the present time, the majority of base stocks, used to formulate lubricating oils, are made from paraffinic crude oils. The literature reports that the naphthenic base stocks are used in lubricant products, which require a good low temperature fluidity but the VI and the oxidation stability are of secondary importance. The paraffinic base stocks are used in lubricant products, which require a high VI and a good oxidation stability (Shell, 2002).

The properties of different conventional solvent-refined base stocks are shown in Table 6.9. The 100N naphthenic base stock was reported to have a low VI of 15 and a low natural pour point of −45°C. The 100N paraffinic base stock, with a similar viscosity, was reported to have a high VI of 100 and a high dewaxed pour point of −18°C (Pirro and Wessol, 2001). The low temperature fluidity of the dewaxed paraffinic base stocks is not as good as the low temperature fluidity of some naphthenic base stocks with very low natural pour points. Bright stocks have much higher viscosity and, according to the literature, it is not possible to measure their viscosity in a Saybolt Universal Viscometer. The viscosity of the bright stocks is measured at 122°F in a Saybolt Furol Viscometer, which has a larger diameter orifice. The viscosity of the bright stocks is reported as Saybolt Furol Seconds (SFS). One hundred and seventy-five bright stocks would have a viscosity of 175 SFS at 122°F (50°C), which is about 700 cSt at 40°C (Phillipps, 1999). However, not every base stock manufacturer follows the same system for measuring and reporting the viscosity of bright stocks. High viscosity 150 SUS bright stock was reported to have also a high VI of 94, a high dewaxed pour point of −9°C, and a high flash point of 316°C (Lynch, 2007). The early literature reports on the deasphalted and dewaxed cylinder

TABLE 6.9
Properties of Different Conventional Solvent-Refined
Base Stocks

Conventional	100N	100N	150 SUS	Cylinder
Base Stocks	**Naphthenic**	**Paraffinic**	**Bright Stock**	**Oil**
KV at 100°F (SUS)	100	100	150	2302
KV at 210°F (SUS)				146.8
KV at 38°C (cSt)	20.5	20.5		
VI	15	100	94	98
Pour point (°C)	−45	−18	−9	—
Flash point (°C)	171	199	316	—

Sources: Pirro, D.M. and Wessol, A.A., *Lubrication Fundamentals*, 2nd ed., Marcel
Dekker Inc., New York, 2001. With permission; Lynch, T.R., *Process Chemistry
of Lubricant Base Stocks*, CRC Press, Boca Raton, FL, 2007. With permission;
Beuther, H. et al., *I&EC Prod. Res. Dev.*, 3, 174, 1964. With permission.

oils having a very high viscosity and a VI of 98 (Beuther, Donaldson, and Henke,
1964). The paraffinic base stocks are available in different viscosities ranging from
light spindle oils to viscous bright stocks.

The properties of different viscosity hydrofinished (HF) paraffinic base stocks are
shown in Table 6.10. The low viscosity HF 90N paraffinic base stock, having a VI of
92 and a dewaxed pour point of −15°C, was reported to contain 0.01 wt% of S and
has a flash point of 190°C. A higher viscosity HF 200N paraffinic base stock, having
a VI of 99 and a dewaxed pour point of −20°C, was reported to contain 0.1 wt% of S
and has a higher flash point of 226°C.

The HF 350N paraffinic base stock, having a VI of 97 and a dewaxed pour point
of −18°C, was reported to contain 0.13 wt% of S and has a higher flash point of
252°C. A high viscosity hydrofined 600N paraffinic base stock, having a VI of 92,

TABLE 6.10
Properties of Different Viscosity HF Paraffinic Base Stocks

Viscosity Grade	HF 90N	HF 200N	HF 350N	HF 600N	HF 650N
KV at 40°C (cSt)	17.4	40.7	65.6	105.9	117.9
KV at 100°C (cSt)	3.7	6.2	8.4	11.3	12.4
VI	92	99	97	92	96
Pour point (°C)	−15	−20	−18	—	−18
Sulfur (wt%)	0.01	0.1	0.13	0.12	0.16
Flash point (°C)	190	226	252	—	272

Sources: Pillon, L.Z. et al., Lubricating oil for inhibiting rust formation, U.S. Patent 5,397,487,
1995; Lubrizol, http://www.lubrizol.com (accessed 2001).

TABLE 6.11

Use of Different Solvent-Refined Base Stocks in Lubricant Products

Naphthenic Oils	Paraffinic Base Stocks	Bright Stocks	Cylinder Oils
Transformer oils	Engine oils	Motor oils	Gear oils
Refrigeration oils	Transmission fluids	Hydraulic oils	Cylinder lubricants
Hydraulic oils	Industrial oils	Gear oils	Journal lubricants
Gear oils	Paper machine oils	Journal lubricants	Greases
Cylinder lubricants	Specialty oils	Cylinder lubricants	
Process oils	Metalworking fluids	Greases	
Rubber process oils	Journal lubricants		
Greases	Greases		

Source: Sequeira, A. Jr., *Lubricant Base Oil and Wax Processing*, Marcel Dekker, Inc., New York, 1994. With permission.

was reported to contain 0.12 wt% of S. Another high viscosity HF 650N paraffinic base stock, having a VI of 96 and a dewaxed pour point of −18°C, was reported to contain 0.16 wt% of S and has a high flash point of 272°C. With an increase in the viscosity of HF paraffinic base stocks, an increase in the S content and the flash point is observed.

The use of different solvent-refined base stocks in lubricant products is shown in Table 6.11. Electrical oils are used in industrial transformers and they require low viscosity oils, having very good low temperature properties. Naphthenic base stocks were reported to be used in transformer oils, refrigeration oils, hydraulic oils, gear oils, cylinder lubricants, rubber process oils, process oils, and greases. For products that operate over a wide temperature range, such as automotive engine oils, low VI naphthenic base stocks are not suitable. Paraffinic base stocks were reported to be used in engine oils, transmission fluids, hydraulic oils, gear oils, turbine oils, paper machine oils, journal lubricants, metalworking fluids, and greases. Bright stocks were reported to be used in motor oils, hydraulic oils, gear oils, cylinder lubricants, journal lubricants, and greases. High viscosity bright stocks and cylinder oils are blended with lower viscosity base stocks to meet the viscosity requirements of different lubricant products. The base stocks, used to formulate lubricants, need to meet various specifications including viscosity, VI, cloud point, pour point, color, color stability, volatility, oxidation stability, flash point, and carbon residue.

The literature reports that a low viscosity base stock, used to formulate the machine oil-cutting fluid, needs to meet additional requirements, such as a low aromatic content, below 10 wt%, and a low sulfur content, to reduce the environmental pollution and the hazard to human health (Men et al., 2001).

The properties of 100N base stocks suitable for use in machine oil-cutting fluid are shown in Table 6.12. The standard 100N mineral base stock, suitable for use in machine oil-cutting fluid, was reported to have a VI above 100, a total acid number (TAN) below 0.05 mg KOH/g, a flash point above 190°C, and a carbon residue below

TABLE 6.12

Properties of 100N Base Stocks Suitable for Use in Machine Oil-Cutting Fluid

Properties	100N Standard	100N Typical
Density at 20°C (g/cm³)	0.8550–0.8700	0.8650
Viscosity at 40°C (mm²/s)	18–23	20
VI	>100	101
TAN (mg KOH/g)	<0.05	0.01
Flash point (°C)	>190	212
Carbon residue (wt%)	<0.02	0.01

Source: Reprinted from Men, G. et al., *Petrol. Sci. Technol.*, 19(9–10), 1251, 2001. With permission.

0.02 wt%. The typical 100N mineral base stock was reported to have a VI of 101, a TAN of 0.01 mg KOH/g, a flash point of 212°C, and a carbon residue of 0.01 wt% (Men et al., 2001). The use of the hydroprocessing can decrease the aromatic and the sulfur contents of petroleum-derived base stocks for use in specialty products. Hydrorefined technical white oils require the use of a second stage hydrorefining step to remove trace aromatics and sulfur contents, and make them suitable for use in food-grade products.

The effect of different crude oils on the paraffinic and naphthenic hydrocarbon contents of hydrorefined white oils is shown in Table 6.13. The hydrorefined white oils, from naphthenic crude oils, were reported to have a low paraffinic content

TABLE 6.13

Effect of Crude Oils on the Hydrocarbon Contents of Hydrorefined White Oils

Crude Oils Hydrorefined	Naphthenic White Oil	Naphthenic White Oil	Paraffinic White Oil	Paraffinic White Oil
KV at 40°C (cSt)	21	76	9	13
Paraffins (%)	13.9	11.8	49.2	42.2
1-Ring naphthenes (%)	23.1	22.0	24.0	26.3
2-Ring naphthenes (%)	21.8	20.3	13.1	15.0
3-Ring naphthenes (%)	16.2	19.6	7.3	7.7
4-Ring naphthenes (%)	14.0	14.3	5.5	6.4
5-Ring naphthenes (%)	7.6	8.0	0.8	2.3
6-Ring naphthenes (%)	3.4	2.8	0.1	0.1
Aromatics (%)	0	0	0	0

Source: Lynch, T.R., *Process Chemistry of Lubricant Base Stocks*, CRC Press, Boca Raton, FL, 2007. With permission.

of 12–14 wt% and no presence of aromatics. The hydrorefined white oils, from paraffinic crude oils, were reported to have a high paraffinic content of 42–50 wt% and no presence of aromatics. The use of different crude oils affects their hydrocarbon composition in terms of the paraffinic and naphthenic contents. While all different hydrorefined white oils, containing no aromatics, meet the requirements for the use in the specialty industrial oils, their solvency properties might vary due to the variation in the naphthenic content, which might affect the solubility of some additives. The use of hydrotreating, hydrocracking, and hydroisomerization also leads to a decrease in the aromatic and heteroatom contents of petroleum-derived base stocks. In many cases, an increase in their VI and a decrease in their volatility were also observed.

The effect of hydroprocessing on the composition, the VI, and the volatility of different viscosity base stocks is shown in Table 6.14. The conventional solvent-refined 150N mineral base stock has typical properties, such as a VI of 96, a dewaxed pour point of −9°C, 34.9 wt% of aromatics, 0.4 wt% of S, 53 ppm of N, and a volatility of 16% off. The hydroprocessed 150N mineral base stock has a VI of 97, a dewaxed pour point of −12°C, no aromatics, only 0.006 wt% of S, 5 ppm of N, and a volatility of 18% off.

Another hydroprocessed 150N mineral base stock has a higher VI of 130, a dewaxed pour point of −12°C, 2 wt% of aromatics, only 0.001 wt% of S, 5 ppm of N, and a lower volatility of 9% off.

The volatility of conventional solvent-refined base stocks increases with an increase in their viscosity. A lower viscosity hydroprocessed 100N base stock has a VI of 125, a dewaxed pour point of −15°C, 6.3 wt% of aromatics, 0.01 wt% of S, 2 ppm of N, and a volatility of 17% off. The use of slack wax hydroisomerization can

---

TABLE 6.14

Effect of Hydroprocessing on the Composition, the VI, and the Volatility of Base Stocks

Processing	Conventional	Hydro	Hydro	Hydro	Hydro
Base stocks	150N	150N	150N	100N	100N
KV at 100°C	5.3 cSt	5.7 cSt	5.6 cSt	4.4 cSt	4.0 cSt
VI	96	97	130	127	144
Pour point (°C)	−9	−12	−12	−15	−18
Saturates (wt%)	65.1	100	98	93.7	100
Aromatics (wt%)	34.9	0	2	6.3	0
Sulfur (wt%)	0.4	0.006	0.001	0.01	0.002
Nitrogen (ppm)	53	5	5	2	2
Volatility (% off)	16	18	9	17	16

Source: Phillipps, R.A., Highly refined mineral oils, in *Synthetic Lubricants and High-Performance Functional Fluids*, 2nd ed., Rudnick, L.R. and Shubkin, R.L., Eds., Marcel Dekker Inc., New York, 1999. With permission.

TABLE 6.15

Use of Hydroprocessed Base Stocks in Different Lubricant Products

Industrial Oils	Specialty Oils	Automotive Oils
Turbine oils	Process oils	Passenger car engine oils
Hydraulic oils	White oils	Heavy-duty engine oils
Compressor oils	Textile oils	Automotive gear oils
Vacuum pump oils	Transformer oils	ATFs
Rolling oils	Spray oils	Tractor fluids
	Cutting oils	Shock absorber fluids
	Quenching oils	Power steering fluids
	Heat transfer oils	Two-stroke oils
		Racing oils

Source: Reproduced from Henderson, H.E., Chemically modified mineral oils, in *Synthetics, Mineral Oils, and Bio-Based Lubricants Chemistry and Technology*, Rudnick, L.R., Ed., CRC Press, Boca Raton, FL, 2005. With permission.

also lead to a lower viscosity XHVI base stock having a VI of 144; practically no aromatics, and S and N contents; and a volatility of 16% off. The literature reports that XHVI base stocks can be manufactured by hydrocracking and isomerization of slack wax or from syngas (Phillipps, 1999).

The use of hydroprocessed base stocks in different lubricant products is shown in Table 6.15. Many different hydroprocessed base stocks are used in different products, including industrial and automotive lubricants. Growing environmental concerns extended the volatility limits to hydraulic, gear box, and shock absorber oils. The standard Noack test temperature of 250°C was lowered to 200°C, 150°C, and even to 100°C to test more volatile lubricating oils under their actual operating conditions (Instrumentation Scientifique de Laboratoire, 2002). The literature reports that the nonconventional processing of petroleum-derived base stocks by hydroisomerization of slack wax accounts for 10%–30% of a total base oil market and is predicted to increase (Calemma et al., 1999). The low temperature, the low-shear-rate viscosity of gear oils, the automatic transmission fluids (ATFs), the torque and the tractor fluids, and the industrial and the automotive hydraulic oils can be determined following the ASTM D 2983 method for low temperature viscosity of lubricants measured by a Brookfield viscometer. The pour point is the lowest temperature at which oil will flow under specific testing conditions, and all lube oil base stocks, except some highly naphthenic ones, must be dewaxed to assure good flow properties of lubricants at ambient temperatures. The use of catalytic dewaxing, known as isodewaxing, in the presence of hydrogen is called hydrodewaxing.

The Brookfield viscosity at −40°C of ATF oils, containing hydrodewaxed base stocks, is shown in Table 6.16. The ATFs are required to perform at very low temperatures and need to meet the severe low temperature Brookfield viscosity specification

TABLE 6.16

Brookfield Viscosity of ATF Oils Containing Hydrodewaxed Base Stocks

ATF Oils	Oil A	Oil B
Hydrodewaxed base stocks		
KV at 100°C (cSt)	4.25	4.62
KV at 40°C (cSt)	19.61	22.01
VI	123	128
Cloud point (°C)	−16	−19
Pour point (°C)	−24	−21
Additive package	Same	Same
Brookfield viscosity at −40°C (cP)	15,720	19,280

Source: Henderson, H.E., Chemically modified mineral oils, in *Synthetics, Mineral Oils, and Bio-Based Lubricants Chemistry and Technology*, Rudnick, L.R., Ed., CRC Press, Boca Raton, FL, 2005. With permission.

at −40°C. The Brookfield viscosity test measures the low temperature viscosity of lubricating oils under low shear conditions. The ATF oil A containing hydrodewaxed base stock, with a VI of 123 and a dewaxed pour point of −24°C, was reported to have a Brookfield viscosity of 15,720 cP at −40°C. Another ATF oil B containing hydrodewaxed base stock, with a higher VI of 128 and a higher dewaxed pour point of −21°C, was reported to have a higher Brookfield viscosity of 19,280 cP at −40°C. While the viscosity and VI are the most important properties of base stocks, low hydrodewaxed pour points are needed to assure good low temperature fluidity of ATF oils.

The Brookfield viscosity at −40°C of ATF oils, containing solvent-dewaxed base stocks, is shown in Table 6.17. The ATF oil #1 containing solvent-dewaxed base stock, with a VI of 136 and a dewaxed pour point of −15°C, was reported to have a Brookfield viscosity of 22,650 cP at −40°C. Another ATF oil #2 containing solvent dewaxed base stock, with a lower VI of 128 and a lower dewaxed pour point of −27°C, was reported to have a higher Brookfield viscosity of 38,150 cP at −40°C.

The Brookfield viscosity test measures the low temperature viscosity of lubricating oils under low shear conditions. The ATF oil #3 containing solvent-dewaxed base stock, with a very high VI of 146 and a dewaxed pour point of −18°C, was reported to be "solid" at −40°C. The preferred molecules for the lube oil manufacture are *iso*-paraffins with a high VI and a low pour point. The solvent dewaxing of XHVI base stocks, with a very high VI and a low volatility, is not sufficient to assure good flow properties of ATF oils at low temperatures. Engine oils require a good low temperature fluidity, a low volatility, and a good oxidation stability. The use of hydroprocessing can produce base stocks with a high VI and a low volatility, but dewaxing is still required to lower their pour points. The solvent dewaxing

TABLE 6.17
Brookfield Viscosity of ATF Oils Containing Solvent Dewaxed Base Stocks

ATF Oils	Oil #1	Oil #2	Oil #3
Solvent-dewaxed base stocks			
KV at 100°C (cSt)	3.56	4.13	3.99
KV at 40°C (cSt)	14.18	18.43	16.39
VI	136	128	146
Cloud point (°C)	−13	−20	−12
Pour point (°C)	−15	−27	−18
Additive package	Same	Same	Same
Brookfield viscosity at −40°C (cP)	22,650	38,150	Solid

Source: Henderson, H.E., Chemically modified mineral oils, in *Synthetics, Mineral Oils, and Bio-Based Lubricants Chemistry and Technology*, Rudnick, L.R., Ed., CRC Press, Boca Raton, FL, 2005. With permission.

and hydrodewaxing of paraffinic base stocks, containing only *iso*-paraffins, and the use of pour point depressants are not always sufficient to meet the low temperature specifications of some lubricant products.

6.3 USE OF SYNTHETIC FLUIDS

While the mineral oils have a dominant share of the lubricating fluid market, the environmental acceptability is another important criterion in selecting lube oil base stock and it includes toxicity, biodegradability, and bioaccumulation (Phillipps, 1999). Biodegradation is the breakdown of a chemical compound by the organisms and the microbes that live in water.

The toxicity requirements of biodegradable oils are different from food-grade lubricants. Food-grade lubricants are nontoxic to human beings but might be toxic to land or aquatic lives.

The literature reports that the lubricant industry must be more concerned with the environmental issues. Biodegradable lubricants require the use of biodegradable oils and biodegradable additives. Development of biodegradable lubricants is difficult due to the poor solubility of hydrocarbons in water and the presence of additives (Betton, 1997). There is also a main unresolved issue of biodegradable oils, which is how to define the biodegradability of the products, in terms of the organisms to be used and the testing procedures (Beitelman, 2004).

The properties and the biodegradability of different chemistry synthetic esters are shown in Table 6.18. According to the literature, for similar molecular weight polyalkylene glycols their pour points can vary depending on the ethylene/propylene oxide ratio (Brown et al., 1997). The first lubricants susceptible to microbial degradation

TABLE 6.18
Properties and Biodegradability of Different Chemistry Synthetic Esters

Synthetic Esters	Diesters	Phthalates	Polyols	Polyoleates
KV at 100°C (cSt)	2–8	4–9	3–6	10–15
KV at 40°C (cSt)	6–46	29–94	14–35	8–95
VI	90–170	40–90	120–130	130–180
Pour point (°C)	−70 to −40	−50 to −30	−60 to −9	−40 to −5
% Biodegradability	75–100	46–88	90–100	80–100

Source: Brown, M. et al., Synthetic base fluids, in *Chemistry and Technology of Lubricants*, 2nd ed., Mortier, R.M. and Orszulik, S.T., Eds., Blackie Academic & Professional, London, U.K., 1997. With permission.

were developed in 1970 for use on outboard engines. These lubricants were based on synthetic esters structurally similar to naturally occurring triglycerides.

The market share of biodegradable products is relatively small and their cost is significantly higher than the cost of mineral oils. Other synthetic fluids, such as polybutenes and polyalfaolefins (PAO), are also used in lubricant products to meet other performance requirements. Synthetic polybutenes and PAO base stocks are available in different viscosity grades.

The properties of different viscosity polybutenes and PAO fluids are shown in Table 6.19. Low viscosity polybutene, with a viscosity of about 3 cSt at 100°C, was reported to have a VI of 92 and a very low pour point of −60°C. With an increase in the viscosity of polybutenes, an increase in the VI and pour point was reported. A very high viscosity polybutene, with a viscosity of 4250 cSt at 100°C, was reported to have a very high VI of 264 and a high pour point of 24°C. An extensive review

TABLE 6.19
Properties of Different Viscosity Polybutenes and PAO Fluids

Viscosity Grade	Polybutene	Polybutene	Polybutene	PAO 4	PAO 10
KV at 100°C (cSt)	3.4	225	4,250	3.8	9.9
KV at 40°C (cSt)	15	7,200	185,000	16.7	64.5
VI	92	125	264	124	137
Pour point (°C)	−60	−7	24	−72	−53

Sources: Brown, M. et al., Synthetic base fluids, in *Chemistry and Technology of Lubricants*, 2nd ed., Mortier, R.M. and Orszulik, S.T., Eds., Blackie Academic & Professional, London, U.K., 1997. With permission; Rudnick, L.R. and Shubkin, R.L., Poly(alfa-olefins), in *Synthetic Lubricants and High-Performance Functional Fluids*, 2nd ed., Rudnick, L.R. and Shubkin, R.L., Eds., Marcel Dekker, Inc., New York, 1999. With permission.

TABLE 6.20
Physical and Chemical Properties of Polybutenes and Their Use in Lubricants

Polybutenes Properties	Lubricant Applications
Viscous grades have high VI	General lubricant products
Depolymerize at high temperatures leaving no residue	Two-stroke oil
UV radiation stable	Metalworking fluid
Good oxidation stability	Gear oil
Colorless	Compressor lubricant
Hydrophobic	Grease
Nontoxic	Wire rope
Can provide tackiness and adhesiveness	Other products

Source: Brown, M. et al., Synthetic base fluids, in *Chemistry and Technology of Lubricants*, 2nd ed., Mortier, R.M. and Orszulik, S.T., Eds., Blackie Academic & Professional, London, U.K., 1997.

of the properties and the use of synthetic polybutenes in lubricants were published (Brown et al., 1997). The PAO base stocks have a high VI, in the range of 124–136, and very low natural pour points, in the range of −53°C to 72°C. At high temperatures, PAO fluids have a very low volatility of 1.8%–11.8% off and a high flash point of 213°C–270°C An extensive review of the properties and the use of synthetic PAO base stocks in lubricants was published (Rudnick and Shubkin, 1999).

The physical and chemical properties of polybutenes and their use in different lubricants are shown in Table 6.20. The synthetic base stocks account for a small percentage of a total base oil market. The main classes of synthetic base stocks include PAO, dibasic acid esters, polyol esters, alkylated aromatics, polyalkylene glycols, and phosphate esters. The synthetic esters have greater swelling tendencies than hydrocarbons, which exclude them from applications where they may contact elastomers designed for use with mineral oils (Rudnick and Shubkin, 1999). The most widely used synthetic base stocks in industrial, automotive, and aircraft applications are PAO base stocks. A relatively low volatility of PAO base stocks relates to a higher flash point that is an important safety feature in many high temperature applications. Also, PAO base stocks have a high VI and a very low natural pour point, which makes them very useful in cold climate applications (Rudnick and Shubkin, 1999).

The use of synthetic fluids in different lubricant products is shown in Table 6.21.

Base stock quality, for use in engine oils, can vary from mineral to partly synthetic and fully synthetic. The technical demands of engine oils can be easily achieved by the use of synthetic base stocks, such as PAO, which have excellent low temperature fluidity and low volatility.

Some engine oils, such as 5W-30 and 10W-30, use hydrocracked or synthetic PAO base stocks to meet the low temperature and low volatility specifications (Gary and Handwerk, 2001). The synthetic engine oils can contain up to 75% of low viscosity

TABLE 6.21

Use of Synthetic Fluids in Different Lubricant Products

Industrial Lubricants	Automotive Lubricants	Aviation
Gas turbine oils	Engine oils	Turbine oils
Hydraulic oils	Gear oils	Piston engine oils
Gear oils	Brake fluids	Hydraulic fluids
Compressor oils	Lubricating greases	Lubricating greases
Metalworking fluids		
Heat transfer oils		
Lubricating greases		

Source: Reproduced from Rudnick, L.R. and Bartz, W.J., Comparison of synthetic, mineral oil, and bio-based lubricant fluids, in *Synthetics, Mineral Oils, and Bio-Based Lubricants Chemistry and Technology*, Rudnick, L.R., Ed., CRC Press, Boca Raton, FL, 2005. With permission.

PAO, up to 20% of synthetic esters, and the reminder is the additive system (Exxon Mobil, 2001). The synthetic base stocks are expensive, but some lubricating oils require the use of synthetic fluids to meet the severe low temperature requirements. The literature reports on the use of inexpensive monomers such as ethylene, propylene, and butene to copolymerize less expensive synthetic base stocks. The laboratory testing in lubricating oils and engine tests indicated a similar performance to commercial PAO (Song, Chiu, and Heilman, 2002). Many synthetic fluids are available; however, their cost versus performance is a key issue when using synthetic oils in the place of petroleum-derived base stocks.

6.4 ADDITIVE SELECTION

A typical lubricant contains petroleum-derived lube oil base stock and additive. The additive is any material added to a base stock to change its property, characteristic, or performance (Lubrizol, 2001). Although many of the base stock properties are modified by the additives, some properties, such as volatility and air entrainment, cannot be modified. The volatility is dependant on selected base stocks, while air entrainment is affected by the chemistry of selected additives. Straight mineral oils contain no additives and include lubricant products, such as thoroughly refined white oils and low-cost once-through lubricants. Straight mineral oils, used in certain aviation piston engines, may contain a pour point depressant (Exxon Mobil, 2002). The quality of a lubricant is affected by the selected base stocks and the additives (Shell, 2002).

Although the viscosity of base stocks decreases with an increase in the temperature, the decrease in the viscosity is smaller in the presence of the VI improver. VI improvers are added to lubricating oils to help maintain viscosity at operating temperatures. Most VI improvers are oil-soluble polymers, which usually exert a greater thickening effect on oil at higher temperatures than lower temperatures.

TABLE 6.22

Chemistry of Additives Used in Lubricant Products

Additive Type	Typical Chemistry
VI improvers	Acrylate polymers
Pour point depressants	Copolymers of polyalkylmethacrylates
Antioxidants	Hindered phenols
	Aromatic amines
	Alkyl sulfides
	Aromatic sulfides
	ZnDTP
Antifoaming agents	Dimethylsilicones (dimethylsiloxanes)
Rust and corrosion inhibitors	Organic acids, esters
	Amino-acid derivatives
	Alkaline compounds
Antiwear additives	ZnDTP
	Zinc dialkyldithiophosphate (ZDDP)
	Tricresylphosphate (TCP)
EP additives	Chlorinated paraffins
	Sulfurized fats, esters
	Zinc dialkyldithiophosphate (ZDDP)
	Molybdenum disulfide (MoS2)
Friction modifiers	Graphite
	Molybdenum disulfide (MoS2)
	Boron nitride (BN)
	Tungsten disulfide (WS2)
	Polytetrafluoroethylene (PTFE)
Detergents	Phenolates, sulfonates, and phosphonates of alkaline elements such as Ca, Mg, and Na
Dispersants	Polyisobutylene succinimides

Source: Kopeliovich, D., *Additives in Lubricating Oils*, http://www.substech. com (accessed 2008).

The efficiency of different acrylic polymers and styrene copolymers in improving the VI of lube oil base stocks was reported to increase with an increase in their molecular weight (Nassar et al., 2005).

The typical chemistry of additives used in lubricant products is shown in Table 6.22. Many lubricating oil additives are highly viscous materials or solids and are used as concentrated solutions in mineral base stocks or petroleum solvents to assure proper mixing. An extensive review of the chemistry of petroleum additives and their use in lubricant products was published (Rudnick, 2003). Silicones are known as the most effective antifoaming agents available in the market. Some original equipment manufacturers require that only nonsilicone antifoaming agents, such as polyacrylate, be used in their plants (Lee and Harris, 2003). The performance of polyoxyalkylene polyol demulsifier in lubricating oil, containing ashless dispersant, was reported to be enhanced by the presence of a free dicarboxylic acid or tetrapropenyl succinic anhydride (Aries and Dunn, 1992). Many chemical molecules, used as additives, are multifunctional. The triazynyl amino ethane derivative was reported as polyfunctional additive effective as a rust inhibitor, an antiwear, an anti-scuffing, and an antioxidant in synthetic lubricating oils (Kelarev et al., 1998). The two-substituted imidazoline derivative was reported as polyfunctional additive effective as a corrosion inhibitor, an antiwear, an anti-scuffing, and an antioxidant in mineral lubricating oils (Kelarev et al., 2000). Some special amine salts of mono- or dialkylphosphoric acid partial esters exhibit anticorrosion properties in addition to their antiwear properties, which make them useful in formulating ashless industrial oils (Braun and Omeis, 2001).

The additive types used in industrial lubricants are shown in Table 6.23. Turbine oils contain antioxidants, rust and corrosion inhibitors, and demulsifiers and antifoaming agents. There are three main categories of turbines, namely, gas, steam, and water. If turbine oils are used in geared turbines, mild, long life, temperature, and oxidation stable EP additives are included (Bock, 2001). Hydraulic oils contain antioxidants, rust inhibitors, antifoaming agents, and additionally antiwear additives. Industrial gear oils contain additives required to impart the resistance to oxidation, good antiwear properties, resistance to foaming, good demulsibility and rust preventive properties, and additionally good EP performance. A modified polyether polymer was reported to improve air separation properties, demulsibility, and the oxidation stability of turbine oils, antiwear hydraulic oils, and industrial gear oils (Ma and Niu, 2000). Some industrial oils contain additionally other "miscellaneous" additives. The oil component of metalworking fluids, which can be mineral,

---

TABLE 6.23
Additive Types Used in Industrial Lubricants

Performance	Surface Active	Miscellaneous
Antioxidants	Antifoaming agents	Oiliness agents
VI improvers	Demulsifiers	Tackiness agents
Pour point depressants	Emulsifiers	Bactericides
Thickeners	Rust inhibitors	Bacteriostats
	Corrosion inhibitors	Fungicides
	Metal deactivators	Dyes
	Metal passivators	Odorants
	Antiwear agents	
	EP agents	

Sources: Sequeira, A. Jr., *Lubricant Base Oil and Wax Processing*, Marcel Dekker, Inc., New York, 1994. With permission; Perez, J.M., Additives for industrial lubricant applications, in *Lubricant Additives Chemistry and Applications*, Rudnick, L.R., Ed., Marcel Dekker, Inc., New York, 2003. With permission.

synthetic, or semisynthetic, is dispersed in water at a concentration of 1–10 wt% and their appearance can vary from a translucent liquid to white milk. Other than emulsifiers, the oil phase also contains an antifoaming agent, a corrosion inhibitor, and an EP additive. Bactericides and fungicides are also included (Denis, Briant, and Hipeaux, 2000). Some derivatives of fatty acids are surface active and are used as additives in industrial lubricants.

The chemistry and nomenclature of the typical fatty acids are shown in Table 6.24. VI improvers and pour point depressants are used to increase the base stock performance. Many different surface active additives are also used in industrial oils. Specialty industrial lubricants that include cylinder oils, process oils, textile oils, and slideway oils require certain types of additives. To prevent the stick-slip phenomena, slideway oils contain polar surface active compounds, mostly fatty acid derivatives, together with antiwear additives and oxidation inhibitors (Kajdas, 1997). Some industrial gear oils were reported to contain VI improvers, antioxidants, metal deactivators, antiwear additives, soluble EP additives, rust inhibitors, pour point depressants, tackifiers, demulsifiers, friction modifiers, antifoaming agents, and antimisting additives (Lee and Harris, 2003). The largest market for additives is in the transportation field, including additives for engines and drivetrains in cars, trucks,

TABLE 6.24
Chemistry and Nomenclature of Typical Fatty Acids

Systematic Name	Nonsystematic Name	Chemistry
Octanoic acid	Caprylic acid (8:0)	$CH_3(CH_2)_6COOH$
Decanoic acid	Capric acid (10:0)	$CH_3(CH_2)_8COOH$
Dodecanoic acid	Lauric acid (12:0)	$CH_3(CH_2)_{10}COOH$
Tetradecanoic acid	Myristic acid (14:0)	$CH_3(CH_2)_{12}COOH$
Hexadecanoic acid	Palmitic acid (16:0)	$CH_3(CH_2)_{14}COOH$
9-Hexadecanoic acid	Palmitoleic acid (16:1)	$CH_3(CH_2)_5CH=CH(CH_2)_7COOH$
Octadecanoic acid	Stearic acid (18:0)	$CH_3(CH_2)_{16}COOH$
9-Octadecanoic acid	Oleic acid (18:1)	$CH_3(CH_2)_7CH=CH(CH_2)_7COOH$
12-Hydroxyoleic acid	Ricinoleic acid	$CH_3(CH_2)_5CH(OH)$ $CH_2CH=CH(CH_2)_7COOH$
9,12-Octadecadienoic acid	Linoleic acid (18:2)	$CH_3(CH_2)_4CH=CHCH_2CH=CH$ $(CH_2)_7COOH$
9,12,15-Octadecatrienoic acid	Linolenic acid (18:3)	$CH_3(CH_2CH=CH)_3(CH_2)_7COOH$
Eicosanoic acid	Arachidic acid (20:0)	$CH_3(CH_2)_{18}COOH$
Docosanoic acid	Behenic acid (22:0)	$CH_3(CH_2)_{20}COOH$
13-Docosenoic acid	Erucic acid (22:1)	$CH_3(CH_2)_7CH=CH(CH_2)_{11}COOH$

Sources: Crawford, J. et al., Miscellaneous additives and vegetable oils, in *Chemistry and Technology of Lubricants*, 2nd ed., Mortier, R.M. and Orszulik, S.T., Eds., Blackie Academic & Professional, London, U.K., 1997; Sigma-Aldrich, *Aldrich Handbook of Fine Chemicals*, St. Louis, MO, 2007–2008.

TABLE 6.25
Additive Types Used in Engine Oils

Performance	Protective	Surface Active
VI improvers	Antioxidants	Antifoaming agents
Pour point depressants	Detergents	Demulsifiers
Seal swell agents	Dispersants	Rust inhibitors
		Corrosion inhibitors
		Metal deactivators
		Antiwear agents
		EP agents
		Friction modifiers

Sources: Sequeira, A. Jr., *Lubricant Base Oil and Wax Processing*,
Marcel Dekker, Inc., New York, 1994; Lubrizol, http://
www.lubrizol.com (accessed 2001).

buses, locomotives, and ships. The greater the number of performance properties required for a lubricant, the more complex the formulation and the more additives are needed. By far, engine oils contain the most additives, including VI improvers. The use of VI improvers is required when blending multigrade engine oils.

The additive types used in engine oils are shown in Table 6.25. Additives used in engine oils to increase the base stock performance are VI modifiers, pour point depressants, and seal swell agents. Antioxidants used in engine oils include hindered phenols and aryl amines. Alkylated diphenylamines have been widely used in engine oils.

Many surface active additives, such as antifoaming agents and demulsifiers, are used to prevent the lubricant degradation. Many different surface active additives, such as rust inhibitors, corrosion inhibitors, antiwear agents, EP agents, and friction modifiers, are used to protect the metal surface. At high operating temperatures, the use of antioxidants and metal deactivators is required. Typical metal deactivators used in engine oils are zinc dithiophosphates (ZnDTP) (Pirro and Wessol, 2001). In engine oils, under high temperature and oxidation conditions, dispersants and detergents are also used to protect the metal surface. The properties of zinc dialkyl dithiophosphates and zinc diaryl dithiophosphates were reported to vary depending on their chemistry, and the ratio of neutral and basic salts used during their production (Evstafev et al., 2001).

The effect of different chemistry ZnDTP additives on lubricant performance is shown in Table 6.26. Some chemicals, used as additives, initially appear to be beneficial for all applications, but this is usually incorrect. Some additives used in lubricant products might be effective in improving a specific performance area, while degrading other properties (Rasberger, 1997). Under high temperature conditions different chemistry ZDDP additives are used in engine oils, including the aryl ZDDP, to achieve the maximum thermal stability (Bardasz and Lamb, 2003). Many additives are used to improve the properties of base stocks, extend their use, and protect the

TABLE 6.26
Effect of Different Chemistry ZnDTP Additives on Lubricant Performance

Properties	Primary ZnDTP	Secondary ZnDTP	Aryl ZnDTP
Thermal stability	Good	Moderate	Very good
Oxidation inhibition	Satisfactory	Good	Moderate
Wear protection	Satisfactory	Good	Bad

Source: Rasberger, M., Oxidative degradation and stabilisation of mineral oil based lubricants, in *Chemistry and Technology of Lubricants*, 2nd ed., Mortier, R.M. and Orszulik, S.T., Eds., Blackie Academic & Professional, London, U.K., 1997. With permission.

metal surface. There are also other additives designed to impart specific properties, such as emulsifiers, tackiness agents, bactericides, and gelling agents for greases (Lubrizol, 2001).

Lubricating greases, processed in grease plants, are gel-like materials created by the dispersion of a thickening agent in a liquid lubricant, usually a metal soap, along with other additives (Singh, 2008). The main ingredient that differentiates the lubricating oil from grease is the presence of a grease thickener.

The type and chemistry of grease thickeners are shown in Table 6.27. Most grease thickeners are soaps comprised of the reaction product of a fatty acid derivative and a metallic hydroxide. Other salts may be combined with the soap thickener to impart higher grease dropping temperature, which is the temperature at which the grease liquefies. A small proportion of lubricating greases are manufactured with nonsoap

TABLE 6.27
Type and Chemistry of Grease Thickeners

Grease Thickener Type	Grease Thickener Chemistry	Grease Type
Soap	Lithium salt of castor oil	
	Calcium stearate (hydrous)	
	Calcium stearate (anhydrous)	
	Lithium complex	Complex grease
	Aluminum complex	Complex grease
	Calcium complex	Complex grease
Nonsoap	Organoclay	
	Polyurea	
	Silica	

Source: Lee, F.L. and Harris, J.W., Long-term trends in industrial lubricant additives, in *Lubricant Additives Chemistry and Applications*, Rudnick, L.R., Ed., Marcel Dekker, Inc., New York, 2003. With permission.

thickeners, such as organoclays or polyurea compounds (Lee and Harris, 2003). The metal salts, such as zinc naphthenates, or amine salts of dinonylnaphthalene sulfonic acid, are also used in lubricating oils, greases, and metalworking fluids. The inclusion of highly alkaline materials helps to neutralize strong acids, as they are formed, greatly reducing the corrosion and the corrosive wear. Salts of the lanolin fatty acid, mostly in combination with sulfonates, are used in the rust preventatives (Braun and Omeis, 2001). Hundreds of different chemical molecules have been found to enhance certain aspects of a lubricant performance and are used as additives. However, some additives can have detrimental side effects if the dosage is excessive or if an interaction with other additives occurs.

6.5   USE OF SURFACTANTS

The literature reports that the incompatibility within the same molecule is the unique property of the surfactant molecule. The surfactants contain a hydrophobic (water-repellent) hydrocarbon and a hydrophilic (water-compatible) head group (Karsa, 1986). The surfactant molecule will change their orientation to place its hydrophobic group in hydrophobic medium, form the adsorbed layer at the interface, and agglomerate into spherical, cylindrical, or lamellar micelles above the critical micelle concentration (CMC). The alkyl benzene sulfonates are classified as anionic, quaternary ammonium salts and imidazolinium salts are classified as cationic, imidazolines are amphotheric while ethoxylates, PEG esters and EO/PO copolymers are classified as nonionic surfactants (Karsa, 1986). According to the literature, ethoxylated alcohols are one of the most widely used classes of nonionic surfactants. They act as emulsifiers, dispersants, and detergents. Their adsorption at the oil/water interface was reported to decrease with an increase in the degree of the ethoxylation and to increase with an increase in the hydrophobe size (Rulison and Falberg, 1997).

The types and chemistry of surfactant molecules are shown in Table 6.28. The literature reports that the main solution behavior affected by the presence or the absence of surfactant molecules can lead to a variation in solubilizing, dispersing, or detergency properties (Karsa, 1986). The addition of amphiphiles is frequently used

TABLE 6.28
Types and Chemistry of Surfactant Molecules

Chemistry of Molecule	Surfactant Types	Examples
Hydrophobe-hydrophile	Anionic	Alkyl benzene sulfonates
Hydrophobe-hydrophile	Cationic	Quaternary ammonium salts
Hydrophobe-hydrophile	Amphoteric	Imidazolines
Hydrophobe-hydrophile	Nonionic	Alkyl phenol ethoxylates

Source: Karsa, D.R., Industrial applications of surfactants—An overview, in Industrial Applications of Surfactants, Karsa, D.R., Ed., The Royal Society of Chemistry, Burlington House, London, U.K., 1986.

TABLE 6.29

Effect of Surfactants on the Solution Behavior and the Interfacial Properties of Liquids

Solution Behavior	Interfacial Properties
Solubility	Foaming or defoaming
Dispersing	Air entrainment
Detergency	Emulsification or demulsification
	Metal wetting or re-wetting

Sources: Karsa, D.R., Industrial applications of surfactants—an overview, in *Industrial Applications of Surfactants*, Karsa, D.R., Ed., The Royal Society of Chemistry, Burlington House, London, U.K., 1986. With permission; Pillon, L.Z., *Interfacial Properties of Petroleum Products*, CRC Press, Boca Raton, FL, 2007. With permission.

to prevent asphaltene precipitation in reservoir rocks and wellbore tubing. It was reported that the adsorption of amphiphiles on the asphaltenes makes them more soluble under the production conditions (Rogel, Contreras, and Leon, 2002). The emulsions can be stabilized by surfactant molecules with an affinity for both phases, oil and water, and adsorb at the interface creating a barrier by forming a film that stabilizes the system (Denis, Briant, and Hipeaux, 2000). The properties of surfactants and their surface activity at the oil/air, oil/water, and oil/metal interfaces can lead to foaming or defoaming, air entrainment, emulsification or demulsification, wetting or nonwetting of the metal surface (Pillon, 2007).

The effect of surfactants on the solution behavior and the interfacial properties of liquids is shown in Table 6.29. The literature reports that oil/metal and oil/solid interfaces are the most important in lubrication because they affect the machine parts and the suspended contaminants (Denis, Briant, and Hipeaux, 2000). A high content of dispersants and detergents is used in engine oils.

Low molecular weight sodium sulfonates are used as emulsifiers and detergents. High molecular weight calcium, magnesium, and sodium sulfonates are used as rust inhibitors (Braun and Omeis, 2001). Many different additives, used in lubricating oils, have the surfactant chemistry. Early literature reports on the surface activity of different derivatives of fatty acids and their effect on the surface tension (ST) of a purified mineral oil known as Nujol. A certain chemistry derivative of fatty acids was found to increase or decrease the ST of mineral oil indicating their surface activity toward the oil/air surface (Bondi, 1951).

The effect of the derivatives of fatty acids on the ST of a purified mineral oil, Nujol, is shown in Table 6.30. Purified mineral oil, such as Nujol, was reported to have an ST of 31 mN/m measured at 20°C and no significant decrease will be observed when measured at 27°C. At the temperature of 27°C, the use of stearic acid, at the treat rate of 0.7%, was reported to actually increase the ST of mineral oil to 32.5 mN/m indicating an interaction with some surface active contaminants

TABLE 6.30
Effect of the Derivatives of Fatty Acids on the ST of Purified Mineral Oil

Purified Mineral Oil (Nujol) Derivatives of Fatty Acids	Temperature (°C)	Neat Liquid ST (mN/m)	Solution ST (mN/m)
Neat Nujol	20	31.0	
Neat stearic acid	27	33.5	
Nujol/stearic acid (0.7%)	27		32.5
Neat lauryl sulfonic acid	27	33.5	
Nujol/lauryl sulfonic acid (1%)	27		25.9
Neat triethanolamine oleate	27	33.5	
Nujol/triethanolamine oleate (9%)	27		31.5

Source: Bondi, A., *Physical Chemistry of Lubricating Oils*, Reinhold Publishing Corporation, New York, 1951. With permission.

and their desorption from the oil/air interface. At the same temperature, the use of lauryl sulfonic acid, at the treat rate of 1%, was found to significantly decrease the ST of mineral oil to 25.9 mN/m indicating its adsorption at the oil/air interface. The use of triethanolamine oleate, at a very high concentration of 9 wt%, was reported to have no significant effect on the ST of highly purified mineral oil. In highly purified mineral oil, stearic acid was reported to increase the ST by causing the desorption of some contaminants, while lauryl sulfonic acid was reported to decrease the ST indicating its adsorption.

The effect of derivatives of fatty acids on the ST of mineral oil is shown in Table 6.31. At 25°C, the mineral oil was reported to have an ST of 31–32 mN/m. At a temperature of 27°C, the use of lauryl sulfonic acid, at the treat rate of 1%, was also

TABLE 6.31
Effect of the Derivatives of Fatty Acids on the ST of Mineral Oil

Mineral Oil Derivatives of Fatty Acids	Temperature (°C)	Neat Liquid ST (mN/m)	Solution ST (mN/m)
Neat mineral oil	25	31–32	
Neat lauryl sulfonic acid	27	33	
Mineral oil/lauryl sulfonic acid (1%)	27		27.8
Neat mineral oil	100	25–26	
Neat sodium stearate	112	25.2	
Mineral oil/sodium stearate (1%)	112		20.8

Source: Bondi, A., *Physical Chemistry of Lubricating Oils*, Reinhold Publishing Corporation, New York, 1951. With permission.

found to decrease the ST of mineral oil to 27.8 mN/m indicating its adsorption at the oil/air interface. With an increase in the temperature from 25°C to 100°C, the ST of mineral oil decreased to 25–26 mN/m. At the temperature of 112°C, the addition of sodium stearate, at the treat rate of 1%, was reported to significantly decrease the ST of mineral oil from 25.2 to 20.8 mN/m indicating its adsorption at the oil/air interface. While a fatty acid, such as stearic acid, did not decrease the ST of mineral oil, its derivative, such as sodium stearate, was found to decrease the ST of mineral oil indicating its adsorption at the oil/air interface. The patent literature reports on the composition of lubricating oils, containing synthetic PAO fluids, and the effect of the additive chemistry on their air release time (Ward, Tipton, and Murray, 1996).

The effect of additives on the air release time of lubricating oils, containing PAO base stocks, is shown in Table 6.32. Lubricating oils, when used in environments in which the oil is subject to mechanical agitation in the presence of a gas, have a tendency to foam. Fluorosilicone antifoaming agents have been suggested for use in some lubricating oils but not for industrial or internal combustion engine crankcase oils. Synthetic lubricating oil #1, containing a PAO blend and a total of 5.6 wt% of many different additives, including dispersant, antioxidant, diluent oil, and seal swell agent, metal deactivator, silicone/fluorosilicone antifoaming blend and red dye, was reported to have an air release time of 3.5 min at 50°C. Another synthetic lubricating oil #2, containing also a PAO blend and the same additives, except a different

TABLE 6.32
Effect of Additives on the Air Release Time of Lubricating Oils

Synthetic Lubricating Oils	Oil #1	Oil #2
Base Stock	PAO Blend	PAO Blend
Succinimide dispersant (wt%)	2.4	2.4
Borated epoxide (wt%)	0.2	0.2
S-containing antioxidant (wt%)	0.5	0.5
Di(para-nonyl phenyl) amine (wt%)	0.65	0.65
Dialkyl hydrogen phosphite (wt%)	0.2	0.2
Alkyl naphthalene (wt%)	0.2	0.2
Diluent oil (wt%)	0.34	0.34
Sulfolane seal swell (wt%)	0.6	0.6
Alkylthiodimercaptothiadiazole (wt%)	0.5	
Dimercaptothiadiazole (wt%)		0.5
Silicone/fluorosilicone blend (ppm)	40	40
Red dye (ppm)	250	250
Kin viscosity at 100°C (cSt)	4.63	4.71
ASTM D 3427 (50°C)		
Air release time (min)	3.5	4.9

Source: Ward, W.C. Jr. et al., Lubrication fluids for reduced air entrainment and improved gear protection, CA Patent 2,184,969, 1996.

TABLE 6.33

Effect of the Derivatives of Fatty Acids on the IFT of Purified Mineral Oil

Purified Mineral Oil (Nujol)	Concentration	Solution at 20°C
Derivatives of Fatty Acids	**(wt%)**	**IFT (mN/m)**
Neat Nujol oil	0	53
Oleic acid	0.25	39
	4.0	25
Oleyl alcohol	0.25	36
	4.0	20
Na oleate	0.25	3
	4.0	2
Glycerol monooleate	0.25	3
	4.0	4

Source: Bondi, A., *Physical Chemistry of Lubricating Oils*, Reinhold Publishing Corporation, New York, 1951. With permission.

chemistry metal deactivator, was reported to have an increased air release time of 4.9 min at 50°C. While silicone and fluorosilicone antifoaming agents are known to increase the air release time, it was actually the use of a different chemistry derivative of thiadiazole that increased the air entrainment of the lubricating oil #2.

The effect of derivatives of fatty acids on the oil/water interfacial tension (IFT) of a purified mineral oil, Nujol, is shown in Table 6.33. Highly purified mineral oil, Nujol, was reported to have an IFT of 53 mN/m measured at 20°C. With an increase in the concentration of the oleic acid from 0.25 to 4.0 wt%, the IFT of Nujol decreased to 25–39 mN/m indicating its adsorption at the oil/water interface. After the addition of 0.25–4.0 wt% of oleyl alcohol, the IFT of Nujol further decreased to 20–36 mN/m also indicating its adsorption at the oil/water interface. With an increase in the concentration of the Na oleate from 0.25 to 4.0 wt%, the IFT of Nujol drastically decreased to 2–3 mN/m indicating an increase in the surface activity of the salts of fatty acids. After the addition of 0.25–4.0 wt% of glycerol monooleate, the IFT of Nujol oil also drastically decreased to 3–4 mN/m indicating an increase in the surface activity of the esters of fatty acids. While different derivatives of fatty acids adsorb at the oil/water interface, some derivatives such as salts and esters were found to be significantly more surface active when used at the same concentration.

The effect of different chemistry derivatives of fatty acids on their spreading coefficient (S) at the oil/water interface of white oil is shown in Table 6.34. When mixed with water, the mineral oil has a negative spreading coefficient (S) of −11.2 mN/m indicating a poor miscibility. The highly purified white oil, when mixed with water, has a negative spreading coefficient (S) of −11.7 mN/m indicating a poor miscibility. However, a fatty acid such as an oleic acid, when mixed with water, has a high spreading coefficient (S) of 25 mN/m indicating a good miscibility.

TABLE 6.34
Effect of the Derivatives of Fatty Acids on Their Spreading on the Water Surface

Oil/Water Interface at 20°C	Oil/Air	Oil/Water	Oil/Water
Derivatives of Fatty Acids	**ST (mN/m)**	**IFT (mN/m)**	**(S) (mN/m)**
Neat mineral oil	31	53.0	−11.2
Neat white oil	32.5	52.0	−11.7
Neat oleic acid	32.5		25
White oil/oleic acid (0.9 wt%)			7.4
White oil/oleyl alcohol (0.9 wt%)			14.4
White oil/glycerol monooleate (0.7 wt%)			37.8

Source: Bondi, A., *Physical Chemistry of Lubricating Oils*, Reinhold Publishing Corporation, New York, 1951. With permission.

The white oil, containing 0.9 wt% of oleic acid, was reported to have a positive spreading coefficient (S) of 7.4 mN/m indicating its improved miscibility with water. The white oil, containing 0.9 wt% of oleyl alcohol, was reported to have an increased spreading coefficient (S) of 14.4 mN/m indicating an increased miscibility with water. The white oil, containing a lower concentration of 0.7 wt% of glycerol monooleate, was reported to have a drastically increased spreading coefficient (S) of 37.8 mN/m due to the presence of additional ester groups indicating a further increase in miscibility with water. Some derivatives of fatty acids, if they have emulsifying properties, can lead to an increase in the emulsion stability.

The effect of Na alkyl sulfate and NaCl on the IFT of purified mineral oil, Nujol, is shown in Table 6.35. The purified mineral oil, Nujol, was found to have a high

TABLE 6.35
Effect of Na Alkyl Sulfate and NaCl on the IFT of Purified Mineral Oil

Purified Mineral Oil (Nujol)	Concentration	Solution at 20°C
(ST of 31 mN/m at 20°C)	**(wt%)**	**IFT (mN/m)**
Neat Nujol oil	0	53.0
Na alkyl sulfate	0.025	15.9
	0.25	6.5
Na alkyl sulfate (0.2% NaCl)	0.025	6.8
	0.25	4.9

Source: Bondi, A., *Physical Chemistry of Lubricating Oils*, Reinhold Publishing Corporation, New York, 1951. With permission.

IFT of 53 mN/m measured at 20°C. With an increase in the concentration of the Na alkyl sulfate from only 0.025–0.25 wt%, the IFT of Nujol oil decreased to 6.5–15.9 mN/m indicating its adsorption at the oil/water interface. With an increase in the concentration of the Na alkyl sulfate from only 0.025–0.25 wt% and the addition of 0.2 wt% NaCl, the IFT of Nujol oil further decreased to 4.9–6.8 mN/m indicating its increased adsorption at the oil/water interface. The presence of salts in the water can further increase the surface activity of some surfactants and affect the lubricant performance. The literature reports that in the presence of calcium ions or other heavy metals, the spreading of the fatty alcohols will not take place and a rigid film at the oil/water interface will form. Calcium ions contained in "hard" water will precipitate soap from the solution and build deposits (Bondi, 1951). Rust preventatives, used for coating metal surfaces with a film that protects against rust formation, are commonly used for the preservation of the equipment in storage. Occasionally rust-preventive oils are required to displace water from a steel surface.

The effect of different additives on the water displacing action and IFT of mineral oil is shown in Table 6.36. Neat mineral oil was reported to have no properties allowing for the water displacing action and the addition of a surface active additive was required. The literature reports that the main components of rust preventatives are petroleum oils, solvents, wax, or asphalt to which a rust inhibitor is added (Pillon, 2007). The addition of additives A and B did not improve the water displacing action of the mineral oil; however, a decrease in the IFT from 49.7 to 8.9–23.6 mN/m was reported indicating their adsorption at the oil/water interface. The addition of additives C, D, and E was reported to improve the water displacing action of mineral oil, and a decrease in IFT from 49.7 to 4.6–20.3 mN/m was also observed. The literature reports that no correlation between the water displacing action and the IFT is observed because once the water reached the metal surface, the IFT is not important (Bondi, 1951). However, the selection of surface active additives, such as rust inhibitors, and their effect on the IFT can affect many other

TABLE 6.36
Effect of Different Additives on the Water Displacing Action and IFT of Mineral Oil

Mineral Oil	Oil/Metal Interface	Oil/Water Interface
	Water Displacing Action	IFT at 20°C (mN/m)
Neat mineral oil	None	49.7
Mineral oil (additive A)	None	23.6
Mineral oil (additive B)	None	8.9
Mineral oil (additive C)	Good	20.3
Mineral oil (additive D)	Good	8.9
Mineral oil (additive E)	Good	4.6

Source: Bondi, A., *Physical Chemistry of Lubricating Oils*, Reinhold Publishing Corporation, New York, 1951. With permission.

aspects of lubricant performance. A decrease in the IFT and the presence of a surface active molecule at the oil/water interface can increase the stability of emulsions. In the presence of water, a very low IFT of a lubricant might lead to an increase in the water solubility.

REFERENCES

Apar Industries, http://www.aparind.com (accessed 2003).

Aries, A. and Dunn, A., *Demulsification of Oils*, PCT Int. Appl., 1992.

ASTM D 2422 standard classification of industrial fluid lubricants by viscosity system, ASTM International, 2007.

Bardasz, E.A. and Lamb, G.D., Additives for crankcase lubricant applications, in *Lubricant Additives Chemistry and Applications*, Rudnick, L.R., Ed., Marcel Dekker, Inc., New York, 2003.

Beitelman, A.D., Environmentally Friendly Lubricants, http://www.wes.army.mil (accessed 2004).

Betton, C.I., Lubricants and their environmental impact, in *Chemistry and Technology of Lubricants*, 2nd ed., Mortier R.M. and Orszulik S.T., Eds., Blackie Academic & Professional, London, U.K., 1997.

Beuther, H., Donaldson, R.E., and Henke, A.M., *I&EC Product Research and Development*, 3(3), 174 (1964).

Bock, V., *Turbine Oils, in Lubricants and Lubrications*, Mang, T. and Dresel, W., Eds., Wiley-VCH GmbH, Weinheim, 2001.

Bondi, A., *Physical Chemistry of Lubricating Oils,* Reinhold Publishing Corporation, New York, 1951.

Braun, J. and Omeis, J., Additives, in *Lubricants and Lubrications*, Mang, T. and Dresel, W., Eds., Wiley-VCH GmbH, Weinheim, 2001.

Brown, M. et al., Synthetic base fluids, in *Chemistry and Technology of Lubricants*, 2nd ed., Mortier, R.M. and Orszulik, S.T., Eds., Blackie Academic & Professional, London, U.K., 1997.

Calemma, V. et al., *Preprints-ACS, Division of Petroleum Chemistry*, 44(3), 241 (1999).

Carter, B.H., Marine lubricants, in *Chemistry and Technology of Lubricants*, 2nd ed., Mortier, R.M. and Orszulik, S.T., Eds., Blackie Academic & Professional, London, U.K., 1997.

Crawford, J., Psaila, A., and Orszulik, S.T., Miscellaneous additives and vegetable oils, in *Chemistry and Technology of Lubricants*, 2nd ed., Mortier, R.M. and Orszulik, S.T., Eds., Blackie Academic & Professional, London, U.K., 1997.

Denis, J., Briant, J., and Hipeaux, J.C., *Lubricant Properties Analysis and Testing*, Editions Technip, Paris, 2000.

Evstafev, V.P. et al., *Chemistry and Technology of Fuels and Oils*, 37(6), 427 (2001).

Exxon Mobil, http://www.exxonmobil.com (accessed 2001).

Gary, J.H. and Handwerk, G.E., *Petroleum Refining, Technology and Economics*, 4th ed., Marcel Dekker, Inc., New York, 2001.

Henderson, H.E., Chemically modified mineral oils, in *Synthetics, Mineral Oils, and Bio-Based Lubricants Chemistry and Technology*, Rudnick, L.R., Ed., CRC Press, Boca Raton, FL, 2005.

Instrumentation Scientifique de Laboratoire, Product News Letter 4602, 2002.

Kajdas, C., Industrial lubricants, in *Chemistry and Technology of Lubricants*, 2nd ed., Mortier, R.M. and Orszulik, S.T., Eds., Blackie Academic & Professional, London, U.K., 1997.

Karsa, D.R., Industrial applications of surfactants—an overview, in *Industrial Applications of Surfactants*, Karsa, D.R., Ed., The Royal Society of Chemistry, Burlington House, London, U.K., 1986.

Kelarev, V.I. et al., *Petroleum Chemistry*, 38(5), 348 (1998).

Kelarev, V.I. et al., *Petroleum Chemistry*, 40(2), 134 (2000).

Kopeliovich, D., Additives in Lubricating Oils, http://www.substech.com (accessed 2008).

Lee, F.L. and Harris, J.W., Long-term trends in industrial lubricant additives, in *Lubricant Additives Chemistry and Applications*, Rudnick, L.R., Ed., Marcel Dekker, Inc., New York, 2003.

Lubrizol, http://www.lubrizol.com (accessed 2001).

Lynch, T.R., *Process Chemistry of Lubricant Base Stocks*, CRC Press, Boca Raton, FL, 2007.

Ma, Z. and Niu, C., *Runhuayou*, 15(4), 19 (2000).

Mang, T., Base oils, in *Lubricants and Lubrications*, Mang, T. and Dresel, W., Eds., Wiley-VCH GmbH, Weinheim, 2001.

Men, G. et al., *Petroleum Science and Technology*, 19(9&10), 1251 (2001).

Mills, A.J., Lindsay, C.M., and Atkinson, D.J., The formulation of automotive lubricants, in *Chemistry and Technology of Lubricants*, 2nd ed., Mortier, R.M. and Orszulik, S.T., Eds., Blackie Academic & Professional, London, U.K., 1997.

Nassar, A.M. et al., *Petroleum Science and Technology*, 23(5–6), 537 (2005).

Perez, J.M., Additives for industrial lubricant applications, in *Lubricant Additives Chemistry and Applications*, Rudnick, L.R., Ed., Marcel Dekker, Inc., New York, 2003.

Phillipps, R.A., Highly refined mineral oils, in *Synthetic Lubricants and High-Performance Functional Fluids*, 2nd ed., Rudnick, L.R. and Shubkin, R.L., Eds., Marcel Dekker Inc., New York, 1999.

Pillon, L.Z., *Interfacial Properties of Petroleum Products*, CRC Press, Boca Raton, FL, 2007.

Pillon, L.Z., Reid, L.E., and Asselin, A.E., Lubricating Oil for Inhibiting Rust Formation, U.S. Patent 5,397,487, 1995.

Pirro, D.M. and Wessol, A.A., *Lubrication Fundamentals*, 2nd ed., Marcel Dekker Inc., New York, 2001.

Placek, D.G., Hydraulics, in chemically modified mineral oils, in *Synthetics, Mineral Oils, and Bio-Based Lubricants Chemistry and Technology*, Rudnick, L.R., Ed., CRC Press, Boca Raton, FL, 2005.

Prince, R.J., Base oils from petroleum, in *Chemistry and Technology of Lubricants*, 2nd ed., Mortier, R.M. and Orszulik, S.T., Eds., Blackie Academic & Professional, London, U.K., 1997.

Rasberger, M., Oxidative degradation and stabilisation of mineral oil based lubricants, in *Chemistry and Technology of Lubricants*, 2nd ed., Mortier, R.M. and Orszulik, S.T., Eds., Blackie Academic & Professional, London, U.K., 1997.

Rogel, E., Contreras, E., and Leon, O., *Petroleum Science and Technology*, 20(7&8), 725 (2002).

Rudnick, L.R., *Lubricant Additives Chemistry and Applications*, Marcel Dekker Inc., New York, 2003.

Rudnick, L.R. and Bartz, W.J., Comparison of synthetic, mineral oil, and bio-based lubricant fluids, in *Synthetics, Mineral Oils, and Bio-Based Lubricants Chemistry and Technology*, Rudnick, L.R., Ed., CRC Press, Boca Raton, FL, 2005.

Rudnick, L.R. and Shubkin, R.L., Poly(alfa-Olefins), in *Synthetic Lubricants and High-Performance Functional Fluids*, 2nd ed., Rudnick, L.R. and Shubkin, R.L., Eds., Marcel Dekker Inc., New York, 1999.

Rulison, C.C. and Falberg, D., *Book of Abstracts*, 214th ACS National Meeting, Las Vegas, NV, September 7–11, 1997.

Russo, J.M. et al., Polymeric Thioheterocyclic Rust and Corrosion Inhibiting Marine Diesel Engine Lubricant Additives, U.S. Patent 5,516,442, 1996.

Sequeira, A. Jr., *Lubricant Base Oil and Wax Processing*, Marcel Dekker, Inc., New York, 1994.

Shell, http://www.shellglobalsolutions.com (accessed 2002).

Sigma-Aldrich, *Handbook of Fine Chemicals*, 2007–2008.

Singh, H., Science in Africa, November 2002, http://www.scienceinafrica.co.za (accessed 2008).

Society of Automotive Engineers, SAE J300 Viscosity Grades for Engine Oils, 2004.

Song, W., Chiu, I.C., and Heilman, W.J., *Lubrication Engineering*, 58(6), 29 (2002).

Ward, W.C. Jr., Tipton, C.D., and Murray, K.A., Lubrication Fluids for Reduced Air Entrainment and Improved Gear Protection, CA Patent 2,184,969, 1996.

7 Effects of Additives on Surface Activities of Turbine Oils

7.1 BASE STOCK AND ADDITIVE PERFORMANCE REQUIREMENTS

Turbine oils are required to have good oxidation stability. The importance of the low temperature properties of industrial oils is dependent on their intended use. The literature reports that there is no need for low pour points of oils to be used inside heated plants or in a continuous service, such as steam turbine (Bock, 2001). Turbine oils are also required to have good interfacial properties, such as resistance to foaming, good air separation properties, good demulsibility, and good rust- and corrosion-preventive properties. High quality turbine oils are formulated from highly refined paraffinic base stocks and have high viscosity-temperature characteristics. Gas turbines for industrial applications require high temperature–operating conditions, and, for particularly high temperature, the use of synthetic base stocks is required (Kajdas, 1997). The early literature reports on the oxidation stabilities of solvent-refined and clay-finished paraffinic base stocks, and the hydrotreated oil used to formulate turbine oils (Beuther, Donaldson, and Henke, 1964). According to the literature, the term "hydrotreated" mineral base stock refers to the removal of undesired components, such as condensed aromatics and polar compounds, by passing the oil over a catalyst in the presence of hydrogen at a temperature ranging between 50°C and 500°C and a pressure ranging between 200 and 4000 psi (Butler and Henderson, 1990).

The oxidation stability of the solvent-refined and hydrotreated oil suitable for use in turbine oils is shown in Table 7.1. The total acid number (TAN) of oil is the weight of potassium hydroxide (KOH) required to neutralize 1 g of oil. The ASTM D 943 oxidation test, known as the Turbine Oil Stability Test (TOST), is usually used to measure the time, in hours, for the oil to reach the TAN value of 2.0 mg KOH/g, which is called the TOST life. Following the ASTM D 943 method, at 95°C, the uninhibited solvent-refined and clay-finished paraffinic base stocks were found to have the TOST life of 90 h. Under the same test conditions, the uninhibited hydrotreated oil was found to have a longer TOST life of 210 h indicating an increase in the oxidation stability. The ASTM D 2272 test method, known as the Rotary Bomb Oxidation Test (RBOT), is used to compare the oxidation stability of steam turbine oils with the same composition in terms of base stocks and additives. According to the early literature, the RBOT test was run at 150°C until a pressure drop of 60 mm inside

TABLE 7.1

Oxidation Stability of the Solvent-Refined Base Stock and the Hydrotreated Oil

Base Stock Processing	Solvent Refined Clay Finishing	Hydrotreated Oil
ASTM D 943 (95°C)		
TAN (mg KOH/g)	2	2
TOST life (h)	90	210
ASTM D 2272 (150°C)		
RBOT life (min)	16	48

Source: Beuther, H. et al., *I&EC Prod. Res. Dev.*, 3(3), 174, 1964.
With permission.

the bomb was measured (Beuther, Donaldson, and Henke, 1964). The uninhibited solvent-refined and clay-finished paraffinic base stocks were reported to have the RBOT life of 16 min. Under the same test conditions, the uninhibited hydrotreated oil was reported to have a longer RBOT life of 48 min.

The performance requirements of the mineral oil blend and the ISO VG 46 turbine oil containing additives are shown in Table 7.2. Hydrofined mineral base stocks have usually a low TAN, below 0.01 mg KOH/g; however, the TAN gradually increases with an increase in the oxidation time. The literature reports that even when using the most refined paraffinic base stocks, antioxidants are still required to increase the oxidation stability of mineral turbine oils and meet the RBOT life specification of a minimum of 300 min (El-Ashry et al., 2006). The oxidation stability of turbine oils is tested following the ASTM D 2272 test, and the test is run until a pressure drop of 25 psi inside the bomb is measured. The literature also reports that while the demulsibility properties of hydrofined mineral oils can vary, the use of antifoaming agents is required to prevent the foaming and the use of rust inhibitor is needed to prevent the rust formation (Pillon, 2007). The literature reports on the surface active properties and oxidation stability of mineral ISO VG 46 oil blends containing different chemistry antioxidants and a pour point depressant (El-Ashry et al., 2006).

The surface active properties and the oxidation stability of mineral ISO VG 46 oil blend containing different antioxidants and a pour point depressant are shown in Table 7.3. The oxidized mineral ISO VG 46 oil blend #1, containing 0.8 wt% of commercial antioxidant, was reported to have an increased RBOT life from 30 to 302 min, and the oil TAN decreased from 4.1 to 1.6 mg KOH/g. Another oxidized mineral ISO VG 46 oil blend #2, containing 0.7 wt% of benzotriazole derivative #1 and 50 ppm of pour point depressant, was found to have a similar RBOT life of 305 min and an increased oil TAN of 2.3 mg KOH/g. The oxidized mineral ISO VG 46 oil blend #3, containing only 0.5 wt% of benzotriazole derivative #2 and 50 ppm of pour point depressant, was found to have an increased RBOT life of 330 min and

TABLE 7.2

Performance Requirements of the Mineral Oil and the ISO VG 46 Turbine Oil

Turbine Oil Performance Requirements	Mineral Oil Blend Properties	ISO VG 46 Oil Specifications
KV at 40°C (cSt)	46.3	41.4–50.6
KV at 100°C (cSt)	6.65	—
VI	94	Min 90
Pour point (°C)	−3	Min −6
Fresh oil TAN (mg KOH/g)	0.015	Max 0.3
Flash point (°C)	188	Min 185
Oxidation stability		
RBOT life (min)	30	Min 300
Oxidized oil TAN (mg KOH/g)	4.1	—
Foaming tendency		
Seq. I at 24°C (mL)	450/0	Max 50/0
Seq. II at 93.5°C (mL)	25/0	0/0
Seq. III at 24°C (mL)	450/0	Max 50/0
Air release time at 50°C (min)	4	Max 6
Demulsibility at 54°C (min)	5	Max 30
Rust protection	None	Yes
Cu corrosion (100°C/3 h)	1a	Max 2

Source: Reprinted from El-Ashry, El-S.H. et al., *Lubr. Sci.*, 18(2), 109, 2006. With permission.

a similar oil TAN of 2.3 mg KOH/g. To increase the RBOT life and decrease the TAN of oxidized mineral turbine oils, the chemistry and treat rates of antioxidants need to be carefully selected.

The chemistry of different additives used in turbine oils and their bench performance testing are shown in Table 7.4. At higher operating temperatures, the use of antioxidants and metal deactivators is required.

Typical turbine oils use paraffinic base stocks and contain additives required to increase their oxidation stability, impart resistance to foaming without air entrainment, assure good water separation properties, and prevent rust and corrosion. Some turbine oils might require the addition of pour point depressants to lower their pour points. The typical antioxidants used in turbine oils are hindered phenols and aromatic amines. Under high-temperature conditions, different chemistry metal deactivators are also used to increase the oxidation stability of turbine oils and protect the metal surface. Such materials include indoles, thiazoles, triazoles, benzotriazoles, and many others. The benzimidazole and the benzotriazole derivatives are used as metal deactivators and corrosion inhibitors. Hydrofined mineral base stocks foam when mixed with air and require the use of antifoaming agents. Turbine oils require

TABLE 7.3

Effect of Antioxidants on the Property and the RBOT Life of ISO VG 46 Oil Blend

Mineral ISO VG 46 Oils	Blend #1	Blend #2	Blend #3
Antioxidants	Commercial	Benzotriazole #1	Benzotriazole #2
Antioxidant (wt%)	0.8	0.7	0.5
Pour point depressant (ppm)	0	50	50
KV at 40°C (cSt)	47.0	47.3	46.8
KV at 100°C (cSt)	6.8	6.8	6.8
VI	96	96	99
Pour point (°C)	−6	−3	−3
Depressed pour point (°C)	—	−6	−6
Flash point (°C)	189	188	188
Fresh oil TAN (mg KOH/g)	0.04	0.02	0.07
Foaming, Seq. I (mL)	190/0	70/0	250/0
Foaming, Seq. II (mL)	20/0	0/0	30/0
Foaming, Seq. III (mL)	340/0	50/0	100/0
Air release time at 50°C (min)	1.4	1.5	1.4
Demulsibility (54°C) (min)	1.4	1.5	1.2
Rust protection	None	None	None
Cu corrosion (100°C/3 h)	1a	1a	1a
RBOT life (min)	302	305	330
Oxidized oil TAN (mg KOH/g)	1.6	2.3	2.3

Source: El-Ashry, El-S.H. et al., *Lubri. Sci.*, 18(2), 109, 2006.

the use of silicone-free antifoaming agents (Mang, 2001). Turbine oils need to have good water separation properties, and demulsifiers are used. To prevent the rust formation, different chemistry rust inhibitors are used. Some turbine oils require the addition of a mild antiwear/EP additive.

7.2 FOAM INHIBITION AND AIR ENTRAINMENT

The literature reports that the excessive foaming of the lubricating oil might lead to inadequate lubrication and mechanical failure. The lubricating oil may no longer be delivered effectively to the moving parts of the equipment as a continuous liquid stream. In addition, foaming may result in the overflow losses of the lubricating oil (Pillon and Asselin, 1998). The agitation of the lubricating oil with air equipment, such as bearings, gears, and pumps, may produce foam and a dispersion of finely divided air bubbles in the oil known as air entrainment. If the residence time in the reservoir is too short to allow the air bubbles to rise to the oil surface, a mixture of air and oil will circulate through the lubricating system (Pirro and Wessol, 2001). Unlike foaming, air entrainment is a dispersion of small air bubbles in oil. One of the key performance requirements of turbine oils is to have a resistance to foaming

TABLE 7.4

Chemistry of Additives Used in Turbine Oils and Their Performance Testing

Additive Function	Additive Chemistry	Performance Testing
Antioxidant	Diaryl amines	ASTM D 943
	Hindered phenols	ASTM D 4310
	Organic sulfides	ASTM D 2272
		ASTM D 1743
		ASTM D 6186
Metal deactivator	Triazoles	
	Benzotriazoles	
	Tolutriazole derivatives	ASTM D 130
	2-Mercaptobenzothiazoles	
Foam inhibitor	PAs	ASTM D 892
Demulsifier	Polyalkoxylated phenols	ASTM D 1401
	Polyalkoxylated PAs	
Rust inhibitor	Alkylsuccinic acids	ASTM D 665
	Imidazoline derivatives	
	Ethoxylated phenols	
Mild antiwear/EP additive	Alkylphosphoric acid ester	ASTM D 4172
	Alkylphosphoric salts	ASTM D 5182

Source: Reproduced from Lee, F.L. and Harris, J.W., Long-term trends in industrial lubricant additives, in *Lubricant Additives Chemistry and Applications*, Rudnick, L.R., Ed., Marcel Dekker, Inc., New York, 2003. With permission.

and a good air separation property. Silicones increase the air release time and are not recommended for use in turbine oils. Polyacrylate (PA) dispersions are used as anti-foaming agents in lubricating oils, which require good air separation properties, such as turbine oils. Hydrocarbons are less flexible than silicones and require polar side chains to be surface active. The presence of ester side chains makes the PA polymer surface active toward the oil/air surface (Pape, 1981).

$$[-CH(COOR)-CH_2-]_n$$

Acrylate copolymers

Many polymer type additives are highly viscous materials and are used as concentrated solutions in diluent oils. The PA antifoaming agent is commercially available in the form of a 40 wt% dispersion in a petroleum solvent, such as kerosene. The surface tension of the 40 wt% PA dispersion was found to be 27.2 mN/m, measured at 24°C, which is below the surface tension of mineral base stocks but higher than the ST of silicones.

The effect of the temperature on the surface activity of the PA (40 wt%) dispersion in mineral 150N base stock is shown in Table 7.5. With an increase in the temperature

TABLE 7.5
Effect of Temperature on the Surface Activity of PA (40 wt%) in Mineral 150N Oil

Antifoaming Agent	Mineral 150N Base Stock		
	ST at 24°C (mN/m)	(S) at 24°C (mN/m)	(S) at 93.5°C (mN/m)
Si 12500	24.0	8.2	5.5
FSi 300	26.2	1.7	3.9
PA (40 wt%)	27.2	5	−1.1

Source: Pillon, L.Z. and Asselin, A.E., Antifoaming agents for lubricating oils, U.S. Patent 5,766,513, 1998.

from 24°C to 93.5°C the efficiency of Si 12500 to spread (*S*) on the oil/air surface of mineral 150N base stock was found to decrease from 8.2 to 5.5 mN/m. With an increase in the temperature, the efficiency of FSi 300 to spread (*S*) on the oil surface of mineral 150N base stock increased from 1.7 to 3.9 mN/m. With an increase in the temperature from 24°C to 93.5°C, the efficiency of PA (40 wt%) to spread (*S*) on the oil/air surface of mineral 150N base stock was found to decrease from 5 mN/m to a negative 1.1 mN/m indicating its surface activity only at low temperatures. The selected antifoaming agent needs to prevent foaming without an increase in the air release time. Following the ASTM D 892 test method, the foaming tendency is the volume of foam measured at 24°C and 93.5°C immediately after the cessation of air flow. The foam stability is the volume of foam measured 10 min after disconnecting the air supply. The ASTM D 3427 test method for air release properties of petroleum oils covers the ability of industrial oils, such as turbine, and hydraulic and gear oils, to separate the entrained air usually at 50°C.

The effect of the PA (40 wt%) dispersion on the foaming tendency and the air release time of different 150N mineral base stocks is shown in Table 7.6. The hydrofined

TABLE 7.6
Effect of PA (40%) on the Foaming and the Air Release Time of Different 150N Base Stocks

Mineral 150N Base Stocks	PA (40 wt%), ppm (Active)	Foam at 24°C, Seq. I (mL)	Foam at 93.5°C, Seq. II (mL)	Air Release Time, at 50°C (min)
150N (phenol) oil	0	345/0	30/0	1.6
	100 (40)	5/0	30/0	2.2
150N (NMP) oil	0	280/0	20/0	2.1
	200 (80)	0/0	20/0	3.2

Source: Pillon, L.Z. and Asselin, A.E., Antifoaming Agents for Lubricating Oils, U.S. Patent 5,766,513, 1998.

150N phenol extracted base stock was found to require 100 ppm of the PA (40 wt%) antifoaming agent, 40 ppm of an active ingredient, to prevent the foaming tendency and only a small increase in the air release time from 1.6 to 2.2 min at 50°C was observed. The hydrofined 150N NMP extracted base stock was found to require a higher treat rate of 200 ppm of the PA (40 wt%) antifoaming agent, 80 ppm of an active ingredient, to prevent the foaming tendency and an increase in the air release time from 2.1 to 3.2 min at 50°C was observed. However, the use of only 20 ppm of the PA (40 wt%), 8 ppm of an active ingredient, was found sufficient to prevent the foaming tendency of 600N mineral base stock indicating an increase in its surface activity due to an increase in the ST of higher viscosity base stock. With an increase in the surface activity of the PA (40 wt%), a significant increase in the air release time of mineral 600N base stock from 9 to 16 min at 50°C was also observed. The mineral ISO VG 32 turbine oils can be formulated by blending paraffinic base stocks with all the required additives, including the PA (40%) antifoaming agent. Higher viscosity mineral ISO VG 46 turbine oils have a higher ST, which leads to an increase in the surface activity of the PA (40 wt%) and lower treat rates are required to prevent foaming.

The effect of the temperature on the surface activity of the PA (40 wt%) in SWI 6 base stock is shown in Table 7.7. SWI 6 base stock has a lower ST when compared to a similar viscosity 150N mineral base stocks and a decrease in the surface activity of antifoaming agent is observed. With an increase in the temperature from 24°C to 93.5°C, the efficiency of Si 12500 to spread on the oil/air surface of SWI 6 base stock was found to drastically decrease from 7 to 1.9 mN/m. With an increase in the temperature, the efficiency of FSi 300 to spread (S) on the oil surface of SWI 6 base stock increased from −0.2 to 1.8 mN/m. The PA antifoaming agent was found to have a negative spreading coefficient of 5.3 mN/m at 24°C and a negative spreading coefficient of 2.0 mN/m at 93.5°C on the oil surface of SWI 6 base stock indicating poor surface activity at high and low temperatures. While silicone and fluorosilicone antifoaming agents are effective in preventing foaming, they increase the air release time of petroleum oils.

The effect of the PA (40%) on the foaming tendency and the air release time of SWI 6 base stock is shown in Table 7.8. Si 12500 was found to be effective in

TABLE 7.7
Effect of Temperature on the Surface Activity of PA (40 wt%) in SWI 6 Base Stock

| Antifoaming Agent | SWI 6 Base Stock | | |
	ST at 24°C (mN/m)	(S) at 24°C (mN/m)	(S) at 93.5°C (mN/m)
Si 12500	24.0	7.0	1.9
FSi 300	26.2	−0.2	1.8
PA (40 wt%)	27.2	−5.3	−2.0

Source: Pillon, L.Z. et al., Method for reducing foaming of lubricating oils, U.S. Patent 6,090,758, 2000.

TABLE 7.8

Effect of PA (40%) on the Foaming and the Air Release Time of SWI 6 Base Stock

Antifoaming Agents	Treat Rate, ppm (Active)	Foam at 24°C, Seq. I (mL)	Foam at 93.5°C, Seq. II (mL)	Air Release, Time at 50°C (min)
Neat SWI 6 oil	0	170/0	25/0	1.0
Si 12500	3	0/0	0/0	2.8
FSi 300	3	35/0	0/0	1.1
PA (40%)	150 (60)	135/0	15/0	1.0

Source: Pillon, L.Z. and Asselin, A.E., Antifoaming agents for lubricating oils, U.S. Patent 5,766,513, 1998.

preventing the foaming of SWI 6 base stock at a very low treat rate of 3 ppm, and the air release time was found to increase from 1 to 2.8 min at 50°C. At the treat rate of 3 ppm, FSi 300 was found to be effective in decreasing the foaming tendency of SWI 6 base stock from 170 to 35 mL at 24°C, but no significant increase in the air release time was observed. At the treat rate of 150 ppm of the PA (40%), which is 60 ppm of an active ingredient, the foaming tendency of SWI 6 base stock only decreased from 170 to 135 mL at 24°C with no increase in the air release time. Due to a very low surface activity of the PA (40%) in SWI 6 base stock, very high treat rates, above 200 ppm, are required to prevent foaming. The use of high treat rates of the PA (40 wt%) in turbine oils, containing SWI 6 base stocks, requires the storage stability testing. In many cases, the addition of mineral base stocks might be required to increase the solvency and the stability of the PA (40 wt%) in some turbine oils. The chemistry and the treat rates of antifoaming agents need to be carefully selected to prevent the foaming of turbine oils without an increase in air entrainment.

7.3   RUST PREVENTION AND DEMULSIBILITY

The patent literature reports on the use of 0.5 wt% of a hindered phenol, 0.03 wt% of diphenylamine, and 0.08 wt% of a metal deactivator, such as triazole derivative, in mineral turbine oils (Butler, Miller, and Nadasdi, 2001). Under high temperature and oxidation conditions, the metal deactivator protects the metal surface and can function as the corrosion inhibitor. Under low-temperature conditions and in the presence of water, the use of rust inhibitors is needed to prevent rusting. The effect of rust inhibitors on the demulsibility properties of steam turbine oils was reported (Luo et al., 1999). The literature reports that the concentration of rust inhibitors needs to be carefully selected to give adequate protection under a variety of mild-to-severe rust regimes but not too high (Mang, 2001). To prevent oil emulsification by water contamination or steam contamination, nonemulsifiable corrosion inhibitors and nonferrous metal passivators are used in turbine oils. Alkylated succinic acids, their partial esters, and half amides are used in industrial oils at low treat rates of 0.01–0.05 wt% to prevent the rusting of the metal surface (Braun and

TABLE 7.9

Effect of EASA on the IFT, the Emulsion Stability, and the Rust-Preventive Properties of White Oil

	White Oil		
	Uninhibited Oil	Inhibited Oil	Inhibited Oil
EASA (wt%)	0	0.05	0.1
IFT at RT (mN/m)	45.1	13.5	9.3
ASTM D 1401 (54°C)			
O/W/E (mL) (after 1 min)	24/39/17	15/23/42	3/7/70
ASTM D 665B (60°C)	Fail	Fail	Pass

Source: Pillon, L.Z. and Asselin, A.E., Lubricating oil having improved rust inhibition and demulsibility, U.S. Patent 5,227,082, 1993.

Omeis, 2001). The ASTM D 1401 method for the water separability of petroleum oils and synthetic fluids requires the mixing of oils with distilled water at 54°C. The rust-preventive properties of lubricating oils are usually tested following the more severe ASTM D 665B rust test in the presence of synthetic water at 60°C. The adsorption of rust inhibitors at the oil/water interface can lead to an increase in the emulsion stability.

The effect of the partially esterified alkylsuccinic acid (EASA) on the IFT, the emulsion stability, and the rust-preventive properties of white oil is shown in Table 7.9. Without the rust inhibitor, the uninhibited white oil was found to form 17 mL of emulsion at 54°C, after standing for 1 min, and was found to have poor rust-preventive properties. After the addition of 0.05 wt% of the partially EASA, the IFT of white oil was found to decrease from 45.1 to 13.5 mN/m indicating its adsorption at the oil/water interface. The inhibited white oil formed 42 mL of emulsion at 54°C, after standing for 1 min, and continued to have poor rust-preventive properties. After the addition of 0.1 wt% of the partially EASA, the IFT of white oil was found to further decrease to 9.3 mN/m indicating its increased adsorption at the oil/water interface. The inhibited white oil formed 70 mL of emulsion at 54°C, after standing for 1 min, but it was found to have good rust-preventive properties and passed the test indicating not even one rust spot on the metal surface.

With an increase in the treat rate of the EASA rust inhibitor, the rust-preventive properties of white oil improved and some increase in the emulsion stability was observed. The processor tensiometer K12 can measure the IFT of oils at different temperatures up to 100°C.

The effect of different chemistry rust inhibitors on the IFT and demulsibility of mineral 150N base stock is shown in Table 7.10. The addition of 0.05 wt% of EASA rust inhibitor decreased the IFT of mineral 150N base stock from 33 to 18 mN/m at 54°C indicating its adsorption at the oil/water interface, but no increase in the emulsion stability was observed. The addition of 0.05 wt% of a basic polyamine (PA) rust inhibitor decreased the IFT of mineral 150N base stock to 14 mN/m at 54°C also indicating its adsorption at the oil/water interface and an increase in the emulsion

TABLE 7.10

Effect of Different Rust Inhibitors on the IFT and Demulsibility of Mineral 150N Oil

Mineral 150N Oil	Neat 150N	EASA	PA	SAA
Rust Inhibitors (wt%)	0	0.05	0.05	0.05
IFT at RT (mN/m)	34	18	15	11
IFT at 54°C (mN/m)	33	18	14	Very low*
ASTM D 1401 (54°C)				
Separation time (min)	10	10	15	35

Source: Pillon, L.Z. and Asselin, A.E., Lubricating oil having improved rust inhibition and demulsibility, U.S. Patent 5,227,082, 1993.

*Lamella broke.

separation time from 10 to 15 min at 54°C, was observed. The addition of 0.05 wt% of basic succinic anhydride amine (SAA) rust inhibitor decreased the IFT of mineral 150N base stock to a very low value at 54°C, which was difficult to measure because the lamella broke, and a drastic increase in the emulsion separation time to 35 min was observed. Depending on their chemistry, some rust inhibitors have emulsifying properties and increase the stability of the oil/water interface. Turbine oils are required to possess the ability to protect the metal surface against rust formation and have good water separation properties.

The effect of different chemistry rust inhibitors on the IFT and demulsibility of SWI 6 base stock is shown in Table 7.11. The addition of 0.05 wt% of the EASA rust inhibitor decreased the IFT of SWI 6 base stock from 43 to 19 mN/m at 54°C indicating its adsorption at the oil/water interface, but no increase in the emulsion stability was observed. Actually, the emulsion separation time decreased from 4 to

TABLE 7.11

Effect of Different Rust Inhibitors on the IFT and Demulsibility of SWI 6 Oil

SWI 6 Base Stock	Neat SWI 6	EASA	PA	SAA
Rust Inhibitors (wt%)	0	0.05	0.05	0.05
IFT at RT (mN/m)	43	20	14	10
IFT at 54°C (mN/m)	42	19	13	Very low*
ASTM D 1401 (54°C)				
Separation time (min)	4	2	10	20

Source: Pillon, L.Z. and Asselin, A.E., Lubricating oil having improved rust inhibition and demulsibility, U.S. Patent 5,227,082, 1993.

*Lamella broke.

2 min. The addition of 0.05 wt% of the PA rust inhibitor decreased the IFT of SWI 6 base stock to 13 mN/m at 54°C also indicating its adsorption at the oil/water interface and an increase in the emulsion separation time from 4 to 10 min at 54°C was observed. The addition of 0.05 wt% of SAA rust inhibitor decreased the IFT of SWI 6 base stock to a very low value at 54°C, which was difficult to measure and a drastic increase in the emulsion separation time to 20 min was observed. Turbine oils are required to separate the emulsion in less than 30 min and have good rust-preventive properties. Mineral turbine oils, containing basic type rust inhibitors, require the addition of demulsifiers to meet the demulsibility specifications. The use of SWI 6 base stock improves the demulsibility properties of turbine oils, containing basic type rust inhibitors, and eliminates the need for demulsifiers.

7.4   EFFECT OF DEMULSIFIER ON FOAMING

In industrial equipment, rusting is the most common type of corrosion. One of the key performance requirements of turbine oils is that they possess the ability to protect the metal surface against rust formation and separate the water. In the case of turbine oils, water contamination and emulsion formation can be a serious problem leading to oil deterioration. The literature reports that the typical rust inhibitors used in most lubricating oils are amine succinates and alkaline earth sulfonates (Pirro and Wessol, 2001). In many cases, the use of corrosion and rust inhibitors leads to an increase in the emulsion stability of turbine oils and demulsifiers must be used. The early literature reports that the air separation properties of turbine oils were found to be degraded by the presence of polymethylsiloxane antifoaming agent and a demulsifier (Smelnitskii and Sergeev, 1970). The patent literature reports that an effective demulsifier for circulating oils was prepared by reacting abietic acid with polyoxypropylenediamine (Frangatos and Davis, 1988). Turbine oils are blended in different viscosity grades. ISO VG 32, 46, and 68 oils are recommended for the lubrication of steam, gas, and hydraulic turbines. The lower viscosity ISO VG 32 and ISO VG 46 turbine oils are used in turbines without transmissions. The higher viscosity ISO VG 68 turbine oil is used for integral gear lubrication due to its increased lubrication properties (Kajdas, 1997).

The turbine oils are required to have good demulsibility properties, and, if the oil inadvertently mixes with the water, the oil and the water must separate. Some demulsifiers might be surface active toward the oil/water and the oil/air interfaces, and might interfere with the performance of PA type antifoaming agents in low viscosity ISO VG 32 turbine oils with a lower ST. An increase in the demulsifier content decreased the emulsion separation time of mineral ISO VG 32 turbine oil, while an increase in the foaming tendency is observed indicating the need to increase the content of the antifoaming agent. An increase in the content of the antifoaming agent would lead to an increase in the air release time of mineral ISO VG 32 turbine oil indicating a need to control the treat rate of the demulsifier. Higher viscosity mineral ISO VG 68 turbine oils require a higher content of the rust inhibitor to prevent the rust formation and require more time to separate emulsions. In higher viscosity ISO VG 68 turbine oils, an increase in the content of the demulsifier did not affect the performance of the antifoaming agent. An increase in the viscosity of mineral ISO

VG 68 turbine oil and an increase in its ST lead to an increase in the surface activity of the antifoaming agent, and a lower treat rate of the antifoaming agent is required to prevent foaming due to its increased surface activity at the oil/air interface.

7.5 OXIDATION RESISTANCE

The oxidation or the thermal breakdown of turbine oils can result in the formation of acids or insoluble solids and render the oil unfit for further use. The oxidation stability of lubricating oils depends on the amount of oxygen contacting the oil, the catalytic effects of metals, and the temperature. The ASTM D 943 test is usually stopped when the TAN of oxidized oil reaches the value of 2, and some industrial oils were reported to take a several thousand hours before the TAN reaches the value of 2 (Pirro and Wessol, 2001). A further improvement in the oxidation stability of turbine oils was reported to be achieved by using API group II base stocks, with a high saturate content above 90 wt% (Lee and Harris, 2003). The literature also reports on the poor performance of some antioxidants in hydrocracked base stocks. After the composition of antioxidants was optimized, the inhibited hydrocracked base stocks were reported to have an improved oxidation stability over the inhibited mineral base stocks (Braun and Omeis, 2001).

The early literature reports on the effect of the solvent-refined and hydrotreated base stocks on the oxidation stability of turbine oils containing the same chemistry additives (Beuther, Donaldson, and Henke, 1964).

The effect of the solvent-refined base stock and the hydrotreated oil on the oxidation stability of turbine oils containing the same chemistry additives is shown in Table 7.12. Following the ASTM D 943 method, the mineral turbine oil, containing the solvent-refined base stock, was found to have the TOST life of 1350 h. Under the same test conditions, the modified turbine oil, containing the hydrotreated oil, was found to have a longer TOST life of above 3000 h. The early literature reports that the

TABLE 7.12
Effect of Different Base Stocks on Oxidation Stability of Turbine Oils

Turbine Oils	Mineral	Modified	Modified
Base Stocks	Solvent Refined	Hydrotreated	Hydrotreated
Additive content	Full dosage	Full dosage	Half dosage
ASTM D 943 (95°C)			
TAN (mg KOH/g)	2	2	2
TOST life (h)	1350	>3000	>3000
ASTM D 2272 (150°C)			
RBOT life (min)	160	386	304

Source: Beuther, H. et al., *I&EC Prod. Res. Dev.*, 3(3), 174, 1964.

ASTM D 2272 test method was run until a pressure drop of 60 mm inside the bomb was measured. The mineral turbine oil containing the solvent-refined base stock was reported to have the RBOT life of 160 min, while under the same test conditions, the modified turbine oil containing the hydrotreated oil was reported to have a longer RBOT life of 386 min (Beuther, Donaldson, and Henke, 1964). Following the ASTM D 943 method and the ASTM D 2275 method, the use of the hydrotreated oil was also reported to decrease the treat rates of the additives that are needed to increase the oxidation resistance of the turbine oils. Another test, the ASTM D 4310 method is used for the determination of sludging and corrosion tendencies of the inhibited mineral oils. An oil sample is contacted with oxygen in the presence of an iron–copper catalyst also at 95°C and after 1000 h, the insoluble material is separated by filtration and is reported as a sludge. The literature reports on the effect of the hydrotreated base stocks on the oxidation resistance of the ISO VG 32 turbine oils, containing 2,6-di-tertiary-butyl-*p*-cresol known as BHT antioxidant and its S-containing derivative, in terms of decreasing the TAN and the sludge formation (Rasberger, 1997).

The effect of the hydrotreated base stock on the oxidation resistance of the ISO VG 32 turbine oils, containing different antioxidants and a corrosion inhibitor, is shown in Table 7.13.

The literature reports that triazine amino derivatives were found effective as corrosion inhibitors in turbine oils (Kelarev et al., 1999). The oxidized mineral ISO VG 32 turbine oil #1A, containing 0.25 wt% of the BHT antioxidant and a corrosion inhibitor, was reported to produce 47 mg of sludge after 1000 h and reached a TAN of 2 after 1100 h. The use of hydrotreated base stocks to formulate ISO VG 32 turbine

TABLE 7.13
Effect of Hydrotreated Base Stock on the TAN and the Sludge of ISO VG 32 Turbine Oils

ISO VG 32 Turbine Oils	Oil #1A	Oil #1B	Oil #2A	Oil #2B
Base Stocks	Mineral	Hydrotreated	Mineral	Hydrotreated
BHT antioxidant (wt%)	0.25	0.25	0	0
BHT–sulfur derivative (wt%)	0	0	0.25	0.25
Corrosion inhibitor (wt%)	0.05	0.05	0.05	0.05
ASTM D 943 (95°C)				
TAN (mg KOH/g)	2	2	2	2
TOST life (h)	1100	2400	2200	4300
ASTM D 4310 (95°C)				
Time (h)	1000	1000	1000	1000
Sludge (mg)	47	39	28	6

Source: Rasberger, M., Oxidative degradation and stabilisation of mineral oil based lubricants, in *Chemistry and Technology of Lubricants*, 2nd ed., Mortier, R.M. and Orszulik, S.T., Eds., Blackie Academic & Professional, London, U.K., 1997.

oil #1B produced less sludge of 39 mg after 1000 h and increased the TOST life to 2400 h. The oxidized mineral ISO VG 32 turbine oil #2A, containing 0.25 wt% of the BHT–sulfur derivative and a corrosion inhibitor, was reported to produce only 28 mg of sludge after 1000 h and reached a TAN of 2 after 2200 h. The use of hydrotreated base stocks to formulate ISO VG 32 turbine oil #2B further reduced the sludge content to 6 mg after 1000 h and drastically increased the TOST life to 4300 h. The selection of rust inhibitors must be made carefully to avoid problems, such as the corrosion of nonferrous metals or the formation of emulsions.

The TAN and the sludge content of the oxidized mineral ISO VG 32 turbine oils containing different antioxidants and a rust inhibitor is shown in Table 7.14. The oxidized mineral ISO VG 32 turbine oil #1, containing 0.25 wt% of the alkylated diphenylamine antioxidant and a rust inhibitor, was reported to produce 172 mg of sludge after 1000 h and reached a TAN of 2 after 2000 h. Another oxidized mineral ISO VG 32 turbine oil #2, containing 0.25 wt% of the organosulfur compound $S(CH_2CH_2-COOR)_2$, was reported to reach a TAN of 2 after only 200 h and produce over 5000 mg of sludge after 1000 h. The oxidized mineral ISO VG 32 turbine oil #3, containing 0.2 wt% of the alkylated diphenylamine and 0.05 wt% of the organosulfur compound $S(CH_2CH_2-COOR)_2$, was reported to produce significantly less sludge of only 89 mg after 1000 h and reached a TAN of 2 after a longer time of 3300 h. A synergism between the alkylated diphenylamine antioxidant and the $S(CH_2CH_2-COOR)_2$ compound, called organosulfur hydroperoxide decomposer, in mineral turbine oils was reported (Rasberger, 1997). The literature reports that the synergistic blend of hindered phenols and diarylamines is used to improve the oxidation stability of many industrial circulating oils, including turbine oils (Migdal, 2003).

TABLE 7.14
TAN and Sludge of Oxidized Mineral ISO VG 32 Turbine Oils

	Mineral ISO VG 32 Turbine Oils		
	Oil #1	Oil #2	Oil #3
Alkylated diphenylamine (wt%)	0.25	0	0.2
$S(CH_2CH_2-COOR)_2$ (wt%)	0	0.25	0.05
Rust inhibitor (wt%)	0.05	0.05	0.05
ASTM D 943 (95°C)			
TAN (mg KOH/g)	2	2	2
TOST life (h)	2000	200	3300
ASTM D 4310 (95°C)			
Time (h)	1000	1000	1000
Sludge (mg)	172	>5000	89

Source: Rasberger, M., Oxidative degradation and stabilisation of mineral oil based lubricants, in *Chemistry and Technology of Lubricants*, 2nd ed., Mortier, R.M. and Orszulik, S.T., Eds., Blackie Academic & Professional, London, U.K., 1997.

TABLE 7.15

Effect of the Treat Rate of the Antioxidant Synergistic Blend on the TOST Life of Turbine Oils

	Mineral Turbine Oils		
	Oil #1	Oil #2	Oil #3
Hindered phenol (wt%)	0.8		0.4
Diphenylamine (wt%)		0.8	0.4
ASTM D 943 (95°C)			
TAN (mg KOH/g)	2	2	2
TOST life (h)	4000	1800	>5000

Source: Migdal, C.A., Antioxidants, in *Lubricant Additives Chemistry and Applications*, Rudnick, L.R., Ed., Marcel Dekker, Inc., New York, 2003. With permission.

The effect of the treat rate of the antioxidant synergistic blend on the TOST life of mineral turbine oils is shown in Table 7.15. The mineral turbine oil #1, containing 0.8 wt% of the hindered phenol type antioxidant, was reported to reach a TAN of 2 after 4000 h, while another mineral turbine oil #2, containing 0.8 wt% of the diphenylamine antioxidant was reported to reach a TAN of 2 after only 1800 h. The mineral turbine oil #3, containing 0.4 wt% of the hindered phenol antioxidant and 0.4 wt% of the diphenylamine antioxidant, was reported to have the TOST life of over 5000 h indicating a drastic increase in the oxidation resistance, in terms of decreasing the formation of acidic by-products by decreasing the treat rate of the antioxidants.

The effect of the synergistic blend of two different chemistry antioxidants on the RBOT life of mineral turbine oils is shown in Table 7.16. The mineral turbine oil

TABLE 7.16

Effect of the Synergistic Blend of Antioxidants on the RBOT Life of Mineral Turbine Oils

	Mineral Turbine Oils		
	Oil #1	Oil #2	Oil #3
Hindered phenol (wt%)	0.5		0.25
Diphenylamine (wt%)		0.5	0.25
ASTM D 2272 (150°C)			
RBOT life (min)	300	600	>700

Source: Migdal, C.A., Antioxidants, in *Lubricant Additives Chemistry and Applications*, Rudnick, L.R., Ed., Marcel Dekker, Inc., New York, 2003. With permission.

#1, containing 0.5 wt% of a hindered phenol type antioxidant, was reported to have the RBOT life of 300 min. Another mineral turbine oil #2, containing 0.5 wt% of a diphenylamine antioxidant, was reported to have a longer RBOT life of 600 min.

The use of 0.25 wt% of a hindered phenol type antioxidant and 0.25 wt% of a diphenylamine antioxidant further increased the RBOT life of mineral turbine oil #3 to over 700 min demonstrating the antioxidant synergism. The literature reports that a synergistic blend of different antioxidants is used to block various oxidation mechanisms and the degree of synergism depends on their chemistry, the base stock, and the oxidation test. The hindered phenolics are excellent antioxidants when used under low-temperature conditions, while the alkylated diphenylamines are excellent antioxidants when used under high-temperature conditions (Migdal, 2003).

The effect of the antioxidant synergistic blend and the metal deactivator on the RBOT life of mineral and modified turbine oil, containing API III base stock, is shown in Table 7.17.

Following the ASTM D 2275 method, the mineral turbine oil, containing 0.5 wt% of the hindered phenol, 0.03 wt% of the diphenylamine, and 0.08 wt% of the triazole derivative, was reported to have the RBOT life of 630 min. At a higher temperature of 150°C, the addition of the metal deactivator was found to increase the RBOT life of mineral turbine oil. However, the modified turbine oil, containing the API group III base stock and the same antioxidant system, was found to have a shorter RBOT life of 450 min indicating a need for different chemistry oxidation inhibitors. The patent literature reports that the use of the phenyl naphthyl amine antioxidant and two different metal deactivators, such as thiadiazole and benzotriazole, was reported to be effective in increasing the RBOT life of turbine oils containing mineral, hydrotreated, hydrocracked, and slack wax hydroisomerate base stocks (Butler, Miller, and Nadasdi, 2001).

The effect of the phenyl naphthyl amine antioxidant and two different metal deactivators on the RBOT life of mineral turbine oils is shown in Table 7.18. With an increase in the phenyl naphthyl amine antioxidant from 0.2% to 0.6 wt% the RBOT life of mineral turbine oils #1–2, containing 0.08 wt% of benzotriazole and 0.01 wt%

TABLE 7.17
Effect of Antioxidant Synergistic Blend and Metal Deactivator on RBOT Life

	Turbine Oils	
	Mineral	Modified
Hindered phenol (wt%)	0.5	0.5
Diphenylamine (wt%)	0.03	0.03
Triazole derivative (wt%)	0.08	0.08
ASTM D 2272 (150°C)		
RBOT life (min)	630	450

Source: Butler, K.D. et al., Industrial oils of enhanced resistance to oxidation, CA Patent 2,395,784, 2001.

TABLE 7.18

Effect of Antioxidant and Metal Deactivators on the RBOT Life of Mineral Turbine Oils

	Mineral Turbine Oils			
	Oil #2	Oil #2	Oil #3	Oil #4
Antioxidant system				
Phenyl naphthyl amine (wt%)	0.2	0.6	0.8	1.0
Benzotriazole (wt%)	0.08	0.08	0.08	0.08
Thiadiazole (wt%)	0.01	0.01	0.01	0.01
Pour point depressant (wt%)	0.1	0.1	0.1	0.1
Rust inhibitor (wt%)	0.08	0.08	0.08	0.08
Demulsifier (wt%)	0.004	0.004	0.004	0.004
Antifoaming agent (wt%)	0.008	0.008	0.008	0.008
ASTM D 2272 (150°C)				
RBOT life (min)	1452	2565	2515	2265

Source: Butler, K.D. et al., Industrial oils of enhanced resistance to oxidation, CA Patent 2,395,784, 2001.

of thiadiazole, drastically increased to 1452–2565 min. However, with a further increase in the phenyl naphthyl amine content to 0.8–1 wt%, the RBOT life of mineral turbine oils #3–4 was found to decrease to 2265–2515 min indicating the need to avoid overtreatment.

The effect of phenyl naphthyl amine antioxidant and two different metal deactivators on the RBOT life of hydrotreated turbine oils is shown in Table 7.19. With an increase in the phenyl naphthyl amine antioxidant from 0.3 to 0.8 wt%, the RBOT life of hydrotreated turbine oils #1–3, containing 0.08 wt% of benzotriazole and 0.01 wt% of thiadiazole, also drastically increased to 1997–3165 min. However, with a further increase in the phenyl naphthyl amine content to 1 wt%, the RBOT life of hydrotreated turbine oil #4 was found to decrease to 2880 min indicating the need to avoid overtreatment.

The effect of the phenyl naphthyl amine antioxidant and two different metal deactivators on the RBOT life of hydrocracked turbine oils is shown in Table 7.20. With an increase in the phenyl naphthyl amine antioxidant from 0.3 to 0.6 wt%, the RBOT life of hydrocracked turbine oils #1–2, containing 0.08 wt% of benzotriazole and 0.01 wt% of thiadiazole, also drastically increased to 2877–3675 min. However, with a further increase in the phenyl naphthyl amine content to 0.8–1 wt%, the RBOT life of hydrocracked turbine oils #3–4 was found to decrease to 3310–3540 min indicating the need to avoid overtreatment.

The effect of base stocks on the RBOT life of turbine oils, containing phenyl naphthyl amine antioxidant and two different metal deactivators, is shown in Table 7.21. The mineral turbine oil containing an optimized antioxidant system, comprising of 0.4 wt% of phenyl naphthyl amine, 0.08 wt% of benzotriazole, and 0.01 wt% of thiadiazole, was reported to have the RBOT life of 1860 min. The hydrotreated turbine

TABLE 7.19

Effect of Antioxidants and Metal Deactivators on the RBOT Life of Hydrotreated Oils

	Hydrotreated Turbine Oils			
	Oil #1	Oil #2	Oil #3	Oil #4
Antioxidant system				
Phenyl naphthyl amine (wt%)	0.3	0.6	0.8	1.0
Benzotriazole (wt%)	0.08	0.08	0.08	0.08
Thiadiazole (wt%)	0.01	0.01	0.01	0.01
Pour point depressant (wt%)	0.1	0.1	0.1	0.1
Rust inhibitor (wt%)	0.08	0.08	0.08	0.08
Demulsifier (wt%)	0.004	0.004	0.004	0.004
Antifoaming agent (wt%)	0.008	0.008	0.008	0.008
ASTM D 2272 (150°C)				
RBOT life (min)	1997	2955	3165	2880

Source: Butler, K.D. et al., Industrial oils of enhanced resistance to oxidation, CA Patent 2,395,784, 2001.

TABLE 7.20

Effect of Antioxidant and Metal Deactivators on the RBOT Life of Hydrocracked Oils

	Hydrocracked Turbine Oils			
	Oil #1	Oil #2	Oil #3	Oil #4
Antioxidant system				
Phenyl naphthyl amine (wt%)	0.3	0.6	0.8	1.0
Benzotriazole (wt%)	0.08	0.08	0.08	0.08
Thiadiazole (wt%)	0.01	0.01	0.01	0.01
Pour point depressant (wt%)	0.1	0.1	0.1	0.1
Rust inhibitor (wt%)	0.08	0.08	0.08	0.08
Demulsifier (wt%)	0.004	0.004	0.004	0.004
Antifoaming agent (wt%)	0.008	0.008	0.008	0.008
ASTM D 2272 (150°C)				
RBOT life (min)	2877	3675	3540	3310

Source: Butler, K.D. et al., Industrial oils of enhanced resistance to oxidation, CA Patent 2,395,784, 2001.

oil containing the same antioxidant system, comprising of 0.4 wt% of phenyl naphthyl amine, 0.08 wt% of benzotriazole, and 0.01 wt% of thiadiazole, was reported to have a longer RBOT life of 2449 min. The hydrocracked turbine oil containing the same antioxidant system, comprising of 0.4 wt% of phenyl naphthyl amine, 0.08 wt%

TABLE 7.21

Effect of Different Base Stocks on the RBOT Life of Turbine Oils

	Turbine Oils			
	Mineral	Hydrotreated	Hydrocracked	Hydroisomerate
Antioxidant system				
Phenyl naphthyl amine (wt%)	0.4	0.4	0.4	0.4
Benzotriazole (wt%)	0.08	0.08	0.08	0.08
Thiadiazole (wt%)	0.01	0.01	0.01	0.01
Pour point depressant (wt%)	0.1	0.1	0.1	0.1
Rust inhibitor (wt%)	0.08	0.08	0.08	0.08
Demulsifier (wt%)	0.004	0.004	0.004	0.004
Antifoaming agent (wt%)	0.008	0.008	0.008	0.008
ASTM D 2272 (150°C)				
RBOT life (min)	1860	2449	3327	4065

Source: Butler, K.D. et al., Industrial oils of enhanced resistance to oxidation, CA Patent 2,395,784, 2001.

of benzotriazole, and 0.01 wt% of thiadiazole, was also reported to have a longer RBOT life of 3327 min. The hydroisomerate turbine oil containing the same antioxidant system, comprising of 0.4 wt% of phenyl naphthyl amine, 0.08 wt% of benzotriazole, and 0.01 wt% of thiadiazole, was reported to have further increased the RBOT life to 4065 min.

The effect of different chemistry antioxidant systems on the RBOT life and the TOST life of turbine oils containing different base stocks is shown in Table 7.22. Following the ASTM D 2275 method, the mineral turbine oil, containing 0.5 wt% of the hindered phenol, 0.03 wt% of the diphenylamine, and 0.08 wt% of the triazole derivative, was reported to have the RBOT life of 630 min. Following the ASTM D 943 method, it was found to reach a TAN of 2 after 5000 h. The hydrocracked turbine oil containing the optimized antioxidant system, comprising of 0.4 wt% of the phenyl naphthyl amine, 0.08 wt% of the benzotriazole, and 0.01 wt% of the thiadiazole, was found to have a long RBOT life of 3120 min. Following the ASTM D 943 test, it was reported to reach a TAN of 2 after 13,660 h. The hydrotreated turbine oil #1 containing the same antioxidant system, comprising of 0.4 wt% of the phenyl naphthyl amine, 0.08 wt% of the benzotriazole, and 0.01 wt% of the thiadiazole, was found to have a shorter RBOT life of 2905 min. However, following the ASTM D 943 test, after a longer time of 14,000 h, it was reported to have a TAN below 2 indicating an increase in the TOST life. Another hydrotreated turbine oil #2 containing the same antioxidant system, comprising of 0.4 wt% of the phenyl naphthyl amine, 0.08 wt% of the benzotriazole, and 0.01 wt% of the thiadiazole, was also found to

TABLE 7.22

Effect of Different Antioxidant Systems on RBOT Life and TOST Life

Turbine Oils	Oil	Oil	Oil #1	Oil #2
Base Stocks	Mineral	Hydrocracked	Hydrotreated	Hydrotreated
Hindered phenol (wt%)	0.5			
Diphenylamine (wt%)	0.03			
Triazole derivative (wt%)	0.08			
Phenyl naphthyl amine (wt%)		0.4	0.4	0.4
Benzotriazole (wt%)		0.08	0.08	0.08
Thiadiazole (wt%)		0.01	0.01	0.01
Pour point depressant (wt%)	0.1	0.1	0.1	0.1
Rust inhibitor (wt%)	0.08	0.08	0.08	0.08
Demulsifier (wt%)	0.004	0.004	0.004	0.004
Antifoaming agent (wt%)	0.008	0.008	0.008	0.008
ASTM D 2272 (150°C)				
RBOT life (min)	630	3,120	2,905	2,430
ASTM D 943 (95°C)				
TAN (mg KOH/g)	2	2	<2	<2
TOST life (h)	5,000	13,660	14,000	16,000

Source: Butler, K.D. et al., Industrial oils of enhanced resistance to oxidation, CA Patent 2,395,784, 2001.

have a significantly shorter RBOT life of 2430 min. However, following the ASTM D 943 test, after a longer time of 16,000 h, it was reported to have a TAN still below 2 indicating a drastic increase in the TOST life.

The ASTM D 5846 test method is used to measure the Universal Oxidation Test (UOT) life of turbine and hydraulic oils. An oil sample is contacted with air at 135°C in the presence of copper and iron metals. The TAN and spot-forming tendency of the oil is measured daily until the oxidation life of the oil has been reached. The oil is considered to be degraded when either the TAN has increased by 0.5 mg KOH/g over that of new oil or when the oil begins to form insoluble solids. The drop of oil placed onto a filter paper shows a clearly defined dark spot surrounded by a ring of clear oil. The part-synthetic ISO VG 32 turbine oil, containing mostly synthetic PAO 6 and a small content of hydrofined 600N mineral base stock, was found to have a shorter UOT life indicating a need for a different chemistry antioxidant system to be used in synthetic oils. An increase in the TAN and the sludge leads to a decrease in the useful life of the lubricating oils and the base stocks, and the chemistry of antioxidants, including other additives and their treat rates, need to be carefully selected to meet the performance requirements of turbine oils.

REFERENCES

Beuther, H., Donaldson, R.E., and Henke, A.M., *I&EC Product Research and Development*, 3(3), 174 (1964).

Bock, V., Turbine oils, in *Lubricants and Lubrications*, Mang T. and Dresel W., Eds., Wiley-VCH GmbH, Weinheim, Germany, 2001.

Braun, J. and Omeis, J., Additives, in *Lubricants and Lubrications*, Mang T. and Dresel W., Eds., Wiley-VCH GmbH, Weinheim, Germany, 2001.

Butler, K.D. and Henderson, H.E., Adsorbent processing to reduce base stock foaming, U.S. Patent 4,600,502, 1990.

Butler, K.D., Miller, A.F., and Nadasdi, T.T., Industrial oils of enhanced resistance to oxidation, CA Patent 2,395,784, 2001.

El-Ashry, El-S.H. et al., *Lubrication Science*, 18(2), 109 (2006).

Frangatos, G. and Davis, R.H., Polyoxyalkylene diamides as lubricant demulsifier additives, U.S. Patent 4,743,387, 1988.

Kajdas, C., Industrial lubricants, in *Chemistry and Technology of Lubricants*, 2nd ed., Mortier, R.M. and Orszulik, S.T., Eds., Blackie Academic & Professional, London, U.K., 1997.

Kelarev, V.I. et al., *Petroleum Chemistry*, 39(3), 196 (1999).

Lee, F.L. and Harris, J.W., Long-term trends in industrial lubricant additives, in *Lubricant Additives Chemistry and Applications*, Rudnick, L.R., Ed., Marcel Dekker, Inc., New York, 2003.

Luo, Y. et al., *Shiyou Lianzhi Yu Huagong*, 30(2), 61 (1999).

Mang, T., Base oils, in *Lubricants and Lubrications*, Mang, T. and Dresel, W., Eds., Wiley-VCH GmbH, Weinheim, Germany, 2001.

Migdal, C.A., Antioxidants, in *Lubricant Additives Chemistry and Applications*, Rudnick, L.R., Ed., Marcel Dekker, Inc., New York, 2003.

Pape, P.G., Silicones: Unique chemicals for petroleum processing, SPE 10089, *56th Annual Fall Technical Conference and Exhibition of SPE*, San Antonio, TX, 1981.

Pillon, L.Z., *Interfacial Properties of Petroleum Products*, CRC Press, Boca Raton, FL, 2007.

Pillon, L.Z. and Asselin, A.E., Lubricating oil having improved rust inhibition and demulsibility, U.S. Patent 5,227,082, 1993.

Pillon, L.Z. and Asselin, A.E., Antifoaming agents for lubricating oils, U.S. Patent 5,766,513, 1998.

Pillon, L.Z., Asselin, A.E., and Vernon, P.D.F., Method for reducing foaming of lubricating oils, U.S. Patent 6,090,758, 2000.

Pirro, D.M. and Wessol, A.A., *Lubrication Fundamentals*, 2nd ed., Marcel Dekker Inc., New York, 2001.

Rasberger, M., Oxidative degradation and stabilisation of mineral oil based lubricants, in *Chemistry and Technology of Lubricants*, 2nd ed., Mortier, R.M. and Orszulik, S.T., Eds., Blackie Academic & Professional, London, U.K., 1997.

Smelnitskii, S.G. and Sergeev, V.A., *Elektronika* (in Russian), 41(8), 26, 1970.

8 Effects of Base Stocks on Surface Activities of Hydraulic Oils

8.1 BASE STOCK AND ADDITIVE PERFORMANCE REQUIREMENTS

The selection of the hydraulic lubricant depends on the hydraulic system operating conditions. Mineral lube oil base stocks provide good lubrication as long as the lubricant film is maintained between moving parts. This kind of hydrodynamic lubrication is mainly dependent on the oil viscosity. When the lubrication film is broken due to high temperature, pressure, and velocity, a chemical compound is needed to react with the metal surface and form a protective coating. This type of lubrication is called "boundary lubrication" (Raab and Hamid, 2003). The literature reports on the influence of lubricant viscosity on the interfacial friction and the surface quality of Pb-coated steel (Deng and Lovell, 2000). Lube oil base stocks need to have good low temperature fluidity and oxidation stability but also good interfacial properties at the oil/air interface, such as resistance to foaming and air entrainment, at the oil/water interface, such as demulsibility, and at the oil/metal interface, such as rust inhibition, corrosion inhibition, and wear prevention. The literature reports that the viscosity and the VI of base stocks affect low temperature fluidity, energy losses, cooling efficiency, and wear protection. Volatility affects oil consumption, oil thickening, and deposit formation. Oxidation stability affects acid formation, metal corrosion, oil thickening, and deposit formation. Solvency affects seal compatibility, engine cleanliness, and the formulation stability (Shell, 2002).

The effect of the base stock properties on the lubricant performance is shown in Table 8.1. Most lubricants are formulated with mineral base stocks because they are relatively low cost, available in a wide viscosity range, and generally good solvents for most additives. In some applications, the base stock selection is based on meeting other specifications such as the low temperature fluidity requirements or the fire-resistance. The synthetic fluids such as polyalfaolefins (PAO), organic esters, glycols, and phosphate esters are only used to meet the specific needs, such as the low temperature fluidity, the oxidation stability, or the fire-resistance where their performance outweighs their cost (Shugarman, 2001). The literature reports that in most industrial lubricants, base stocks are selected based on the viscosity requirements, the oxidation stability, the fire-resistance, and the biodegradability. They carry the load in hydrodynamic lubrication, remove heat and debris from friction and wear, and help to seal out contaminants (Shugarman, 2001). Hydraulic oils need to minimize

TABLE 8.1
Effect of Base Stock Properties on
Lubricant Performance

Base Stock Properties	Effect on Lubricant Performance
Viscosity and VI	Low temperature fluidity
	Energy losses
	Wear protection
	Cooling efficiency
Volatility	Oil consumption
	Oil thickening
	Deposit formation
Oxidation stability	Acid formation
	Metal corrosion
	Oil thickening
	Deposit formation
Interfacial properties	Foaming
	Air release
	Emulsification
	Metal protection
Solvency	Additive solubility
	Formulation stability
	Seal compatibility
	Engine cleanliness

Sources: Shell, http://www.shellglobalsolutions.com (accessed 2002). With permission; Pillon, L.Z., *Interfacial Properties of Petroleum Products*, CRC Press, Boca Raton, FL, 2007. With permission.

friction, wear, and deposits, flush away wear particles and contamination, remove heat, and protect metal surfaces from rust and corrosion (Pirro and Wessol, 2001). The typical applications, where hydraulic oils are used, require the proper viscosity, VI, oxidation stability, foam resistance, good air and water separation properties, good rust and wear protection, and compatibility.

The typical and some specific application requirements of hydraulic oils are shown in Table 8.2. The hydraulic oils need to meet the surface activity requirements at the oil/air, the oil/water, and the oil/metal interface. The foaming of hydraulic oils can cause pressure drops leading to poor lubrication of hydraulic valve tappets, and reduce the cooling effect. The literature reports that when air is introduced into oil under pressure, there is no foaming because the air concentration is low. However, the presence of entrained air in hydraulic systems is of concern because it makes oil more compressible and can to lead to cavitations in the pump (Denis, Briant, and Hipeaux, 2000). The literature reports that water might be present in hydraulic oil

TABLE 8.2

Typical and Some Specific Performance Requirements of Hydraulic Oils

Typical Application Requirements	Specific Application Requirements
Viscosity	Soluble oils
VI	High water content fluids
Oxidation stability	Fire resistant fluids
Foam resistance	Environmental performance
Good air release properties	
Good water separation properties	
Rust protection	
Wear protection	
Compatibility	

Source: Pirro, D.M. and Wessol, A.A., *Lubrication Fundamentals*, 2nd ed., Marcel Dekker Inc., New York, 2001. With permission.

because of water leaks in heat exchangers, washdown procedures, or condensation of atmospheric moisture. Most condensation occurs above the oil level in reservoirs as machines cool during idle or shutdown periods (Pirro and Wessol, 2001). During shutdown, when surfaces that are normally covered with oil, may be unprotected and subject to water condensation. During operation, the water may be broken up into droplets and mixed with oil forming an emulsion that may be drawn into the pump (Pirro and Wessol, 2001). To prevent water contamination, hydraulic oils have to separate the water to prevent the formation of rust. Water and oxygen can cause rusting of ferrous surfaces in hydraulic systems in operations that experience high humidity and temperature changes within the reservoirs (Pirro and Wessol, 2001).

The chemistry of additives used in hydraulic oils including different bench performance tests is shown in Table 8.3. The additives commonly used in hydraulic oils are VI improvers, antioxidants, metal deactivators, antiwear additives, rust inhibitors, pour point depressants, demulsifiers, and antifoaming agents. Hydraulic oils need to have a high VI to minimize viscosity change over a wide temperature range to provide long service life. The literature reports that the olefin copolymers (OCP) and polyalkymethacrylates (PAMA) chemistry VI improvers are used in hydraulic oils (Braun and Omeis, 2001). Hydraulic oils need to use shear stable VI improvers. The silicone-free antifoaming agents were reports used in metalworking processes cutting fluids and hydraulic fluids (Bock, 2001). Depending on the specific application, additional additives are used. For marketing, identification or leak detection purposes some lubricants contain dyes. Most of these compounds are solids suspended in mineral oils or solved in aromatic solvents. In some cases, to avoid the coloration of lubricating oil, fluorescent dyes are used to detect leaks under UV light (Braun and Omeis, 2001). In terms of their chemistry, the hydraulic lubricants can be divided into four groups, such as mineral oils, synthetic oils, water-based fluids, and polymer solutions (Kajdas, 1997).

TABLE 8.3

Chemistry of Additives and Performance Testing of Hydraulic Oils

Additive Function	Additive Chemistry	Performance Testing
Shear stable VI improvers	Poly(alkyl methacrylates)	ASTM D 5621
	OCP	
	Styrene diene copolymers	
Pour point depressants	Poly(alkyl methacrylates)	ASTM D 97
		ASTM D 5949
Antioxidants	Diaryl amines	ASTM D 2272
	Hindered phenols	ASTM D 1743
	Phenothiazine	ASTM D 4310
	Metal dialkyldithiocarbamates	ASTM D 6186
	Ashless dialkyldithiocarbamates	
	Metal dialkyldithiophosphates	
Metal deactivators	Thiadiazoles	ASTM D 130
	Benzotriazoles	
	Tolutriazole derivative	
	2-Mercaptobenzothiazoles	
Foam inhibitors	Polydimethylsiloxanes	ASTM D 892
	Polyacrylates	
Rust inhibitors	Alkylsuccinic acid derivatives	ASTM D 665
	Metal sulfonates	
	Ethoxylated phenols	
Corrosion inhibitors	Fatty amines	ASTM D 130
	Imidazoline derivatives	
	Ammonium sulfonates	
	Salts of phosphate esters and amines	
Demulsifiers	Polyalkoxylated phenols	ASTM D 1401
	Polyalkoxylated polyols	ASTM D 2711
	Polyalkoxylated polyamines	
Antiwear additives	Alkylphosphoric acid ester	ASTM D 4172
	Alkylphosphoric salts	ASTM D 2882
	Dialkyldithiophosphates	ASTM D 5182
	Metal dialkyldithiocarbamates	
	Dithiophosphate esters	

Source: Lee, F.L. and Harris, J.W., Long-term trends in industrial lubricant additives, in *Lubricant Additives Chemistry and Applications*, Rudnick, L.R., Ed., Marcel Dekker, Inc., New York, 2003. With permission.

The different types of mineral hydraulic lubricants and their general characteristics are shown in Table 8.4. The straight mineral oils were reported to be products of the HH group. HL mineral type hydraulic oils are formulated to have excellent oxidation stability, good rust- and corrosion-preventive properties, and resistance to foaming (Kajdas, 1997). HL type hydraulic oils are recommended for use as fluid media

TABLE 8.4

Types and Characteristics of Different Hydraulic Lubricants

Hydraulic Lubricants	Hydraulic Fluid Chemistry	General Characteristics
Group I	Mineral oil	Noninhibited
Group I	Mineral oil	Rust and oxidation inhibited
Group I	HL mineral type	Improved antiwear properties
Group I	HL mineral type	Improved VI
Group I	HM mineral type	Improved VI
Group I	HM mineral type	With antistic-slip properties
Group II	Synthetic fluid	No fire resistant properties
Group II	Synthetic fluid	Fire resistant
Group III	Water-in-oil emulsion	Fire resistant
Group III	Oil-in-water emulsion	Fire resistant
Group IV	Polymer solution	Fire resistant

Source: Kajdas, C., Industrial lubricants, in *Chemistry and Technology of Lubricants*, 2nd ed., Mortier R.M. and Orszulik S.T., Eds., Blackie Academic & Professional, London, U.K., 1997. With permission.

in hydraulic systems and hydraulic pumps under moderate operating conditions. HL type hydraulic oils are also recommended for circulation, splash, bath, and ring oiling systems of plain and antifriction bearings, gears of industrial machinery, chain drives, and crankcase lubrication (Apar Industries, 2003). The addition of the antiwear additive leads to HM products while the addition of the VI improvers leads to HV group of the mineral hydraulic oils. The antiwear type hydraulic oils containing antioxidants, rust inhibitors, antiwear additives, pour point depressants, and antifoaming agents are recommended for use as fluid media for hydraulic systems and hydraulic pumps under severe operating conditions (Apar Industries, 2003). The addition of an antistick-slip additive or friction modifier can lead to a different type of HR or HG group of mineral hydraulic oils (Kajdas, 1997). Depending on the specific performance requirements, the mineral base stocks, the synthetic fluids, or the vegetable oils are used to formulate hydraulic oils. A key issue when using synthetic base stocks in place of mineral oils is cost versus performance.

8.2   SOLVENCY EFFECT OF NAPHTHENIC OILS

The literature reports that the most important physical characteristic of hydraulic oil is its viscosity. Too low a viscosity can lead to excessive wear and leakage. Too high a viscosity can result in excessive heating, higher energy consumption, and a lower mechanical efficiency (Pirro and Wessol, 2001). The hydraulic systems normally require oil with the viscosity range of 32–68 cSt at 40°C. In some cases, when start-up and operating temperatures are lower, lower viscosity ISO VG 22 oil might be required. In some other cases, when start-up and operating temperatures

TABLE 8.5

VI and Typical Characteristics of Petroleum-Derived Base Stocks

Petroleum-Derived Base Stocks	Mineral VI < 20	Mineral VI 20–85	Mineral VI 85–110	Modified VI > 140
Low temperature fluidity	Very good	Poor	Poor	Poor
Volatility	High	Low	Low	Very low
Oxidation stability	Poor	Good	Good	Good
Solvency	Very good	Good	Good	Poor

Sources: Shell, http://www.shellglobalsolutions.com (accessed 2002). With permission; Pillon, L.Z., *Interfacial Properties of Petroleum Products*, CRC Press, Boca Raton, FL, 2007. With permission.

are higher, higher viscosity ISO VG 100 oil might be used (Pirro and Wessol, 2001). The petroleum-derived base stocks are being refined to meet various quality requirements of lubricating oils and are classified as low viscosity index (LVI), medium VI (MVI), and high VI (HVI) (Shell, 2002). The use of nonconventional refining can also lead to extra high VI (XHVI) base stocks.

The VIs and the typical characteristics of different petroleum-derived base stocks are shown in Table 8.5. LVI base stocks having VI < 20, are characterized as having very good low temperature fluidity, high volatility, poor oxidation stability, and very good solvency. MVI mineral base stocks having VI of 20–85, are characterized as having poor low temperature fluidity, low volatility, good oxidation stability, and good solvency. High VI mineral base stocks having VI of 85–110 are characterized as having poor low temperature fluidity, low volatility, good oxidation stability, and good solvency (Shell, 2002). The mineral base stocks are being refined to meet various quality requirements of lubricating oils from different crude oils and the naphthenic oils are known to have a low VI while the paraffinic base stocks have a high VI. The XHVI base stocks, produced by the hydroisomerization of slack wax have VI > 140, and are characterized as having poor low temperature fluidity, very low volatility, good oxidation stability, and poor solvency properties (Pillon, 2007).

Some hydraulic oils use re-refined base stocks. The used lube oil re-refining is a multistep operation, which can involve a caustic treatment and vacuum distillation needed to recover lube oil molecules from used lubricating oil. After the pretreatment and the thin-film distillation, the oil is hydrotreated under mild conditions. A final distillation step yields base stocks having a different viscosity (Betton, 1997). To protect the catalyst, the used lubricating oils need to meet the maximum limit requirements for contaminants and have a low level of ashless additives. The ashless additives are usually used at low levels in industrial lubricating oils. The VI modifiers are the only additives used at high levels but these are polymers that contain carbon, hydrogen, oxygen, and no metals (Butler, 1996). The used lubricating oils that can be hydrotreated are the oils used in low severity lubricating applications,

TABLE 8.6
Typical Properties of 150N and 500N
Re-Refined Oils

	Re-Refined Oils	
	150N	500N
KV at 100°C (cSt)	5 ± 0.2	11 ± 0.4
VI	90–110	90–110
Pour point (°C)	−9	−9
TAN (mg KOH/g)	0.1	0.1
Water (ppm)	Max. 50	Max. 50
Volatility (% loss)	Max. 18	Max. 6
Flash point (°C)	Min. 210	Min. 230

Source: Betton, C.I., Lubricants and their environmental impact, in *Chemistry and Technology of Lubricants*, 2nd ed., Mortier, R.M. and Orszulik, S.T., Eds., Blackie Academic & Professional, London, U.K., 1997. With permission.

such as turbine systems, compressors, electrical insulating equipment, transformers, hydraulic systems, paper manufacturing machines, and other machinery where there is no metal to metal contact (Stein and Bowden, 2004).

The typical properties of different viscosity 150N and 500N re-refined oils are shown in Table 8.6. Re-refined base stocks must meet the typical properties requirements, such as viscosity, VI, pour point, volatility, flash point, TAN, and water content, but also other specifications, such as color, odor, metal content, sulfated ash content, and chlorine content (Betton, 1997). The mineral 150N re-refined oil is required to have a kinematic viscosity of 5 cSt at 100°C, a VI in the range of 90–110, a pour point of −9°C, a TAN of 0.1 mg KOH/g, a max water content of 50 ppm, a volatility of max 18% off, and a flash point of min 210°C. Higher viscosity mineral 500N refined oil is required to have a kinematic viscosity of 11 cSt at 100°C, a VI in the range of 90–110, a pour point of −9°C, a TAN of 0.1 mg KOH/g, a max water content of 50 ppm, a volatility of max 6% off, and a flash point of min 230°C. Some re-refined base stocks have a dark color or a specific odor, which might indicate a poor quality and not acceptable for use in some products. Some used oils contain polycyclic aromatic hydrocarbons (PAH), mostly due to combustion processes, and their content must also be controlled (Betton, 1997). The most effective re-refining technology, in terms of producing high quality base stocks is known as the distillation/hydrotreatment. Re-refined base oils, produced from used lubricants, account for a small percentage of a total base oil market.

Some hydraulic oils are formulated for use in hydraulic systems, which operate outside at low winter temperatures and high ambient summer temperatures. In applications where the base stock selection is based on meeting the low temperature fluidity requirements at the min cost, the naphthenic oils are used. Usually VI improvers and commercial antiwear additive packages are used to formulate hydraulic oils. In some cases, the addition of pour point depressant might be required. The Brookfield viscosity test measures the low temperature viscosity of lubricating oils, such as hydraulic fluids and gear oils, under low shear conditions. To meet the Brookfield viscosity requirements at −40°C, base stocks need to have very low pour points. Silicones are the most surface-active antifoaming agents and many refineries use Si 12500 diluted in kerosene (10 wt%) to improve their dispersion in lubricating oils. To prevent air entrainment, the silicone antifoaming agent is used below 1 ppm of an active ingredient. The ASTM D 892 method for foaming characteristics of lubricating oils covers the determination of their foaming tendency and stability at 24°C and at 93.5°C. The ASTM D 892 Option A method requires the use of a high-speed blender to assure a good dispersion of an antifoaming agent.

The foaming tendency of the antiwear ISO VG 22 hydraulic oils containing different low viscosity naphthenic oils and the diluted Si 12500 antifoaming agent is shown in Table 8.7. The antiwear mineral ISO VG 22 hydraulic oil #1 containing low viscosity naphthenic oil, VI improver, antiwear additive package, pour point depressant, and no antifoaming agent was found to have a high foaming tendency of 150–200 mL at 24°C. The silicone antifoaming agents are used below 1 ppm of an active ingredient to prevent air entrainment of hydraulic oils. The antiwear ISO VG 22 hydraulic oil #2 containing 0.0005 vol% of Si 12500 (10 wt%), 0.5 ppm of an active ingredient, was also found to have a high foaming tendency of 150–200 mL at 24°C. The use of the ASTM D 892 method Option A is useful when testing lubricating oils containing antifoaming agents and having an increased tendency to foam. The ASTM D 892 Option A method requires the use of a high-speed blender to assure a good dispersion of an antifoaming agent; however, the antiwear ISO VG 22 hydraulic oil #2 continued to foam. The antiwear ISO VG 22 hydraulic oil #3 containing 0.003 vol% of Si 12500 (10 wt%), 3 ppm of an active ingredient was also found to have a high foaming tendency of 150–200 mL at 24°C indicating that even an increased the treat rate of Si 12500 is not effective in preventing foaming.

The foaming tendency of the antiwear ISO VG 32 hydraulic oils containing different naphthenic oils and diluted Si 12500 is shown in Table 8.8. The higher

TABLE 8.7
Foaming Tendency of Antiwear ISO VG 22 Hydraulic Oils

	Antiwear ISO VG 22 Hydraulic Oils		
	Naphthenic #1	Naphthenic #2	Naphthenic #3
Si 12500 (10 wt%) (vol%)	0	0.0005	0.003
ASTM D 892			
Foaming tendency 24°C (mL)	150–200	150–200	150–200
ASTM D 892 Option A			
Foaming tendency 24°C (mL)	150–200	150–200	150–200

TABLE 8.8
Foaming Tendency of Antiwear ISO VG 32 Hydraulic Oils

	Antiwear ISO VG 32 Hydraulic Oils		
	Naphthenic #1	Naphthenic #2	Naphthenic #3
Si 12500 (10 wt%) (vol%)	0	0.0005	0.003
ASTM D 892			
Foaming tendency at 24°C (mL)	250–300	250–300	250–300
ASTM D 892 Option A			
Foaming tendency at 24°C (mL)	250–300	250–300	250–300

viscosity antiwear ISO VG 32 hydraulic oil #1 containing a low viscosity naphthenic oil, a VI improver, an antiwear additive package, a pour point depressant, and no antifoaming agent was found to have a higher foaming tendency of 250–300 mL at 24°C. The antiwear ISO VG 32 hydraulic oil #2 containing 0.0005 vol% of Si 12500 (10 wt%), 0.5 ppm of an active ingredient was also found to have a high foaming tendency of 250–300 mL at 24°C. After the use of the ASTM D 892 method Option A, the antiwear ISO VG 32 hydraulic oil #2 continued to foam. The antiwear ISO VG 32 hydraulic oil #3 containing 0.003 vol% of Si 12500 (10 wt%), 3 ppm of an active ingredient, was also found to have a high foaming tendency of 250–300 mL at 24°C indicating that even an increased treat rate of Si 12500 is not effective in the prevention of foaming. The literature reports that if the process fluid dissolves the silicone antifoaming agent or has a lower surface tension than silicone, an increase in foaming might be observed (ProQuest, 1996). While good solubility is usually a requirement for petroleum additives, the silicone antifoaming agents require a good dispersion and not solubility in order to be surface active and to prevent foaming.

The literature reports that the insolubility of antifoaming agents is not necessity but highly desirable. The insoluble antifoaming agents tend to concentrate at available interfaces that increase their effectiveness at low concentrations. An antifoaming agent, which has a low surface tension can become a foam promoter when soluble in a foaming medium (Rome, 2001).

With an increase in the viscosity of mineral hydraulic oils, their depressed pour points increase and their ST increase leading to an increase in the surface activity of antifoaming agents. The use of less than 1 ppm of the silicone antifoaming agent is sufficient to prevent the foaming of high viscosity mineral ISO VG 100 hydraulic oils. Fresh hydraulic oils must have the ability to quickly separate the water and higher viscosity oils require more time to separate the emulsion due to the viscosity effect. The early review of commercial railroad–car journal box oils, produced from mineral refined and re-refined oils, found them to have good demulsibility and rust-preventive properties (Strigner et al., 1980). With an increase in the viscosity of antiwear mineral hydraulic oils, the emulsion separation time gradually increases but good rust-preventive properties are observed.

The patent literature reports that while the use of silicones is not always effective in the prevention of foaming, it does always increase the air release time of mineral oils (Pillon and Asselin, 1998).

The literature reports that the removal of asphaltenes, aromatic compounds, and the silicone antifoaming agents from hydraulic oils by the percolation through clay improved their air separation properties (Fomina, Chesnokov, and Fuks, 1987). The introduction of air into the hydraulic system can change the compressibility, which will result in poor system performance, particularly during the production of close-tolerance parts (Pirro and Wessol, 2001). Different bench tests are used to test the air separation properties of hydraulic oils. The air separation properties of hydraulic oils are usually tested following the ASTM D 3427 method for air release properties of petroleum oils. Another test, NF 48-614, is used to test the air separation proper-ties of fire resistant hydraulic oils (Denis, Briant, and Hipeaux, 2000). The literature reports that the copper or copper-containing materials are often used in hydraulic components and copper pipes are used in cooling systems. The poor copper corro-sion protective properties of hydraulic oils can cause an entire hydraulic system to fail (Bock, 2001).

The air entrainment and the surface activity of mineral ISO VG 46 hydraulic oils containing different ash content are shown in Table 8.9. Hydraulic oils are tested for 3 h at 100°C following the ASTM D 130 method and the copper strip is rated as slightly discolored (rating 1), moderately discolored (rating 2), heavily discolored (rating 3), and a dark discoloration (rating 4) indicating corrosion. The mineral ISO VG 46 hydraulic oil #1 containing an ash content of 0.08 wt%, was reported to have the air release time of 6 min at 50°C, it required 10 min to separate the emulsion at 54°C, its corrosion protection was rated 2, and FM wear was 0.32 mm. The mineral

TABLE 8.9

Surface Activity of ISO VG 46 Hydraulic Oils Containing Different Ash Content

	ISO VG 46 Hydraulic Fluids	
	Mineral Oil #1	Mineral Oil #2
Ash content (wt%)	0.08	0.14
KV at 40°C (cSt)	42.4	43.0
VI	97	90
Solid point (°C)	−15	−15
TAN (mg KOH/g)	0.5	0.9
Color (CST units)	2.5	3.5
ASTM D 3427 (50°C)		
Air release time (min)	6	11
ASTM D 1401 (54°C)		
Oil/water/emulsion (mL)	41/39/0	41/39/0
Separation time (min)	10	45
Corrosion (2 wt% of water)	2	4
FM wear (mm)	0.32	0.45

Source: Radchenko, L.A. and Chesnokov, A.A., *Chem. Technol. Fuels Oils*, 39(3), 122, 2003.

ISO VG 46 oils containing 3 ppm of silicone antifoaming agent, active ingredient have the air release time of about 5–6 min measured at 50°C. Another mineral ISO VG 46 hydraulic oil #2 containing a higher ash content of 0.14 wt% was reported to have a longer air release time of 11 min at 50°C, it required a longer time of 45 min to separate the emulsion, its corrosion protection was rated 4, and FM wear was 0.45 mm. The literature also reports on the filterability of mineral hydraulic oils. It was reported that their colloidal stability decreases with an increase in the temperature and causes the clogging of filters (Radchenko and Chesnokov, 2003). With an increase in the ash content of mineral ISO VG 46 hydraulic oils, an increase in the air release time and the emulsion separation time was reported, as well as an increase in the corrosion and wear.

8.3   WEAR PREVENTION AND OXIDATION RESISTANCE OF PARAFFINIC OILS

The hydraulic oils contain additives required to impart the resistance to oxidation, good water separation, good rust-preventive properties, good antiwear properties, and resistance to foaming without air entrainment. However, after long service periods, sludge is formed and the demulsibility properties of hydraulic oils degrade due to the presence of the surface-active oxidation by-products capable of adsorbing at the oil/water interface and having emulsifying properties that can lead to rusting. An increase in the TAN of the oxidized hydraulic oils indicates an increase in the acid type oxidation by-products that lead to an increase in the corrosion. Rust and corrosion lead to an increase in wear. The literature reports that some oils form excessive sludge and deposit on the catalyst wires without a significant increase in the neutralization number (Pirro and Wessol, 2001). The literature also reports that there is a need to upgrade the oxidation resistance, the air separation properties, and the demulsibility properties of mineral hydraulic oils (Lee and Harris, 2003).

The many different tests used to measure the oxidation stability of hydraulic oils are shown in Table 8.10. While there are many different oxidation tests listed, there are basically two types of oxidation bench tests. One type of the oxidation test is the ASTM D 2272 known as the rotary bomb oxidation test (RBOT) that measures the induction time, which is the time from the beginning of the oxidation to the drop of the oxygen pressure and indicates the aging time of the lubricant at 150°C. The test is run until a pressure drop of 25 psi inside the bomb is measured (Braun and Omeis, 2001).

The effect of different petroleum-derived base stocks on the RBOT life of ISO VG 32 hydraulic oils containing the same additive package is shown in Table 8.11. With a decrease in the aromatic and the S and N contents of the petroleum-derived base stocks from API I to API III group, the RBOT life of ISO VG 32 hydraulic oils increased from 105 to 620 min. Another type of oxidation test is the ASTM D 943 also known as the turbine oxidation stability test (TOST), which is used to measure the condition of the lubricant after the specific defined test period by measuring several aging parameters such as the acid content, the TAN, the viscosity change, and the sludge formation at 95°C (Braun and Omeis, 2001). The test

TABLE 8.10

Different Tests Used to Measure the Oxidation Stability of Hydraulic Oils

Test Method	Test Conditions	Test Results
ASTM D 943 (TOST)	Oil/air/metals /water at 95°C	TAN and sludge
ASTM D 2272 (RBOT)	Oil/oxygen/copper/water at 150°C	Time for a 25 psi pressure drop
ASTM D 4310	Oil/oxygen/metals/water at 95°C	Sludge
ASTM E 537 (DSC)	Oil/air in metal pan	Time for oxidation onset
DIN 51554 (Baader)	Oil/air/metals at 110°C	TAN and viscosity increase
DIN 51373	Oil/air/copper/iron at 120°C	TAN and viscosity increase
Cincinnati machine P-68, P-69, P-70	Oil/ copper/steel at 135°C	Metals color change, sludge
FTMS 5308.6 (Federal)	Oil/air /five metals at 175°C	TAN and viscosity increase
Penn state micro-oxidation test (PSMO)	Oil/air at 225°C	Deposits and weight loss
FTMS 3462 Panel coker	Oil sprayed on steel surface at 300°C	Deposit formation

Source: Reproduced from Placek, D.G., Hydraulics, in *Synthetics, Mineral Oils, and Bio-Based Lubricants Chemistry and Technology*, Rudnick, L.R., Ed., CRC Press, Boca Raton, FL, 2005. With permission.

TABLE 8.11

Effect of Base Stocks on Oxidation Resistance of ISO VG 32 Hydraulic Oils

ISO VG 32 Hydraulic Oils Base Stock Category	Mineral API I	Modified API II	Modified API III
Additives	Same	Same	Same
ASTM D 2272 (150°C)			
RBOT life (min)	105	400	620

Source: Henderson, H.E., Chemically modified mineral oils, in *Synthetics, Mineral Oils, and Bio-Based Lubricants Chemistry and Technology*, Rudnick, L.R., Ed., CRC Press, Boca Raton, FL, 2005. With permission.

is usually stopped when the oxidized oil reaches the TAN of 2 mg KOH/g. The literature reports on the effect of different chemistry antiwear additives on the TAN and the sludge of oxidized antiwear type hydraulic oils (Rasberger, 1997). The ASTM D 4310 test method is used for the determination of the sludging and the corrosion tendencies of inhibited mineral oils at 95°C and the presence of

TABLE 8.12

Effect of ZnDTP Additives on TAN and Sludge of Mineral Hydraulic Oils

Antiwear Hydraulic Oils	Mineral #1	Mineral #2
Additives (Total)	0.9–1.2 wt%	0.9–1.2 wt%
ZnDTP type	ZnDTP ($P \sim 0.035\%$)	ZnDTP ($P \sim 0.1\%$)
ASTM D 943 (95°C)		
TAN (mg KOH/g)	2	2
TOST life (h)	2500–3400	1500–1800
ASTM D 4310 (95°C)		
Time (h)	1000	1000
Sludge (mg)	45–90	130–430

Source: Rasberger, M., Oxidative degradation and stabilisation of mineral oil based lubricants, in *Chemistry and Technology of Lubricants*, 2nd ed., Mortier, R.M. and Orszulik, S.T., Eds., Blackie Academic & Professional, London, U.K., 1997. With permission.

iron–copper catalyst. After 1000 h, the insoluble material is separated by filtration and is reported as the sludge.

The effects of different chemistry ZnDTP additives on the TAN and the sludge content of oxidized antiwear mineral hydraulic oils are shown in Table 8.12. The antiwear mineral hydraulic oil #1 containing ZnDTP additive having 0.035% of P was reported to form 45–90 mg of sludge after 1000 h and reached a TAN of 2 after 2500–3400 h. Another antiwear mineral hydraulic oil #2 containing ZnDTP additive having 0.1% of P was reported to form more sludge of 130–430 mg after 1000 h and reached a TAN of 2 after only 1500–1800 h. The variation in the chemistry of ZnDTP additives and their P content was reported to affect the sludge and the TAN of oxidized antiwear mineral hydraulic oils.

The effects of different ashless additive packages on the TAN and the sludge of oxidized antiwear mineral hydraulic oils are shown in Table 8.13. The antiwear mineral hydraulic oil #1 containing the normal grade ashless additive package was reported to form 140–320 mg of sludge after 1000 h and reached a TAN of 2 after 1800–2800 h. Another antiwear mineral hydraulic oil #2 containing the premium grade ashless additive package was reported to form only 35–90 mg of sludge after 1000 h and reached a TAN of 2 after 2800–3800 h. The variation in the chemistry of different ashless additive packages was also reported to affect the sludge and the TAN of oxidized antiwear mineral hydraulic oils. A modified ASTM D 943B procedure is used to determine the TAN and the sludge content of hydraulic oils after 1000 h of oxidation at 95°C. After oxidation at 95°C for 1000 h, the antiwear modified ISO VG 32 hydraulic oil containing API III base stock was found to have a TAN below 0.5 mg KOH/g and formed 20 mg of sludge.

TABLE 8.13

Effect of Ashless Additives on TAN and Sludge of Oxidized Hydraulic Oils

Antiwear Hydraulic Oils	Mineral Oil #1	Mineral Oil #2
Additives (Total)	0.9–1.2 wt%	0.9–1.2 wt%
Ashless additives	Normal grade	Premium grade
ASTM D 943 (95°C)		
TAN (mg KOH/g)	2	2
TOST life (h)	1800–2800	2800–3800
ASTM D 4310 (95°C)		
Time (h)	1000	1000
Sludge (mg)	140–320	35–90

Source: Rasberger, M., Oxidative degradation and stabilisation of mineral oil based lubricants, in *Chemistry and Technology of Lubricants*, 2nd ed., Mortier, R.M. and Orszulik, S.T., Eds., Blackie Academic & Professional, London, U.K., 1997. With permission.

8.4 PERFORMANCE OF SYNTHETIC FLUIDS

Some hydraulic oils are formulated for use in hydraulic systems, which operate outside at low winter temperatures and high ambient summer temperatures. The paraffinic base stocks have a high VI but are solvent dewaxed to −12°C or −15°C that makes them not suitable for severe low temperature applications. Some low viscosity naphthenic oils were found to have very low natural pour points of −39°C to −57°C indicating excellent low temperature fluidity below −40°C but have a low VI. The synthetic PAO fluids have high VI and very low natural pour points. Some synthetic esters have also high VI, very low natural pour points, and are fluids at −40°C (Lakes, 1999). Synthetic hydraulic fluids having a high VI and low pour points, are an alternative for severe low and high temperature applications (Pirro and Wessol, 2001). The ASTM D 4871 guide for the universal oxidation/thermal stability test apparatus covers the temperature range from 50°C to 375°C in the presence of air, oxygen, nitrogen, or other gases. The stability of oils may be measured in the presence or absence of water or soluble or insoluble catalysts. Following the ASTM D 4871 test at 170°C, the universal oxidation test (UOT) life was measured after the oxidized oils reached the TAN of 2 mg KOH/g.

The effect of different petroleum-derived base stocks and the synthetic PAO on the UOT life of ISO VG 32 hydraulic oils containing the same additive package is shown in Table 8.14.

With a decrease in the aromatic and the S and N contents of the petroleum-derived base stocks from API I to API III group, the UOT life of ISO VG 32 hydraulic oils increased from 100 to 145 h. The synthetic ISO VG 32 hydraulic oil containing PAO base stock was reported to have an increased UOT life of 148 h. The synthetic PAO

TABLE 8.14

Effect of Mineral Base Stocks and PAO on UOT Life of ISO VG 32 Hydraulic Oils

ISO VG 32 Hydraulic Oils	Mineral	Mineral	Mineral	PAO
Base Stock Category	API I	API II	API III	API IV
ASTM D 4871 (170°C)				
TAN (mg KOH/mg)	2	2	2	2
UOT life (h)	100	115	145	148

Source: Henderson, H.E., Chemically modified mineral oils, in *Synthetics, Mineral Oils, and Bio-Based Lubricants Chemistry and Technology*, Rudnick, L.R., Ed., CRC Press, Boca Raton, FL, 2005. With permission.

fluids have a resistance to foaming and the literature reports that the use of synthetic base stocks could solve the foaming problems of industrial machines (Hubmann and Lanik, 1988). The use of synthetic base stocks was also reported to improve the air separation properties and the demulsibility of lubricating oils (Jelitte and Beyer, 1994). The antiwear part-synthetic ISO VG 32 hydraulic oils containing some low viscosity naphthenic oils, PAO 4, VI improvers, antiwear additive packages, and pour point depressants were found to have depressed pour points below −54°C and a very low Brookfield viscosity at −40°C. However, the antiwear part-synthetic ISO VG 32 hydraulic oils were found to have a high foaming tendency at 24°C indicating the need for an antifoaming agent. After the addition of 0.0005 vol% of diluted Si 12500 (10 wt%), 0.5 ppm of active ingredient, no decrease in the foaming tendency was observed indicating the solvency effect of naphthenic oils. The antiwear part-synthetic ISO VG 32 hydraulic oils containing naphthenic oils and PAO 4 were found to require only 10 min to separate the emulsion and have good rust-preventive properties.

The patent literature reports on poor rust-preventive properties of full synthetic oils containing PAO 2 and polyolester, and the need to use rust inhibitors and acid scavengers (Nadasdi and Hayter, 1999).

The rust-preventive properties of synthetic lubricating oils containing the rust inhibitors and the acid scavengers are shown in Table 8.15. The synthetic lubricating oil containing 2 vol% of PAO 2 about 95 vol% of polyolester and an antiwear additive package was reported to have poor rust-preventive properties. The additive package was reported to contain an antioxidant, a copper passivator, an antiwear agent, an EP additive, and an antifoaming agent. The addition of the polyamine rust inhibitor Hitec 536, effective in preventing the formation of rust in mineral lubricating oils, was reported not effective in preventing the formation of rust in synthetic oil #1. The addition of the sulfonate rust inhibitor Nasul EDS, also effective in preventing the formation of rust in mineral lubricating oils, was reported not effective in preventing the formation of rust in synthetic oil #2. The addition of the acid scavenger #1, Additin RC 4220, was found not effective in improving the performance of Hitec 536

TABLE 8.15

Use of Rust Inhibitors and Acid Scavengers in Synthetic Lubricating Oils

	Synthetic Oils				
	Oil #1	Oil #2	Oil #3	Oil #4	Oil #5
PAO 2 (vol%)	2	2	2	2	2
Polyolester (vol%)	~95	~95	~95	~95	~95
Antiwear/EP package (vol%)	2.7	2.7	2.7	2.7	2.7
Polyamine rust inhibitor (vol%)	0.5		0.5		0.5
Sulfonate rust inhibitor (vol%)		0.5		0.5	
Acid scavenger #1 (vol%)			0.2	0.2	
Acid scavenger #2 (vol%)					0.2
ASTM D 665B rust test	Fail	Fail	Fail	Fail	Pass

Source: Nadasdi, T.T. and Hayter, W.N., Method for producing lubricating oils with anti-rust properties, CA Patent 2,344,324, 1999.

and Nasul EDS rust inhibitors in synthetic oils #3–4. The use of the polyamine rust inhibitor, Hitec 536, and the addition of the carbodiimide acid scavenger #2, Additin RC 8500, was found effective in preventing the formation of rust in synthetic oil #5.

The TAN and the sludge of the oxidized synthetic hydraulic oils are shown in Table 8.16. After 1000 h of oxidation at 95°C, the synthetic hydraulic oil #1 containing 20 vol% of PAO 2, 76 vol% of polyolester, the synergistic blend of hindered

TABLE 8.16

TAN and Sludge of Oxidized Synthetic Hydraulic Oils

	Synthetic Hydraulic Oils		
	Oil #1	Oil #2	Oil #3
PAO 2 (vol%)	20	0	0
Polyolester (vol%)	76	96	76
Diisodecyladipate ester (vol%)	0	0	20
Hindered phenol (vol%)	0.3	0.3	0.3
Diphenylamine (vol%)	0.3	0.3	0.3
Antiwear additive package	2.7	2.7	2.7
Acid scavenger	0.5	0.5	0.5
ASTM D 943 (95°C)	1000 h	1000 h	1000 h
TAN (mg KOH/g)	0.03	0.07	0.02
Sludge (mg)	548	3517	2571

Source: Nadasdi, T.T., Polar oil based industrial oils with enhanced sludge performance, CA Patent 2,344,323, 1999.

phenol and diphenylamine antioxidants, the antiwear additive package, and the acid scavenger was reported to have a very low TAN of only 0.03 mg KOH/g but a high sludge content of 548 mg. Another synthetic hydraulic oil #2 containing 96 vol% of polyolester, the synergistic blend of hindered phenol and diphenylamine antioxidants, the antiwear additive package, and the acid scavenger was reported to have a low TAN of 0.07 mg KOH/g and a very high sludge content of 3517 mg. The synthetic hydraulic oil #3 containing 76 vol% of polyolester, 20 vol% of another ester, the synergistic blend of hindered phenol and diphenylamine antioxidants, the antiwear additive package, and the acid scavenger was reported to have a very low TAN of only 0.02 mg KOH/g but a high sludge content of 2571 mg. The patent literature reports that the analysis of the sludge indicated the presence of iron oxide that would indicate the formation of rust (Nadasdi, 1999). Food-grade hydraulic oils require the use of food-grade base oils such as petroleum-derived white oil, synthetic PAO or polyalkylglycols (PAG), and food-grade approved additives. Hydrorefined white oils require the use of second stage hydrorefining step to remove trace aromatics and make them suitable for use in food-grade products.

The chemistry of additives approved for the use in the food-grade lubricants is shown in Table 8.17. The typical VI improvers used in food-grade lubricants are

TABLE 8.17
Chemistry of Food-Grade Additives

Food-Grade Additive Types	Food-Grade Additive Chemistry
Phenol type antioxidants	Hindered phenols
	Hindered thiophenols
	Hindered 4-hydroxy acid esters
	Hindered 4-thiobenzoic acid esters
	Dithioesters
	Hindered alkylene esters
Amine type antioxidants	Aromatic amines
	Phenyl naphthyl amines
	Diphenyl amines
	N,N-dialkyl phenylene diames
Phosphite type antioxidants	Tris(2,4-di-*tert*-butyl phenyl) phosphite
Metal deactivator	Aryl triazole derivatives
Rust inhibitors	Fatty acids
	Fatty acid esters
Corrosion inhibitors	Amine derivative of fatty acids
Antiwear and EP agents	Amine phosphates
	TPPT
Antifoaming agent	Polydimethylsiloxane

Source: Raab, M.J. and Hamid, S., Additives for food-grade lubricant applications, in *Lubricant Additives Chemistry and Applications*, Rudnick, L.R., Ed., Marcel Dekker, Inc., New York, 2003. With permission.

polyisobutylenes and other high molecular weight polymers. The general classes of antioxidants used in lubricating oils are phenols, amines, and phosphates and many of them are approved as food-grade additives. The food-grade phenols include the hindered phenols, the hindered thiophenols, and the phenols, such as the 2,6-di-*tert*-butyl-*p*-cresol, and the butylated hydroxy toluene (BHT). The food-grade amines include the aminic antioxidants, such as phenyl naphthyl amine and diphenyl amine, and the aromatic amines, such as *N*-phenyl-alpha-naphthylamine and di-*tert*-octyl diphenyl amine (Raab and Hamid, 2003). The metal deactivator, such as the aryl triazole derivative, is used in food-grade lubricant products. The typical compounds used as the rust inhibitors in food-grade lubricants are the fatty acids and their esters. The food-grade esters of sorbitan, glycerol, and other polyhydric alcohols and poly-alkylene glycols as well as the esters from fatty alcohols alkoxylated with alkylene oxides or the sorbitan alkoxylated with alkylene oxides or the succinic acid (Raab and Hamid, 2003). The use of the fatty acids and the amines to form the succinic anhy-dride amine derivatives can be used as a rust and the corrosion inhibitor (Raab and Hamid, 2003). The literature reports that the antiwear and the extreme pressure (EP) agents are organic compounds containing P and S, such as amine phosphates and tri-phenyl phosphorothionate (TPPT) that form metal salts at high temperatures (Raab and Hamid, 2003).

The properties of mineral and the synthetic food-grade hydraulic oils are shown in Table 8.18. The mineral food-grade ISO VG 46 hydraulic oil contain-ing petroleum-derived white oil was reported to have a VI of 85 and a pour point of −20°C. The literature reports that the typical synthetic fluids, such as PAO and PAG, are used in the food-grade synthetic hydraulic fluids that require good oxida-tion stability (Raab, 2005). The use of synthetic fluid, such as PAO, has drastically increased the VI to 133 and decreased the pour point to −50°C. The use of another synthetic fluid, such as PAG, has increased the VI to 102 and increased the pour point to −15°C.

TABLE 8.18
Properties of Mineral and Synthetic Food-Grade ISO VG 46 Hydraulic Oils

Food-Grade Hydraulic Fluids	ISO VG 46 Oil Mineral White Oil	ISO VG 46 Oil Synthetic PAO	ISO VG 46 Oil Synthetic PAG
KV at 40°C (cSt)	47.8	42.3	46.0
KV at 100°C (cSt)	6.7	7.2	6.8
VI	85	133	102
Pour point (°C)	−20	−50	−15
Flash point (°C)	204	254	185
Pump test (104°C)	Pass	Pass	Pass

Source: Reproduced from Raab, M.J., Synthetic-based food-grade lubricants and greases, in *Synthetics, Mineral Oils, and Bio-Based Lubricants Chemistry and Technology*, Rudnick, L.R., Ed., CRC Press, Boca Raton, FL, 2005. With permission.

TABLE 8.19

Recommended Operating Temperature Ranges of Different Hydraulic Oils

Hydraulic Fluids	Short Service (°C)	Long Service (°C)
Mineral oil	135–149	93–121
Synthetic PAO	288	149–232
Synthetic diesters	204	177
Synthetic polyglycols	191–204	163–177

Source: Reproduced from Placek, D.G., Hydraulics, in *Synthetics, Mineral Oils, and Bio-Based Lubricants Chemistry and Technology*, Rudnick, L.R., Ed., CRC Press, Boca Raton, FL, 2005. With permission.

The effect of petroleum-derived mineral base stocks and different synthetic fluids on the recommended operating temperature ranges of hydraulic oils are shown in Table 8.19. Hydraulic oils need to have good oxidation stability but also good rust-preventive properties. The performance of surface-active additives, such as rust inhibitors, varies depending on the selected base stocks and the lubricant solvency properties. Rust inhibitors need to be in the monomer form to adsorb on the metal surface. The literature reports on the use of carbodiimide acid scavenger technology (Duncan, Reyes-Gavilan, and Constantini, 2002). The patent literature reports that the use of acid scavengers improves the rust-preventive properties of synthetic oils; however, during the oxidation, the formation of iron oxide and an increase in the sludge was observed (Nadasdi, 1999). When the acidic by-products are formed, the acid scavengers might interact with them leading to a very low TAN of oxidized synthetic oils but also poor rust-preventive properties causing the formation of iron oxide and an increase in the sludge. Both, mineral and synthetic fluids have their advantages and disadvantages that can be called inferior properties.

The list of some inferior properties of mineral oils and some inferior properties of synthetic fluids are shown in Table 8.20. When using synthetic oil in place of petroleum-derived base stock, the surface activity requirements of some lubricants and their performance need to be carefully compared. The recent literature reports that high quality mineral oils are used in most applications and were reports to have higher pressure–viscosity coefficients than common synthetic fluids and thus allowing for greater film thickness at given operating temperatures (Potteiger, 2007). In some cases, synthetic fluids perform better in machines subjected to low temperatures because they have high VI and low pour points. In addition, synthetic fluids might be preferable because many synthetic fluids have an increased thermal and oxidation stability. According to the literature, some customers may encounter problems when switching from mineral products to biodegradable oils because some oils are not compatible. The use of polyglycol-based products may require a total evacuation of the system in addition to one or two flushings (Beitelman, 2004).

TABLE 8.20

Some Inferior Properties of Mineral Oils and Synthetic Fluids

Some Inferior Properties of Mineral Oils	Some Inferior Properties of Synthetic Fluids
VI	Hydrolytic behavior
Low temperature fluidity	Toxicological behavior
Volatility	Compatibility with seals
Thermal stability	Additive solubility
Oxidation stability	Rust and corrosion behavior
Biodegradability	Price

Source: Rudnick, L.R. and Bartz, W.J., Comparison of synthetic, mineral oil, and bio-based lubricant fluids, in *Synthetics, Mineral Oils, and Bio-Based Lubricants Chemistry and Technology*, Rudnick, L.R., Ed., CRC Press, Boca Raton, FL, 2005. With permission.

The literature reports that certain properties of mineral base stocks are superior to synthetic fluids and only some properties of synthetic fluids are superior to mineral base stocks.

8.5 USE OF BIODEGRADABLE OILS

There is an increasing worldwide trend toward the use of products for industrial, commercial, and even household applications that have a minimal impact on the environment. Some industrial oils can escape to the environment due to accidents such as a gearbox seal failure or high pressure hydraulic line fracture. To decrease the effect of used lubricants on the environment, the use of readily biodegradable lubricants is needed. The highest volume of environmentally friendly lubricants is being used as hydraulic fluids. The literature reports on the use of soybean oils in tractor transmission hydraulic fluids, industrial hydraulic fluids for process and machinery applications, food-grade hydraulic fluids, and many other specialty lubricants and greases (Honary, 2001). The chemistry and the type of additives used in lubricants are dependent on the base fluid and end use of the lubricant. The purpose of the additive is to impart those properties to the lubricant that are essential to its function but are not present in the base fluid. The recent literature reports that the additives have generally low aquatic toxicity when studied using standard test methodology in fish, daphnia, and single-celled algae. The most toxic additives were reported to be zinc-containing antiwear/antioxidant molecules (Betton, 2008).

The properties, the biodegradability, and the cost of mineral oils, synthetic esters, polyethylene glycols, and crop oils are shown in Table 8.21. Many synthetic biodegradable fluids and natural crop oils are available; however, their cost is higher. Biodegradable lubricants require the use of biodegradable oils and

TABLE 8.21

Biodegradability and Cost of Mineral Oils, Synthetic Fluids, and Crop Oils

Lubricant Type Oils Properties	Mineral Base Stocks	Synthetic Esters	Polyethylene Glycols	Crop Oils
Low temperature limit (°C)	−20 to −30	−30 to −40	−30	−25
Seal compatibility	Good	Limited	Limited	Good
Varnish compatibility	Good	Good	Limited	Good
Water solubility	Low	Low	Compatible	Low
Biodegradability (%)	20	10–90	77–99	99
Cost comparison	1	6	3–4	2–3

Source: Beitelman, A.D., Environmentally friendly lubricants, http://www.wes.army.mil (accessed 2004).

biodegradable additives. The biodegradable hydraulic fluids can be formulated by blending different viscosity synthetic polyolesters and the approved additives. The patent literature reports that the biodegradable hydraulic fluids containing synthetic polyolesters, require the use of approved additives, such as sulfurized fatty acid ester EP agent, alkylated phenyl naphthyl amine antioxidant, triazole and dimercaptothiadiazole metal deactivators, aryl phosphate antiwear agent, and succinimide rust inhibitor (Nadasdi, 1999). The mineral oils are relatively resistant to the oxidation and require only 0.5 wt% of an antioxidant. The antioxidant performance in synthetic fluids is different from the petroleum-derived lubricants and often 2–3 wt% is required (Shanks et al., 2006). The synthetic fluids require antioxidants and there are already questions about the use of aromatic amines, phenols, and Zn-containing additives in these new fluids, called "bio-lubes" (Shanks et al., 2006).

The TAN and the sludge of oxidized biodegradable synthetic hydraulic fluids containing different polyolesters are shown in Table 8.22. After oxidation at 95°C for 336 h, the biodegradable synthetic hydraulic fluid #1 containing two different polyolesters and the antiwear additive package was reported to have a high TAN of 2.5 mg KOH/g and already formed sludge. After the addition of the acid scavenger, the biodegradable synthetic hydraulic fluid #2 was oxidized for 1000 h at 95°C and found to have a TAN < 2 mg KOH/g and a high sludge content of 2548 mg. After the addition of the synergistic blend of the hindered phenol and the diphenylamine antioxidants, the biodegradable synthetic hydraulic fluid #3 was oxidized for 1000 h at 95°C and was found to have a TAN < 2 mg KOH/g and but an even higher sludge content of 2980 mg. The literature reports that hydraulic fluids containing biodegradable polyol esters and ashless additives outperform vegetable oils in terms of hydrolytic and oxidation stability (Duncan, Reyes-Gavilan, and Constantini, 2002). Most biodegradable lubricating oils use rapeseed oil (known as canola oil), other crop oils, polyglycols, and synthetic esters (Beitelman, 2004).

TABLE 8.22

TAN and Sludge of Oxidized Biodegradable
Synthetic Hydraulic Fluids

	Biodegradable Hydraulic Oils		
	Oil #1	Oil #2	Oil #3
Polyolester #1 (%)	77	77	77
Polyolester #2 (%)	20	20	20
Antiwear package (%)	2.7	2.7	2.7
Hindered phenol (%)	0	0	0.3
Diphenylamine (%)	0	0	0.3
Acid scavenger (%)	0	0.5	0.5
ASTM D 943 (95°C)	336 h	1000 h	1000 h
TAN (mg KOH/g)	2.5	<2	<2
Sludge (mg)	Trace	2548	2980

Source: Nadasdi, T.T., Polar oil based industrial oils with enhanced
sludge performance, CA Patent 2,344,323, 1999.

TABLE 8.23

Oxidation Resistance of Rapeseed Oil and Rapeseed
Oil-Based Hydraulic Fluid

	Neat	ISO VG 32
Biodegradable Oil	**Rapeseed Oil**	**Hydraulic Oil**
	Composition (%)	
Rapeseed oil	100	>99
Alkylated diphenylamine	0	0.3
Antiwear agent	0	0.3
Metal deactivator	0	0.05
Non-silicone antifoaming agent	0	0.005
DIN 51 544/3 (95°C for 3 days)		
Viscosity increase at 40°C (%)	75	12
TAN increase (mg KOH/g)	0.9	0.3

Source: Rasberger, M., Oxidative degradation and stabilisation of mineral
oil based lubricants, in *Chemistry and Technology of Lubricants*,
2nd ed., Mortier, R.M. and Orszulik, S.T., Eds., Blackie Academic
& Professional, London, U.K., 1997. With permission.

The oxidation resistance of the biodegradable rapeseed oil and the hydraulic fluid
containing a rapeseed oil and additives are shown in Table 8.23. After oxidation at
95°C for 3 days, the viscosity of the oxidized rapeseed oil increased by 75% and
the TAN increased by 0.9. After oxidation at 95°C for 3 days, the viscosity of the

oxidized rapeseed oil-based hydraulic oil increased by 12% and the TAN increased by 0.3. Hydraulic oils need to meet various performance requirements and use base stocks ranging from mineral oils, hydroprocessed oils, and synthetic fluids to vegetable oils. The literature reports that the biodegradable lubricants based on the rapeseed oil have been developed for the use in chain saws and the hydraulic systems of snowmobiles (Beitelman, 2004). The literature also reported that the oxidized hydraulic fluids containing rapeseed oil were found to show large changes in the viscosity, the TAN, and the corrosion (Placek, 2005). While the lubricant industry must be more concerned with the environmental issues and develop new products using natural and synthetic esters, new chemistry ashless additives need to be developed to improve their performance (Shanks et al., 2006). The vegetable oil-based hydraulic fluids have a poor oxidation stability when compared to petroleum-based fluids or hydraulic oils containing synthetic esters.

REFERENCES

Apar Industries, http://www.aparind.com (accessed October 2003).

Beitelman, A.D., Environmentally friendly lubricants, http://www.wes.army.mil (accessed 2004).

Betton, C.I., Lubricants and their environmental impact, in *Chemistry and Technology of Lubricants*, 2nd ed., Mortier, R.M. and Orszulik, S.T., Eds., Blackie Academic & Professional, London, U.K., 1997.

Betton, C.I., Lubricants, in *Environmental Technology in the Oil Industry*, Orszulik S.T., Ed., 2nd ed., Springer, Berlin, 2008.

Bock, W., Hydraulic oils, in *Lubricants and Lubrications*, Mang, T. and Dresel, W., Eds., Wiley-VCH GmbH, Weinheim, 2001.

Braun, J. and Omeis, J., Additives, in *Lubricants and Lubrications*, Mang, T. and Dresel, W., Eds., Wiley-VCH GmbH, Weinheim, 2001.

Butler, K.D., Process to produce lube oil base stock by low severity hydrotreating of used industrial circulating oils, CA Patent 2,170,138, 1996.

Deng, Z. and Lovell, M.R., *Wear*, 244(1,2), 41 (2000).

Denis, J., Briant, J., and Hipeaux, J.C., *Lubricant Properties Analysis and Testing*, Editions Technip, Paris, 2000.

Duncan, C., Reyes-Gavilan, J., and Constantini, D., *Lubrication Engineering*, 58(9), 18 (2002).

Fomina, O.N., Chesnokov, A.A., and Fuks, I.G., *Russian Journal of Applied Chemistry*, 1, 16 (1987).

Henderson, H.E., Chemically modified mineral oils, in *Synthetics, Mineral Oils, and Bio-Based Lubricants Chemistry and Technology*, Rudnick, L.R., Ed., CRC Press, Boca Raton, FL, 2005.

Honary, L.A.T., *Machinery Lubrication Magazine*, Issue September 2001, http://www. machinerylubrication.com (accessed 2008).

Hubmann, A. and Lanik, A., *Tribologie und Schmierungestechnik* (in German), 35(3), 138 (1988).

Jelitte, R. and Beyer, C., *Synthetic Lubrication*, 10(4), 285 (1994).

Kajdas, C., Industrial lubricants, in *Chemistry and Technology of Lubricants*, 2nd ed., Mortier, R.M. and Orszulik, S.T., Eds., Blackie Academic & Professional, London, U.K., 1997.

Lakes, S.C., Automotive crankcase oils, in *Synthetic Lubricants and High-Performance Functional Fluids*, 2nd ed., Marcel Dekker Inc., New York, 1999.

Lee, F.L. and Harris, J.W., Long-term trends in industrial lubricant additives, in *Lubricant Additives*, Rudnick, L.R., Ed., Marcel Dekker, Inc., New York, 2003.

Nadasdi, T.T., Polar oil based industrial oils with enhanced sludge performance, CA Patent 2,344,323, 1999.

Nadasdi, T.T. and Hayter, W.N., Method for producing lubricating oils with anti-rust properties, CA Patent 2,344,324, 1999.

Pillon, L.Z., *Interfacial Properties of Petroleum Products*, CRC Press, Boca Raton, FL, 2007.

Pillon, L.Z. and Asselin, A.E., Antifoaming agents for lubricating oils, U.S. Patent 5,766,513, 1998.

Pirro, D.M. and Wessol, A.A., *Lubrication Fundamentals*, 2nd ed., Marcel Dekker Inc., New York, 2001.

Placek, D.G., Hydraulics, in *Synthetics, Mineral Oils, and Bio-Based Lubricants Chemistry and Technology*, Rudnick, L.R., Ed., CRC Press, Boca Raton, FL, 2005.

Potteiger, J., Machinery lubrication, Issues Nov/Dec 2007, http://www.machinerylubrication. com (accessed 2008).

ProQuest Document ID 10552585, *Chemical Engineering*, 103(12), 99 (1996).

Raab, M.J., Synthetic-based food-grade lubricants and greases, in *Synthetics, Mineral Oils, and Bio-Based Lubricants Chemistry and Technology*, Rudnick, L.R., Ed., CRC Press, Boca Raton, FL, 2005.

Raab, M.J. and Hamid, S., Additives for food-grade lubricant applications, in *Lubricant Additives Chemistry and Applications*, Rudnick, L.R., Ed., Marcel Dekker, Inc., New York, 2003.

Radchenko, L.A. and Chesnokov, A.A., *Chemistry and Technology of Fuels and Oils*, 39(3), 122 (2003).

Rasberger, M., Oxidative degradation and stabilisation of mineral oil based lubricants, in *Chemistry and Technology of Lubricants*, 2nd ed., Mortier, R.M. and Orszulik, S.T., Eds., Blackie Academic & Professional, London, U.K., 1997.

Rome, F.C., *Hydrocarbon Asia*, 42 (May/June 2001).

Rudnick, L.R. and Bartz, W.J., Comparison of synthetic, mineral oil, and bio-based lubricant fluids, in *Synthetics, Mineral Oils, and Bio-Based Lubricants Chemistry and Technology*, Rudnick, L.R., Ed., CRC Press, Boca Raton, FL, 2005.

Shanks, D. et al., *Lubrication Science*, 18, 87 (2006).

Shell, http://www.shellglobalsolutions.com (accessed 2002).

Shugarman, A.L., Machinery lubrication magazine, Issue September 2001, http://www. machinerylubrication.com (accessed 2008).

Stein, W.H. and Bowden, R.W., *Machinery Lubrication Magazine*, Issue July 2004, http:// www.machinerylubrication.com (accessed 2008).

Strigner, P.L. et al., *Lubrication Engineering*, 36(7), 390 (1980).

9 Surface Activity of Engine Oils

9.1 BASE STOCK AND ADDITIVE PERFORMANCE REQUIREMENTS

The literature reports that the main function of engine oil is to extend the life of moving parts operating under different conditions of temperature, speed, and pressure. At low temperatures, the lubricant is expected to flow to prevent the moving parts from oil starvation. At high temperatures, the lubricant needs to keep the moving parts apart to prevent wear (Bardasz and Lamb, 2003). Engine oils require base stocks, which have good low temperature fluidity, a low volatility, and a high oxidation resistance. Petroleum-derived base stocks are mixtures of different molecular weight hydrocarbons and their volatility decreases with an increase in their viscosity. At a constant viscosity, their volatility decreases with an increase in their viscosity index (VI). The hydrocracking of large and heavy molecules leads to lower viscosity base stocks and some hydrocracked base stocks can have a high volatility (Phillipps, 1999). The literature reports that in 1985, the European carmakers instituted a max 15% Noack volatility specification for engine oils to minimize the oil consumption in engines operating at high speeds and high temperatures. In 2001, the volatility limits were tightened to a max 13% off for the European synthetic engine oils (Instrumentation Scientifique de Laboratoire, 2002). The volatility can be measured at 250°C and reported as the Noack volatility or estimated using the gas chromatography (GC) technique. The low viscosity solvent-refined base stocks have a high volatility and do not meet the volatility requirements of low viscosity engine oils.

The effect of the hydroprocessing on the VI and the volatility of different petroleum-derived base stocks are shown in Table 9.1. The solvent-refined 100N mineral base stock, produced using conventional refining and classified as API group I, has typical properties such as a high VI (HVI) of 92, a dewaxed pour point of −18°C, 18.2 wt% of aromatics, 0.46 wt% of S, 17 ppm of N, and a high volatility of 32% off. The hydrocracked 100N mineral base stock, classified as API group II has a conventional HVI of 96, a dewaxed pour point of −12°C, only 1.1 wt% of aromatics, only 0.003 wt% of S, 5 ppm of N, and a high volatility of 31% off. The use of hydroprocessing can lead to base stocks having a very high VI (VHVI) of 128, a dewaxed pour point of −27°C, 7.7 wt% of aromatics, 0.08 wt% of S, 2 ppm of N, and a lower volatility of 16% off. The use slack wax hydroisomerization can lead to API group III base stocks having an extra high (XHVI) of 144, a dewaxed pour point of −18°C, no aromatics, only 0.002 wt% of S, 2 ppm of N, and a lower volatility of 11% off. With an increase in the viscosity of petroleum-derived base stocks, their volatility decreases. The use of slack wax hydroisomerization can also lead to a higher

TABLE 9.1

Effect of Hydroprocessing on the VI and the Volatility of Petroleum-Derived Base Stocks

API Category	API I	API II	API III	API III	API III
Base Stocks	**HVI**	**HVI**	**VHVI**	**XHVI 5**	**XHVI 6**
KV at 100°C (cSt)	3.81	4.01	3.99	5.14	5.8
VI	92	96	128	146	142
Pour point (°C)	−18	−12	−27	−18	−21
Saturates (wt%)	81.8	98.9	92.3	100	100
Aromatics (wt%)	18.2	1.1	7.7	0	0
Sulfur (wt%)	0.46	0.003	0.08	0.002	0.001
Nitrogen (ppm)	17	5	2	2	0
Volatility (% off)	32	31	16	11	9

Source: Phillipps, R.A., Highly refined mineral oils, in *Synthetic Lubricants and High-Performance Functional Fluids*, 2nd ed., Rudnick L.R. and Shubkin R.L., Eds., Marcel Dekker Inc., New York, 1999. With permission.

viscosity XHVI 6 base stock having a VI of 142, a dewaxed pour point of −21°C, practically no aromatics, only 0.001 wt% of S, no N content, and a low volatility of 9% off. Despite the use of base stocks having a HVI, additives such as VI improvers are needed when formulating multigrade engine oils.

The chemistry of typical polymers used as VI improvers in engine oils is shown in Table 9.2. The literature reports that there are five basic core technologies

TABLE 9.2

Chemistry of Polymers Used as VI Improvers in Engine Oils

Polymer Chemistry	Polymer Type	VI Improver Name
Polyalkylmethacrylate	Linear copolymer	PMA
Olefin copolymer	Linear copolymer	OCP
Hydrogenated styrene–butadiene	Linear tapered block copolymer	HSB
Hydrogenated styrene–isoprene	Linear A–B block copolymer	HIS
Styrene–ester copolymer	Linear copolymer	SPE
Hydrogenated radial polyisoprene	Star polymer	HRI
Polyisobutylene	Linear homopolymer	PIB

Sources: MacAlpine, G.A. et al., Wax isomerate having a reduced pour point, CA Patent 2,054,096, 1991. With permission; Stambaugh, R.L., Viscosity index improvers and thickeners, in *Chemistry and Technology of Lubricants*, 2nd ed., Mortier R.M. and Orszulik S.T., Eds., Blackie Academic & Professional, London, U.K., 1997. With permission.

of VI improvers used in engine oils, which include polymethacrylates (PMA), poly(ethylene-*co*-propylene), and its modifications (olefin copolymers or OCP), hydrogenated poly(styrene-*co*-butadiene) (HSB), hydrogenated poly(styrene-*co*-isoprene) (HIS), and esterified poly(styrene-*co*-maleic anhydride) (SPE). Other VI improvers such as polyisoprene and polyisobutylene were also reported used in some lubricating oils (Stambaugh, 1997). There are also combinations of PMA/OCP systems used as VI improvers in engine oils (MacAlpine, Halls, and Asselin, 1991). With a decrease in the temperature, the kinematic viscosity of base stocks gradually increases and some VI improvers affect the low temperature fluidity of engine oils. The Cold Cranking Stimulator (CCS) test measures the apparent viscosity of the oil at low temperatures and at high shear rates. Under high shear conditions, the use of PMA, OCP, and HSB type VI improvers was reported to have a less effect on the CCS viscosity, which allows increasing the content of higher viscosity base stocks leading to a decrease in the volatility of engine oils. The use of polyisobutylene was reported to increase the CCS viscosity and require the use of lighter viscosity base stocks leading to an increase in the volatility of engine oils (Stambaugh, 1997). Many other additives are used to improve the properties of base stocks, extend their use, and protect the metal surface.

The composition and the additive chemistry of 5W-20 engine oil containing 100N hydrocracked base stock are shown in Table 9.3. The patent literature reports on the composition of 5W-20 engine oil, which contained 79 vol% of hydrocracked 100N base stock, 6.1 vol% of OCP type VI improver, 0.3 vol% of phenolic type antioxidant, 0.2 vol% of corrosion inhibitor, 1 vol% of two different ZDDP additives, 10.5 vol% of dispersant, 3 vol% of two different detergents, and 0.006 vol% of a demulsifier (Gao et al., 2001). Detergents are described in terms of the metal ratio, the soap content, the sulfate ash, and the total base number (TBN). The commercial "neutral" sulfonates were reported to have the TBN of 30 or less while the "basic" detergents

TABLE 9.3
Composition and Additive Chemistry of 5W-20 Engine Oil

5W-20 Engine Oil	Additive Chemistry	Content (wt%)
Hydrocracked 100N base stock		79.034
VI improver	OCP type	6.1
Antioxidant	Phenolic type	0.3
Corrosion inhibitor	—	0.2
Antioxidant/antiwear	Primary ZDDP	0.69
Antioxidant/antiwear	Secondary ZDDP	0.36
Dispersant	Borated PIBSA-PAM	10.5
Detergent	Ca salicylate (165 TBN)	2.25
Detergent	Mg salicylate (345 TBN)	0.56
Demulsifier	—	0.006

Source: Gao, J. et al., Antioxidant additive composition for lubricating oils, CA Patent 2,397,885, 2001.

have the TBN of 200–500 (Pirro and Wessol, 2001). The addition of an antifoaming agent is also required to meet the foaming tendency specification of engine oils. The VI improvers and the dispersants are typically the highest treat additives used in engine oils. The commercial additive packages containing antioxidants, antiwear additives, dispersants, detergents, pour point depressants, and antifoaming agents are usually used to formulate engine oils and are called detergent inhibitor (DI) additive package. In some cases, the addition of pour point depressants is also required to meet the depressed pour point specifications. The patent literature reports on the raffinate hydroconversion (RHC) process and the performance of RHC base stocks in engine oils. To meet the low temperature fluidity and the volatility requirements of 5W engine oils, the RHC base stocks are required to have a kinematic viscosity in the range of 4–5 cSt at 100°C, a Noack volatility of a max 15% off, and a CCS viscosity below 2500 cP at −25°C (Crosthwait, May, and Deane, 2001).

The low temperature fluidity and the volatility of 5W-20 engine oils containing RHC base stocks, VI improver, and different additive packages are shown in Table 9.4. The 5W-20 engine oil #1 containing an RHC base stock, a VI improver, and an additive package A was reported to have a volatility of 14.8% off and the CCS viscosity of 1990 cP at −25°C, which further increased to 3730 cP when tested at −30°C. Another 5W-20 engine oil #2 containing a different RHC base stock, a VI improver, and a different additive package B was reported to have an increased volatility of 15.3% off, a higher CCS viscosity of 2250 cP at −25°C, which further increased to 4490 cP at −30°C. To ensure that the engine oils have good low temperature fluidity, other performance requirements such as the lubricant pumpability are tested. The viscosity measured by the Mini Rotary Viscometer (MRV) is used to determine the lubricant pumpability at low temperatures. The formulated 5 W engine oils need to meet the kinematic viscosity at 100°C, the CCS viscosity at −25°C, and the MRV viscosity at −35°C specifications. Some VI improvers are known to be unstable under high shear conditions leading the fully formulated oils

TABLE 9.4
Low Temperature Fluidity and Volatility of 5W-20 Oils Containing RHC Base Stocks

5W-20 Engine Oils	Oil #1	Oil #2
KV of RHC base stocks	4.5 cSt at 100°C	4.7 cSt at 100°C
Additives	Package A	Package B
KV at 100°C (cSt)	8.3	8.4
Volatility (% off)	14.8	15.3
CCS at −25°C (cP)	1,990	2,250
CCS at −30°C (cP)	3,730	4,490
MRV at −35°C (cP)	28,100	16,400
HTHS at 150°C (cP)	2.6	2.75

Source: Crosthwait, H.K. et al., Method for optimizing the fuel economy of lubricant base stock, CA Patent 2,390,229, 2001.

to fail to retain the original viscosity grade. The ASTM D 3945 test method for shear stability of polymer-containing fluids is used to measure the degradation in the kinematic viscosity of the fluid under the test conditions. The reduction in viscosity is reported as % loss of the initial kinematic viscosity and is a measure of the shear stability of the polymer-containing fluids. The 10W-30 engine oils have a higher kinematic viscosity at 100°C and need to meet the CCS viscosity requirement at a higher temperature of −20°C. The patent literature reports that to meet the performance requirements of 10W engine oils, the RHC base stocks require a kinematic viscosity in the range of 5–6 cSt at 100°C, a Noack volatility below 15% off, and a CCS viscosity below 2500 cP at −20°C (Crosthwait, May, and Deane, 2001).

The low temperature fluidity and the volatility of 10W-30 engine oils containing RHC base stocks, VI improvers, and different additive packages are shown in Table 9.5. The 10W-30 engine oil #1 containing an RHC base stock, a VI improver, and the additive package A was reported to have a lower volatility of 10.1% off and a CCS viscosity of 1910 cP at −20°C, which further increased to 3840 cP when tested at −25°C. Another 10W-30 engine oil #2 containing the same RHC base stock, a VI improver, and a different additive package B was reported to have a similar volatility of 10.4% off but a higher CCS viscosity of 2110 cP at −20°C, which further increased to 4170 cP at −25°C. The 10W-30 engine oil #3 containing another RHC base stock, a VI improver, and another additive package C was reported to have an increased volatility of 11.4% off and a CCS viscosity of 2000 cP at −20°C, which further increased to 3960 cP when tested at −25°C. According to the literature, the volume of hydrocracked base stocks is affected by their VI and an increase in the VI, decreases the volume. The shear stable VI improvers have a low molecular weight and are inherently poorer thickeners. The use of high molecular weight VI improvers increases their thickening effect leading to a decrease in their treat rates but decreases their shear stability (Stambaugh, 1997).

TABLE 9.5
Low Temperature Fluidity and Volatility of 10W-30 Oils Containing RHC Oils

10W-30 Engine Oils	Oil #1	Oil #2	Oil #3
KV of RHC Base Stocks	5.3 cSt at 100°C	5.3 cSt at 100°C	5.6 cSt at 100°C
Additives	Package A	Package B	Package C
KV at 100°C (cSt)	10.7	10.7	10.3
Volatility (% off)	10.1	11.4	10.4
CCS at −20°C (cP)	1,910	2,000	2,110
CCS at −25°C (cP)	3,840	3,960	4,170
MRV at −30°C (cP)	11,460	12,200	21,000
HTHS at 150°C (cP)	3.22	3.33	3.18

Source: Crosthwait, H.K. et al., Method for optimizing the fuel economy of lubricant base stock, CA Patent 2,390,229, 2001.

TABLE 9.6

Effect of Base Stocks and VI Improver on CCS Viscosity of Heavy-Duty Engine Oils

Engine Oils	15W-40 Oil #1	15W-40 Oil #2	10W-40 Oil
Base Stocks	API I	API II	API III
VI improver (wt%)	8.5	6.8	8.7
KV at 100°C (cSt)	14.8	15.1	14.6
KV at 40°C (cSt)	107	112	95.8
VI	142	140	154
CCS at −15°C (cP)	2670	2980	1810
CCS at −20°C (cP)	—	—	2920
HTHS at 150°C (cP)	4.15	4.33	4.18

Source: Reproduced from Henderson, H.E., Chemically modified mineral oils, in *Synthetics, Mineral Oils, and Bio-Based Lubricants Chemistry and Technology*, Rudnick, L.R., Ed., CRC Press, Boca Raton, FL, 2005. With permission.

The effects of different base stocks and the VI improver content on the CCS viscosity of 15W-40 and 10W-40 heavy-duty engine oils are shown in Table 9.6. The 15W-40 engine oil #1 containing the API I group base stock, 8.5 wt% of VI improver, and the additive package was reported to have a CCS viscosity of 2670 cP at −15°C. Another 15W-40 engine oil #2 containing the API II group base stock, only 6.8 vol% of VI improver, and the additive package was reports to have a higher CCS viscosity of 2980 cP at −15°C. The literature reports that the use of the API II group base stock and a decrease in the VI improver content from 8.5% to 6.8 vol%, leads to an increase in the high temperature high shear stability (HTHS) but no improvement in the CCS viscosity at −15°C of 15W-40 oil is observed (Henderson, 2005a). The 10W-40 engine oil containing API III group base stock having a higher VI, 8.7 wt% of VI improver, and the additive package was reported to have a lower CCS viscosity of 1810 cP at −15°C, which further increased to 2920 cP when tested at −20°C. The use of the API III group base stock having a HVI, decreased the CCS viscosity at −15°C and thus allowing to meet the more severe CCS viscosity requirements at −20°C of 10W-40 engine oils but, with an increase in the VI improver content to 8.7 vol%, some decrease in the high shear stability (HTHS) viscosity was reported (Henderson, 2005a). The literature reports that to meet very severe low temperature requirements of 0W-30 engine oils, the gas to liquid (GTL) base stocks having a VI of 126–128, a low volatility of about 10% off, and very low pour points of −43°C to −45°C were required (Henderson, 2005b).

The composition and the low temperature fluidity of 0W-30 engine oils containing GTL base stocks are shown in Table 9.7. One of the critical low temperature requirements of engine oils is to meet the CCS viscosity specifications measured at the temperature of −5°C to −30°C. The CCS viscosity relates to the cold starting

TABLE 9.7

Low Temperature Fluidity of 0W-30 Engine Oils Containing GTL Base Stocks

	GTL Base Stocks	
	Oil #1	Oil #2
KV at 100°C (cSt)	4.23	4.24
VI	126	128
Pour point (°C)	−45	−43
0W-30 engine oils		
GTL content (vol%)	82.6	63.9
VI improver (%)	7.2	21
Additive package (%)	10	15
Pour point depressant (%)	0.2	0.2
KV at 100°C (cSt)	10.1	11.7
Volatility (% off)	10.3	10.5
CCS at −35°C (cP)	4,297	5,623
MRV at −40°C (cP)	16,400	19,000

Source: Henderson, H.E., Gas to liquids, in *Synthetics, Mineral Oils, and Bio-Based Lubricants Chemistry and Technology*, Rudnick, L.R., Ed., CRC Press, Boca Raton, FL, 2005. With permission.

characteristics of engine oils and should be as low as possible (Mills, Lindsay, and Atkinson, 1997). The selected base stocks and the additives affect the volatility and the low temperature fluidity of engine oils and need to be carefully selected. To meet the volatility requirements of engine oils, base stocks having a higher VI are required. The volatility requirement of a max 15% off can be met by using 105 VI base stocks in SAE 15W-40 engine oil commonly used in Europe. Lower viscosity grade engine oils require higher VI base stocks to meet the volatility requirement of SAE 5W-30 and 10W-30 oils used in North America (Crosthwait, May, and Deane, 2001). The patent literature also reports that to assure good low temperature fluidity of engine oils, 5 W engine oils are required to have a gel index of max 12 and required the use of other additives such as sorbitan esters (Koenitzer and May, 2001). The ASTM D 5133 method for low temperature, low shear rate, viscosity/temperature dependence of lubricating oils is used to measure the Scanning Brookfield Gel Index.

The effect of different sorbitan esters on the gel index and the depressed pour point of mineral 5W-30 engine oils are shown in Table 9.8. The mineral 5W-30 engine oil #1 containing 80.22 vol% of mineral 100N base stock, 10.56 vol% of dialkylfumarate-vinylacetate (DFVA) VI improver, and 9.22 vol% of additive package was reported to be clear, have a high gel index of 15.2, and a depressed pour point of −39°C. The use of 0.2 wt% of polyoxyethylene(4)sorbitan monostearate (Tween

TABLE 9.8

Effect of Sorbitan Esters on Gel Index and Depressed Pour Point of 5W-30 Oils

	Mineral 5W-30 Engine Oils		
	Oil #1	Oil #2	Oil #3
Mineral 100N base stock (vol%)	80.22	80.22	80.22
DFVA VI improver (%)	10.56	10.56	10.56
Additive package (%)	9.22	9.22	9.22
Sorbitan ester (Tween 61) (%)	0	0.2	0
Sorbitan ester (Tween 65) (%)	0	0	0.2
Appearance	Clear	Clear	Clear
Gel index	15.2	8.7	6.5
Depressed pour point (°C)	−39	—	−36

Source: Koenitzer, B.A. and May, C.J., Reducing low temperature scanning Brookfield gel index value in engine oils, CA Patent 2,385,419, 2001.

61) was reported effective in decreasing the gel index of mineral 5W-30 engine oil #2 to 8.7. The use of 0.2 wt% of polyoxyethylene(20)sorbitan tristearate (Tween 65), was found to be more effective in decreasing the gel index to 6.5 but the depressed pour point was reported to increase to −36°C.

The additives affect the low temperature fluidity of engine oils. The use of different chemistry VI improvers can affect the CCS viscosity of engine oils while the use of surfactants, such as sorbitan esters, can affect their gel index and their depressed pour point. To meet the volatility, the low temperature fluidity and the shear stability viscosity requirements of engine oils at the minimum cost, the base stocks, the VI improvers, and the additive packages need to be carefully selected.

9.2 POUR POINT DEPRESSION

The literature reports that the pour point of winter-grade engine oil must be low enough to permit the oil to be dispensed readily and to flow to the pump suction in the engine at the lowest anticipated ambient temperature (Pirro and Wessol, 2001). The use of pour point depressants was found to be the most effective in thinner oils, such as SAE 10, SAE 20, and SAE 30. Only a small effect of pour point depressants in higher viscosity oils, such as SAE 50, was observed (Crawford, Psaila, and Orszulik, 1997). Most mineral oils contain some wax that will crystallize at low temperatures. The literature reports that the high viscosity mineral base stocks have pour points higher than their cloud points and, at low temperatures, they cease to flow because their viscosity becomes too high rather than because of the wax formation (Prince, 1997). The pour point depressants were developed to interfere with

TABLE 9.9
Effect of Pour Point Depressants on the DSC Onset Wax
Crystallization of SWI 6 Oil

Solvent-Dewaxed SWI 6 Base Stock	DSC Onset Wax Crystallization (°C)	DSC Wax Content (−20°C to −40°C) (wt%)	Pour Point (°C)
Neat SWI 6 base stock	−25	9.1	−21 (dewaxed)
FVA (0.5 wt%)	−25.7	8.9	−30
PMA #1 (0.5 wt%)	−26.2	8.2	−33
PMA #2 (0.5 wt%)	−25.4	7.6	−30
PMA #2/FVA (0.5/0.5%)	−26	8.2	−33
PMA #2/PMA #1 (0.5/0.5%)	−26.2	8	−36

Source: Reprinted from Pillon, L.Z., *Petrol. Sci. Technol.*, 20(1–2), 101, 2002. With permission.

the growth of the crystal structure and were reports not effective in depressing the viscosity-limited pour points (Pirro and Wessol, 2001). The literature reports on the effect of the fumarate-vinylacetate (FVA) and the polyalkylmethylacrylates (PMA) pour point depressants on the onset wax crystallization temperature of solvent-dewaxed SWI 6 base stock measured using the differential scanning calorimetric (DSC) technique (Pillon, 2002).

The effect of different pour point depressants on the DSC onset wax crystallization temperature and the pour point depression of solvent-dewaxed SWI 6 base stock are shown in Table 9.9. The DSC analysis of solvent-dewaxed SWI 6 base stock having a HVI of 146 and a low volatility of 9% off, showed the onset wax crystallization at −25°C, which is below the dewaxed pour point of −21°C indicating the viscosity-limited pour point. In the temperature range from −20°C to −40°C, SWI 6 base stock was found to contain 9.1 wt% of wax, which would contain low branching density *iso*-paraffins since no *n*-paraffins are found in solvent-dewaxed base stocks. In the presence of 0.5 wt% of FVA pour point depressant, the onset wax crystallization temperature was found to decrease to only −25.7°C and the wax precipitation, in the temperature range from −20°C to −40°C, was suppressed to only 8.9 wt%. The pour point of SWI 6 base stock containing FVA was depressed to −30°C. In the presence of 0.5 wt% of PMA pour point depressants #1–2, the onset wax crystallization temperature decreased to −25.4°C or −26.2°C and the wax precipitation, in the temperature range from −20°C to −40°C, was suppressed to 7.6–8.2 wt%. The pour point of SWI 6 base stock containing PMA was depressed to −30°C or −33°C. In the presence of the PMA/FVA blend, the onset wax crystallization temperature was found to decrease to −26°C and the wax precipitation, in the temperature range from −20°C to −40°C, was suppressed to 8.2 wt%. The pour point of SWI 6 base stock containing the PMA/FVA blend was depressed to −33°C. In the presence of two different PMA/PMA pour point depressants, the onset wax crystallization temperature (range from −20°C to −40°C) was suppressed to 8 wt%. The pour point of SWI 6 base stock containing the

TABLE 9.10
Effect of Pour Point Depressants on the Onset Wax Crystallization of 10W-40 Oils

10W-40 Engine Oils (VII/Additive Package)	DSC Onset Wax Crystallization (°C)	DSC Wax Content (−20°C to −40°C) (wt%)	Pour Point (°C)
No pour point depressant	−17	6.2	−36
PMA #1/FVA (0.5/0.5%)	−20.6	8.3	−39
PMA #1/ PMA #2 (0.5/0.5%)	−21.6	6.6	−42

Source: Reprinted from Pillon, L.Z., *Petrol. Sci. Technol.*, 20(1–2), 101, 2002. With permission.

PMA/PMA blend, was further depressed to −36°C. The 10W-40 engine oils can be formulated by blending a SWI 6 base stock, a 600N mineral base stock, a VI improver, and the additive package.

The effect of pour point depressants on the DSC onset wax crystallization and the wax content of 10W-40 oils containing SWI 6 and mineral 600N base stocks are shown in Table 9.10. The 10W-40 engine oil containing SWI 6 base stock, mineral 600N base stock, VI improver, and the additive package was found to have the DSC onset wax crystallization temperature of −17°C and the wax content of 6.2 wt% in the temperature range of −20°C to −40°C. Without the pour point depressant, the pour point of 10W-40 engine oil was depressed to −36°C. After the addition of PMA/FVA (0.5/0.5%) pour point depressants, the DSC onset wax crystallization temperature was found to decrease to −20.6°C but the wax content, in the temperature range from −20°C to −40°C actually increased to 8.3 wt%. The pour point of 10W-40 engine oil containing PMA/FVA (0.5/0.5%) pour point depressants was further depressed to −39°C. After the addition of PMA #1/PMA #2 (0.5/0.5%), the DSC onset wax crystallization temperature was found to further decrease to −21.6°C and the wax content, in the temperature range from −20°C to −40°C further decreased to 6.6 wt%. The pour point of 10W-40 engine oil containing PMA #1/PMA #2 (0.5/0.5%) was further depressed to −42°C. Traditionally, the CCS viscosity of engine oils was targeted to meet the CCS viscosity of max 32.5P at −25°C for 5W engine oils and max 32.5P at −20°C for 10 W engine oils (Crosthwait, May, and Deane, 2001). To meet the CCS viscosity requirements of 10W-40 engine oils, the use of 57.7 vol% of SWI 6 having a VI of 146 and a dewaxed pour point of −21°C, 20 vol% of mineral 600N mineral base stock, about 10 vol% of VI improver, and 13 vol% of additive package was required (MacAlpine, Halls, and Asselin, 1991).

The CCS viscosity and the depressed pour points of 10W-40 engine oils containing SWI 6 and mineral 600N base stocks are shown in Table 9.11. Pour point depressants are used in lubricating oils at low treat rates, usually below 0.5 wt%. The 10W-40 engine oil #1 containing 0.4 wt% of PMA pour point depressant was found to have a depressed pour point of only −36°C. The patent literature reports that a mixture of two different PMA polymers having average molecular weights of

TABLE 9.11

CCS Viscosity and Depressed Pour Points of 10W-40 Engine Oils

	10W-40 Engine Oils			
	Oil #1	Oil #2	Oil #3	Oil #4
SWI 6 base stock (vol%)	57.7	57.7	57.7	57.7
Mineral 600N base stock (vol%)	20	20	20	20
OCP VI improver (%)	9.3	8.8	6.3	0
Additive package (%)	12.6	12.6	12.6	12.6
PMA (MW 92,000) (%)	0.4	0.4	0.4	0.4
PMA (MW 43,000) (%)	0	0.5	3	9.3
CCS viscosity at −20°C (P)	Max 32.5	Max 32.5	Max 32.5	Max 32.5
Depressed pour point (°C)	−36	−42	−42	−45

Source: MacAlpine, G.A. et al., Wax isomerate having a reduced pour point, CA Patent 2,054,096, 1991.

92,000 and 43,000 was found to be the most effective in depressing the pour point of 10W-40 engine oils containing SWI 6 and mineral 600N base stocks (MacAlpine, Halls, and Asselin, 1991). The 10W-40 engine oils #2–3 containing 0.4% of PMA polymer having a molecular weight of 92,000 and 0.5%–3% of PMA polymer having a lower molecular weight of 43,000 were found to have a lower depressed pour point of −42°C. The 10W-40 engine oils #2–3 containing 0.4% of PMA polymer having a molecular weight of 92,000 and 0.5%–3% of PMA polymer having a lower molecular weight of 43,000 were found to have a lower depressed pour point of −42°C.

Another 10W-40 engine oil #4 containing no OCP VI improver but 0.4% of PMA polymer having a molecular weight of 92,000, and 9.3% of PMA polymer having a low molecular weight of 43,000 was found to have a further depressed pour point of −45°C. The molecular weight of PMA type VI improvers can usually vary from 5,000 to 100,000 but some PMA VI improvers were found to have a very high molecular weight, above 500,000, measured using gel permeation chromatography (GPC) (Pillon and Asselin, 1993).

The effect of high molecular weight PMA VI improvers on pour point depression of 10W-40 engine oils, containing SWI 6 and 600N mineral base stocks, is shown in Table 9.12. The 10W-40 engine oil #1 containing 5.8 vol% of PMA VI improver having a high molecular weight of 500,000 and no pour point depressant was found to have a depressed pour point of −30°C. Another 10W-40 engine oil #2 containing 5.8 vol% of different PMA VI improver having a high molecular weight of 511,000 and no pour point depressant was found to have a lower depressed pour point of −33°C. The 10W-40 engine oil #3 containing 5.8 vol% of PMA VI improver having a high molecular weight of 600,000 and no pour point depressant was found to have a further decreased depressed pour point of −42°C. Some additives are multifunctional and the PMA type VI improvers can also function as pour point depressants.

TABLE 9.12

Effect of High Molecular Weight PMA VI Improvers on Pour Points of 10W-40 Oils

	10W-40 Engine Oils		
	Oil #1	Oil #2	Oil #3
SWI 6 base stock (vol%)	61.2	61.2	61.2
Mineral 600N base stock (vol%)	20.4	20.4	20.4
PMA VI I (MW 500,000) (%)	5.8	0	0
PMA VI I (MW 511,000) (%)	0	5.8	0
PMA VI I (MW 600,000) (%)	0	0	5.8
Additive package (%)	12.6	12.6	12.6
Pour point depressant (%)	0	0	0
Depressed pour point (°C)	−30	−33	−42

Source: Pillon L.Z. and Asselin A.E., Wax isomerate having a reduced pour point, U.S. Patent 5,229,029, 1993.

The use of PMA VI improvers having a high molecular weight of 500,000–600,000 decreased the content of VI improver; however, the shear stability requirement of engine oils is another major factor affecting the VI improver choice.

The depressed pour points of part-synthetic 0W-40 engine oils containing different VI improvers are shown in Table 9.13. The part-synthetic 0W-40 oil #1 containing mineral base stock, PAO 4, ester, OCP VI improver, another PMA VI improver #1, and the additive package A was found to have a depressed pour point of −51°C and the shear stability of 12.51 cSt and thus meeting the specification of a min 12.5 cSt. Another part-synthetic 0W-40 oil #2 containing OCP VI improver and another PMA VI improver #2 was also found to have a depressed pour point of −51°C but the shear stability viscosity of 12.43 cSt and thus not meeting the specification. The part-synthetic 0W-40 oil #3 containing only PMA VI improver #2 and a different additive package B was reported to have a lower depressed pour point of −54°C but shear stability viscosity of only 11.66 cSt and thus not meeting the specification. Another part-synthetic 0W-40 oil #4 containing OCP VI improver, PMA VI improver #2 with the addition of polyisoprene VI improver was reported to have a depressed pour point of −54°C and an increased shear stability viscosity of 12.98 cSt and thus meeting the specifications of a min 12.5 cSt.

The effect of the mineral base stock on the volatility and the CCS viscosity of part-synthetic 0W-40 engine oils are shown in Table 9.14. The part-synthetic 0W-40 engine oil #1 containing 26.76 vol% of mineral base stock, 36 vol% of PAO 4, 10 vol% of ester, two different VI improvers, and the additive package A was reported to have a volatility of 17.8% off and a CCS viscosity of 2810 cP at −30°C. Another part-synthetic 0W-40 engine oil #2 containing 28.76 vol% of mineral base stock, 34 vol% of PAO 4, 10 vol% of ester, two different VI improvers, and the additive package A was reported to have a higher volatility of 18.5% off and a higher CCS

TABLE 9.13

Depressed Pour Points of Different Part-Synthetic 0W-40 Engine Oils

	Part-Synthetic 0W-40 Oils			
	Oil #1	Oil #2	Oil #3	Oil #4
Mineral base stock (vol%)	26.76	26.34	29.38	26.63
Synthetic PAO 4 (vol%)	36	36.84	33	30.09
Synthetic ester (vol%)	10	11	15	16.56
OCP VI improver (vol%)	8	7.3		7.3
PMA VI improver #1 (vol%)	5.8			
PMA VI improver #2 (vol%)		6	10.1	6
Polyisoprene VI improver (vol%)				5.2
Additive package A (%)	13.44	13.44		
Additive package B (%)			12.52	12.52
KV at 100°C (cSt)	15.61	15.47	15.8	15.78
Pour point (°C)	−51	−51	−54	−54
CCS at −30°C (cP)	2,810	2,860	2,950	2,870
MRV at −40°C (cP)	16,600	—	14,500	18,500
Shear stability viscosity (cSt)	12.51	12.43	11.66	12.98

Source: Gao, J.Z., A partly synthetic multigrade crankcase lubricant, CA Patent 2,263,562, 1999.

TABLE 9.14

Volatility and CCS Viscosity of Part-Synthetic 0W-40 Engine Oils

	Part-Synthetic 0W-40 Engine Oils	
	Oil #1	Oil #2
Mineral base stock (vol%)	26.76	28.76
Synthetic PAO 4 (vol%)	36	34
Synthetic ester (vol%)	10	10
OCP VI improver (vol%)	8	8
PMA VI improver (vol%)	5.8	5.8
Additive package A (vol%)	13.44	13.44
KV at 100°C (cSt)	15.61	15.71
Volatility (% off)	17.8	18.5
CCS viscosity at −30°C (cP)	2810	2870

Source: Gao, J.Z., A partly synthetic multigrade crankcase lubricant, CA Patent 2,263,562, 1999.

viscosity of 2870 cP at −30°C. With a small increase in the content of mineral base stock and a decrease in PAO 4, an increase in the volatility and the CCS viscosity at −30°C is observed. There are three basic types of base stocks used in engine oils, which are mineral, modified mineral such as hydrocracked and hydroisomerized slack wax, and synthetic oils. If the performance is not achieved with the mineral or the modified mineral base stocks, synthetic base stocks such as PAOs are used (Prince, 1997). The literature reports that with a good choice of base stocks and a proper choice of pour point depressant, engine oils can be formulated with many different chemistry VI improvers (Stambaugh, 1997). Engine oils need to perform at low and high temperatures and require base stocks having a good low temperature fluidity, a low volatility, and a good oxidation stability. The synthetic PAO base stocks are frequently mixed with mineral oils to decrease their volatility and improve their low temperature fluidity and the oxidation stability.

9.3   HIGH-TEMPERATURE FOAMING

Foaming of engine oils can result in insufficient lubrication by preventing oil from reaching surfaces that are in need of lubrication (Pillon and Asselin, 1998). The crankcase oils used in the internal combustion engines represent a typical application in which the lubricating oil is subjected to a severe mechanical agitation that results in foaming. For an antifoaming agent to be surface active, their surface tension (ST) needs to be lower, by several units of mN/m, than that of foaming fluid (Ross, 1987). The performance of antifoaming agents in engine oils was reported to be affected by the properties of base stocks, their treat rates, and the blending temperature (Wang and Zhu, 2000). Different chemistry VI improvers are used in engine oils and the literature reports on their shear stability and the thickening efficiency (Braun and Omeis, 2001). Antifoaming agents must be effective over a fairly wide temperature range of from below freezing to above the boiling point of water. The fluorosilicone antifoaming agents were reported to be more effective than silicones in the prevention of foaming of automatic transmission fluids (ATF) at high temperatures up to 160°C (Takakura, 1999).

The foaming test requirements of engine oils are shown in Table 9.15. The foaming characteristics of engine oils are measured following the ASTM D 892 foam test at the temperature of 24°C (Seq. I), at 93.5°C (Seq. II), and again at 24°C (Seq. III). The engine oils need to meet the foaming tendency specification of a max 10 mL at 24°C and a max 50 mL at 93.5°C. The ASTM D 6082 foam test method describes the procedure for determining the foaming characteristics of lubricating oils, specifically transmission fluids and motor oils at 150°C (Seq. IV). No specific performance targets for the foaming tendency of engine oil at 150°C have been established. The most important classes of additives used in engine oils are VI improvers, antioxidants, antiwear additives, dispersants, detergents, pour point depressants, and antifoaming agents. The early literature reports on the effect of the temperature on the ST of different mineral lubricating oils (Bondi, 1951).

The effect of the temperature on the ST of different mineral lubricating oils is shown in Table 9.16. With an increase in the temperature from 25°C to 200°C, the ST of mineral lubricating oil #1 decreased from 32.3 to 19.5 mN/m. With an

TABLE 9.15
Foaming Test Requirements of Engine Oils

	Engine Oils		
	5W-30	10W-30	15W-40
ASTM D 892 foam test (option A)	Max	Max	Max
Seq. I at 24°C (mL)	10/0	10/0	10/0
Seq. II at 93.5°C (mL)	50/0	50/0	50/0
Seq. III at 24°C (mL)	10/0	10/0	10/0
ASTM D 6082 foam test			
Seq. IV at 150°C (mL)	Report	Report	Report

Source: American Petroleum Institute, *Engine Oil Licensing and Certification System,* API 1509, 15th ed., American Petroleum Institute, Washington, DC, April 2002.

TABLE 9.16
Effect of Temperature on ST of Different Mineral Lubricating Oils

Temperature (°C)	Lubricating Oils		
	Mineral Oil #1	Mineral Oil #2	Mineral Oil #3
		ST (mN/m)	
25	32.3	31.0	31.6
100	26.3	25.1	25.6
200	19.5	18.3	19.1

Source: Bondi, A., *Physical Chemistry of Lubricating Oils*, Reinhold Publishing Corporation, New York, 1951.

increase in the temperature from 25°C to 200°C, the ST of another mineral lubricating oil #2 decreased from 31 to 18.3 mN/m. With an increase in the temperature from 25°C to 200°C, the ST of mineral lubricating oil #3 decreased from 31.6 to 19.1 mN/m. With an increase in the temperature and a decrease in the ST of lubricating oils, a decrease in the surface activity of antifoaming agents will be observed. At 25°C, different mineral lubricating oils were reported to have an ST in a range of 31–32.3 mN/m, which decreased to 25.1–26.3 mN/m at 100°C. The VI improvers and the dispersants are typically the highest treat additives used in engine oils. The use of XHVI base stocks having a VHVI over 145, decreases the need for VI improvers in engine oils (Phillipps, 1999). Antifoaming agents are usually present in commercial additive packages and some engine oils require the addition of a pour point depressant.

TABLE 9.17

Foaming Tendency of Mineral and Modified 10W-40 Engine Oils

	10W-40 Engine Oils	
	Mineral Oil	Modified Oil
Mineral 100N base stock (vol%)	13	0
Mineral 150N base stock (vol%)	59	35
SWI 6 base stock (vol%)	0	47
VI improver (%)	16	6
Additive package (%)	12	13
Pour point depressant (%)	0.3	0
ASTM D 892 foam test		
Seq. I at 24°C (mL)	0/0	0/0
Seq. II at 93.5°C (mL)	130/0	230/0

Source: Pillon, L.Z. et al., Method for reducing foaming of lubricating oils, U.S. Patent 6,090,758, 2000.

The foaming tendency of the mineral and the modified 10W-40 engine oils containing the same VI improver and the same commercial additive package is shown in Table 9.17. The mineral 10W-40 engine oil containing 13 vol% of mineral 100N base stock, 59 vol% of mineral 150N base stock, 16 vol% of a VI improver, 12 vol% of a commercial additive package, and 0.3 wt% of a pour point depressant was found to have a total resistance to foaming at 24°C and a high foaming tendency of 130 mL at 93.5°C. With an increase in the temperature, a decrease in the surface activity of antifoaming agent is observed. The modified 10W-40 engine oil containing only 35 vol% of mineral 150N base stock, 47 vol% of SWI 6 base stock, only 6 vol% of a VI improver, and 13 vol% of a commercial additive package was also found to have a total resistance to foaming at 24°C and a higher foaming tendency of 230 mL at 93.5°C. The use of SWI 6 base stock having a lower ST, leads to a further decrease in the surface activity of an antifoaming agent at a higher temperature. Silicone antifoaming agents are usually effective at the treat rate of 2–3 ppm of an active ingredient in mineral and SWI 6 base stocks.

The effect of low viscosity fluorosilicone, FSi 300, on the foaming tendency of the modified 10W-40 engine oils is shown in Table 9.18. The modified 10W-40 engine oil #1 containing 5 ppm of undiluted FSi 300 was found effective in the prevention of foaming at 24°C; however, a severe foaming tendency of 240 mL at 93.5°C was observed. The modified 10W-40 engine oil #2 containing 10 ppm of undiluted FSi 300 was found effective in the prevention of foaming at 24°C; however, a more severe foaming tendency of 290 mL at 93.5°C was observed. With an increase in the treat rate of low viscosity FSi 300, an increase in the foaming tendency is observed indicating its dissolution in the modified 10W-40 engine oil at 93.5°C and acting as a foam promoter.

TABLE 9.18
Performance of Low Viscosity FSi 300 in Modified 10W-40 Engine Oils

	10W-40 Engine Oils	
	Modified Oil #1	Modified Oil #2
Mineral 150N base stock (vol%)	59	35
SWI 6 base stock (vol%)	47	47
VI improver (%)	6	6
Additive package (%)	13	13
Undiluted FSi 300 (ppm)	5	10
ASTM D 892 foam test		
Seq. I at 24°C (mL)	0/0	0/0
Seq. II at 93.5°C (mL)	240/0	290/0

Source: Pillon, L.Z. et al., Method for reducing foaming of lubricating oils, U.S. Patent 6,090,758, 2000.

TABLE 9.19
Performance of Low Viscosity Si 350 in Modified 10W-40 Engine Oils

	10W-40 Engine Oils	
	Modified Oil #1	Modified Oil #2
Mineral 150N base stock (vol%)	59	35
SWI 6 base stock (vol%)	47	47
VI improver (%)	6	6
Additive package (%)	13	13
Undiluted Si 350 (ppm)	5	10
ASTM D 892 foam test		
Seq. I at 24°C (mL)	0/0	0/0
Seq. II at 93.5°C (mL)	240/0	260/0

Source: Pillon, L.Z. et al., Method for reducing foaming of lubricating oils, U.S. Patent 6,090,758, 2000.

The effect of low viscosity silicone, Si 350, on the foaming tendency of the modified 10W-40 engine oils is shown in Table 9.19. The modified 10W-40 engine oil #1 containing 5 ppm of undiluted Si 350 was found effective in the prevention of foaming at 24°C; however, a severe foaming tendency of 240 mL at 93.5°C was observed. The modified 10W-40 engine oil #2 containing 10 ppm of undiluted Si 350 was found effective in the prevention of foaming at 24°C; however, a more severe foaming tendency of 260 mL at 93.5°C was observed. With an increase in the treat rate of low viscosity Si 350, an increase in the foaming tendency was observed also indicating its dissolution in the modified 10W-40 engine oil at 93.5°C and acting as a foam promoter. While silicone antifoaming agents can be added directly to engine oil, they are more often diluted with a solvent such as kerosene, solvent naphtha, xylene, and toluene and a typical solution contains 10 wt% of Si 12500 to improve its dispersion in oil (Pillon, Vernon, and Asselin, 2000).

The effect of diluted Si 12500 (10 wt%) on the foaming tendency of the mineral and the modified 10W-40 engine oils is shown in Table 9.20. The use of 0.001 vol% of diluted Si 12500 (10 wt%), 10 ppm of an active ingredient was found effective in decreasing the foaming tendency of mineral 10W-40 oil from 130 to 35 mL at 93.5°C. However, the use of 0.001 vol% of diluted Si 12500 (10 wt%), also 10 ppm of active ingredient decreased the foaming tendency of the modified 10W-40 oil from 230 to only 65 mL at 93.5°C, indicating a need for a more effective antifoaming agent. Engine oils need to meet the foaming tendency specification of a max 50 mL at 93.5°C and a higher viscosity Si 12500 was found more effective in preventing a high temperature foaming; however, at a very high treat rate.

The effect of high viscosity Si 60000 on the foaming tendency of the modified 10W-40 engine oils is shown in Table 9.21. The modified 10W-40 engine oil #1 containing 5 ppm of undiluted Si 60000 was found to have a total resistance to foaming at 24°C and only decreased the foaming tendency from 230 to 165 mL at 93.5°C. The modified 10W-40 engine oil #2 containing 10 ppm of undiluted Si 60000 was found to have a total resistance to foaming at 24°C and was found more effective in decreasing the foaming tendency to 80 mL at 93.5°C. The literature reports that the coking process requires temperatures as high as 400°C–500°C and only very high viscosity silicones, such as Si 100000, were reported effective in the prevention of foaming

TABLE 9.20

Performance of Diluted Si 12500 in Mineral and Modified 10W-40 Engine Oils

	10W-40 Engine Oils	
	Mineral Oil	Modified Oil
Mineral 100N base stock (vol%)	13	0
Mineral 150N base stock (vol%)	59	35
SWI 6 base stock (vol%)	0	47
VI improver (%)	16	6
Additive package (%)	12	13
Pour point depressant (%)	0.3	0
Si 12500 (10 wt%) (vol%)	0.001	0.001
ASTM D 892 foam test		
Seq. I at 24°C (mL)	0/0	0/0
Seq. II at 93.5°C (mL)	35/0	65/0

Source: Pillon, L.Z. et al., Method for reducing foaming of lubricating oils, U.S. Patent 6,090,758, 2000.

TABLE 9.21
Performance of High Viscosity Si 60000 in Modified 10W-40 Engine Oils

	10W-40 Engine Oils	
	Modified Oil #1	Modified Oil #2
Mineral 150N base stock (vol%)	59	35
SWI 6 base stock (vol%)	47	47
VI improver (%)	6	6
Additive package (%)	13	13
Undiluted Si 60000 (ppm)	5	10
ASTM D 892 foam test		
Seq. I at 24°C (mL)	0/0	0/0
Seq. II at 93.5°C (mL)	165/0	80/0

Source: Pillon, L.Z. et al., Method for reducing foaming of lubricating oils, U.S. Patent 6,090,758, 2000.

during the coking process (Pillon, 2007). To improve the dispersion of high viscosity silicones in engine oils, Si 60000, was used as 5 wt% dispersion in kerosene while the very high viscosity Si 100000 was used as a 1 wt% dispersion in kerosene.

The effect of diluted Si 60000 (5 wt%) and Si 100000 (1 wt%) on the foaming tendency of the modified 10W-40 engine oils is shown in Table 9.22. The modified

TABLE 9.22
Performance of Diluted High Viscosity Si 60000 and Si 100000 in 10W-40 Oils

	Modified 10W-40 Engine Oils		
	Oil #1	Oil #2	Oil #3
Mineral 150N base stock (vol%)	59	35	35
SWI 6 base stock (vol%)	47	47	47
VI improver (%)	6	6	6
Additive package (%)	13	13	13
Si 60000 (5 wt%) (vol%)	0.005	0.02	
Si 100000 (1 wt%) (vol%)			0.1
ASTM D 892 foam test			
Seq. I at 24°C (mL)	0/0	0/0	0/0
Seq. II at 93.5°C (mL)	40/0	30/0	10/0

Source: Pillon, L.Z. et al., Method for reducing foaming of lubricating oils, U.S. Patent 6,090,758, 2000.

10W-40 engine oil #1 containing 0.005 vol% of diluted Si 60000 (5 wt%), 5 ppm of an active ingredient was found effective in decreasing the foaming tendency to 40 mL at 93.5°C and thus meeting the specification. With an increase in the treat rate of diluted Si 60000 (5 wt%) from 5 to 10 ppm of an active ingredient, a further decrease in the foaming tendency of modified 10W-40 engine oil #2 to 30 mL at 93.5°C was observed. The modified 10W-40 engine oil #3 containing 0.1 vol% of diluted Si 100000 (1 wt%), 10 ppm of an active ingredient was found to further decrease the foaming tendency to 10 mL at 93.5°C. To prevent high temperature foaming of engine oils, very high viscosity silicone-antifoaming agents need to be used. With an increase in the temperature, the surface activity of the antifoaming agents decreases due to a decrease in the ST of engine oils. At high temperatures, higher viscosity silicones have a higher efficiency to spread on the oil surface. To assure their dispersion in engine oils, a very low concentration in kerosene is needed and their storage stability needs to be tested.

9.4 AIR ENTRAINMENT

The air entrainment was reported to be affected by the temperature and engine oils were reported to entrain more air at higher temperatures (Kubo and Lillywhite, 1994). The air entrainment can decrease the engine life and increase the fuel usage due to reduced lubrication caused by the enhanced oil oxidation (Pillon and Asselin, 1998). The selection of antifoaming agent for use in lubricating oils will depend on the application. To prevent air entrainment, turbine oils require the use of non-silicone antifoaming agents. Silicone antifoaming agents have been used to prevent foaming in a wide range of automotive lubricating oils, including the diesel and the crankcase oils. The most effective antifoaming agents are silicones and are known to increase the air entrainment of lubricant products. The typical silicone antifoaming agent, Si 12500, is used at treat rate of 3 ppm to prevent the foaming of engine oils. The commercial additive packages usually contain antioxidants, antiwear additives, dispersants, detergents, and antifoaming agents; however, their content and their chemistry can vary. Many additives used in engine oils are highly surface active or have surfactant chemistry and can adsorb at the oil/air interface leading to an increase in the air release time.

The mineral oil containing only a VI improver was found to have a high Seq. I and Seq. III foaming tendency measured at 24°C and a low Seq. II foaming tendency measured at 93.5°C. The air release time of 3 min measured at 50°C indicated no significant effect of the VI improver on air entrainment of mineral base stocks. The mineral 10W-30 engine oil #1 containing a VI improver and the additive package A was found to have a total resistance to foaming at 24°C but an increased foaming tendency at 93.5°C. The addition of the additive package A increased the air release time of 10W-30 engine oil #1 to 17 min at 50°C. Another mineral 10W-30 engine oil #2 containing a VI improver and the additive package B was also found to have a total resistance to foaming at 24°C and a lower foaming tendency at 93.5°C. The addition of the additive package B further increased in the air release time of 10W-30 engine oil #2 to 20 min at 50°C indicating a higher content of antifoaming agent or a higher content of other surface-active additives capable of adsorbing at the

TABLE 9.23
Surfactant Chemistry of Dispersants and Detergents Used in Engine Oils

Surfactant Types	Chemistry of Molecule	Examples
Dispersants		"PIBSA"
Hydrophobe–hydrophile	Hydrocarbon tail—solubilizer	Polyisobutenyl
	O, P and/or N head—polar	Succinic anhydrides
Detergents		Neutral or overbased sulfonates
Hydrophobe–hydrophile	Hydrocarbon tail—solubilizer	Alkyl aromatic hydrocarbon
	Metal head—polar	Ca, Mg, or Na sulfonates

Source: Colyer, C.C. and Gergel, W.C., Detergents and dispersants, in *Chemistry and Technology of Lubricants*, 2nd ed., Mortier, R.M. and Orszulik, S.T., Eds., Blackie Academic & Professional, London, U.K., 1997.

oil/air interface. The dispersants and the detergents used in engine oils are surface active because they have surfactant chemistry. Some engine oils contain overbased detergents in the form of small particles of inorganic carbonates.

The surfactant chemistry of different dispersants and detergents used in engine oils is shown in Table 9.23. The literature reports that the typical molecule used as dispersant is a surfactant containing a hydrocarbon tail, the oleophilic group, and a "polar head" containing oxygen, phosphorus, or nitrogen derivative. The oleophilic group enables the molecule to be fully soluble in the base stock while the "polar head" is used to adsorb particles of sludge and prevent them from settling out on critical parts of the engine (Colyer and Gergel, 1997). The dispersants consist of a long hydrocarbon backbone, typically polyisobutene of a molecular weight > 900, and a succinimide head group formed by reaction with a polyamine. Alkylsuccinimides, alkylsuccinic esters, and the Mannich reaction products are typical dispersant type molecules (Colyer and Gergel, 1997). The literature reports that the detergent molecules are also surfactants containing the oleophilic group and the "polar head" which contains metal cations and is used to neutralize corrosive acids (Colyer and Gergel, 1997). The surface-active dispersants and detergents are used at high treat rates in engine oils and their adsorption at the oil/air interface can lead to a decrease in the surface activity of antifoaming agents and an increase in the air entrainment.

The effect of the temperature on the surface activity of different viscosity silicone antifoaming agents in neat SWI 6 base stock and modified 10W-40 engine oil containing a VI improver and a commercial additive package is shown in Table 9.24. With an increase in the temperature from 24°C to 93.5°C, the efficiency of low viscosity Si 350 to spread (S) on the oil surface of SWI 6 base stock was found to drastically decrease from 8.9 to 0.9 mN/m. In the presence of a VI improver and the additive package, the efficiency of Si 350 to spread on the oil surface of modified 10W-40 engine oil further decreased to 0.6 mN/m at 93.5°C. With an increase in the temperature, the efficiency of higher viscosity Si 12500 to spread

TABLE 9.24

Effect of Temperature on Surface Activity of Silicones in 10W-40 Engine Oil

Different Silicones	Neat SWI 6	Neat SWI 6	Modified 10W-40 Oil
Spreading (S) (mN/m)	(S) at 24°C	(S) at 93.5°C	(S) at 93.5°C
Si 350	8.9	0.9	0.6
Si 12500	6.6	1.9	1.6
Si 60000	7.9	7.1	2.9
Si 100000	7.7	6.1	3.7

Source: Pillon, L.Z. et al., Method for reducing foaming of lubricating oils, U.S. Patent 6,090,758, 2000.

(S) on the oil surface of SWI 6 base stock was also found to drastically decrease from 6.6 to 1.9 mN/m. In the presence of a VI improver and the additive package, the efficiency of Si 12500 to spread on the oil surface of modified 10W-40 engine oil further decreased to 1.6 mN/m at 93.5°C. With an increase in the temperature, the efficiency of high viscosity Si 60000 to spread (S) on the oil surface of SWI 6 base stock was found to decrease from 7.9 to 7.1 mN/m. In the presence of a VI improver and the additive package, the efficiency of Si 60000 to spread on the oil surface of modified 10W-40 engine oil further decreased to 2.9 mN/m at 93.5°C. With an increase in the temperature, the efficiency of a very high viscosity Si 100000 to spread (S) on the oil surface of SWI 6 base stock was also found to decrease from 7.7 to 6.1 mN/m. In the presence of a VI improver and the additive package, the efficiency of Si 100000 to spread on the oil surface of modified 10W-40 engine oil further decreased to 3.7 mN/m at 93.5°C. In some engine oils, the adsorption of other surface-active additives present in the additive packages can decrease the spreading of the silicone antifoaming agents on the oil surface leading to a severe foaming at high temperatures and also lead to a severe air entrainment.

9.5 OXIDATION RESISTANCE AND WEAR

The literature reports on different oxidation by-products of engine oils. Under low temperature oxidation conditions, peroxides, alcohols, aldehydes, ketones, and water are formed. Under high temperature oxidation conditions, the volatility increases, acids are produced, the oil viscosity increases, and sludge and varnish are formed. The soot is the oxidation by-product and the literature reports that wear in diesel engines is accelerated by the soot (Bardasz and Lamb, 2003). The lubricant needs to prevent the damage to engine caused by contamination. Many different antioxidants have been commercialized to improve the oxidation resistance of engine oils; however, the use of phenol derivatives and organic amines as antioxidants showed only limited effectiveness. The aromatic diamines were reported to be preferably used with the metal salts and a metal deactivator. The test data indicated a low viscosity increase, slow sludge build up, and a low corrosion (Migdal, 2003). The patent literature reports on the effect of the diphenylamine derivative and the molybdenum (Mo) dithiocarbamate on the oxidative thickening of the Mobil 500N base stock containing over 97 wt% of the saturate content, and the mineral 600N base stock containing 80 wt% of saturate and 20 wt% of aromatic content. The oils were oxidized at 165°C under a mixed nitrogen/air flow in the presence of 40 ppm of a soluble iron used as the catalyst (Gao, 2001).

The effect of different chemistry antioxidants on the oxidative thickening of different base stocks containing polyisobutenylsuccinimide dispersant is shown in Table 9.25. After 48 h of oxidation at 165°C and in the presence of Fe used as the catalyst, the viscosity of Mobil 500N base stock containing 6 wt% polyisobutenylsuccinimide dispersant and 1 wt% of octylated/butylated diphenylamine antioxidant was found to increase by 181.1%. After oxidation under the same conditions, the viscosity of mineral 600N base stock also containing 6 wt% polyisobutenylsuccinimide dispersant and 1 wt% of octylated/butylated diphenylamine antioxidant was found to increase by only 64.3%. After 48 h of oxidation at 165°C and in the presence of Fe used as the catalyst, the viscosity of Mobil 500N base stock containing 6 wt% polyisobutenylsuccinimide dispersant, 0.5 wt% of octylated/butylated diphenylamine, and 0.5 wt% of Mo dithiocarbamate was found to increase by only 1.7%. After oxidation under the same conditions, the viscosity of mineral 600N base stock also containing 6 wt% polyisobutenylsuccinimide dispersant, 0.5 wt% of octylated/butylated diphenylamine, and 0.5 wt% of Mo dithiocarbamate was found to increase by only 3.7%. The use of the diphenylamine derivatives is not effective in decreasing the oxidative thickening and the addition of the metal containing antioxidants

TABLE 9.25

Effect of Antioxidants on Oxidative Thickening of Different Base Stocks

	Oil Blends			
	Mobil (500N)	Mineral (600N)	Mobil (500N)	Mineral (600N)
	Composition (wt%)			
Base stock	93	93	93	93
Dispersant	6	6	6	6
Diphenylamine	1	1	0.5	0.5
Mo dithiocarbamate	0	0	0.5	0.5
KV at 100°C (cSt)	12.31	13.18	12.38	13.06
Oxidation (165°C/48 h)				
KV at 100°C (cSt)	34.6	21.66	12.59	15.54
Viscosity increase (%)	181.1	64.3	1.7	3.7

Source: Gao, J., Antioxidant additive composition for lubricating oils, CA Patent 2,397,885, 2001.

such as Mo dithiocarbamate is required. The patent literature reports on the effect of the addition of Mo dithiocarbamate on the oxidation resistance of 5W-30 engine oils containing a VI improver, a commercial additive package, and a pour point depressant (Gao, 2001).

The effect of Mo dithiocarbamate on the residue formation of oxidized 5W-30 engine oil containing a commercial additive package is shown in Table 9.26. Different tests are used to measure deposits of oxidized engine oils that need to meet the deposit specification of a max 40 mg. The TEOST-33 is used to measure deposits, as filtered residue, and on a heated rod. Another TEOST-MHT-2 oxidation test, similar to TEOST-33, was reported to use 8 g of oil sample at a temperature of 285°C for 24 h. After oxidation at 285°C for 24 h, the 5W-30 engine oil #1 containing a commercial additive package was reported to produce 5.2 mg of filtered residue and 70 mg of heated rod residue. After oxidation under the same conditions, the 5W-30 engine oil #2 containing the same commercial additive package and an additionally 1 wt% of Mo dithiocarbamate was reported to produce only 0.9 mg of filtered residue and only 26.4 mg of heated rod residue and thus meeting the deposit specification of a max 40 mg. The oxidation of engine oils leads to the formation of the residue and the use of different antioxidants and dispersants can only delay the formation of deposits and reduce the rate at which they accumulate on the metal surface. The literature reports on the effect of mineral and PAO-based engine oils containing the same additive package on their oxidative thickening. The oils were tested using the hot oil oxidation test at 160°C in the presence of iron and copper catalysts for 5 days (Rudnick and Shubkin, 1999).

TABLE 9.26
Effect of Mo Dithiocarbamate on Residue of Oxidized 5W-30 Engine Oil

	5W-30 Engine Oils	
	Oil #1	Oil #2
Base stock (vol%)	81.51	81.51
VI improver (%)	10	10
Additive package (%)	8.29	8.29
Pour point depressant (%)	0.2	0.2
Mo dithiocarbamate (%)	0	1
Oxidation (285°C/24 h)		
Filtered residue (mg)	5.2	0.9
Heated rod residue (mg)	70	26.4

Source: Gao, J. et al., Antioxidant additive composition for lubricating oils, CA Patent 2,397,885, 2001.

TABLE 9.27
Effect of Mineral and PAO Base Stocks on Oxidative Thickening of Engine Oils

	Fresh Engine Oils	
	Mineral-Based Oil	PAO-Based Oil
Kinematic viscosity	95 cSt at 40°C	94 cSt at 40°C
Hot oil oxidation test (160°C/5 days)		
KV at 40°C (cSt)	146.3	96.8
Viscosity increase (%)	54	3

Source: Reproduced from Rudnick, L.R. and Shubkin, R.L., Poly(alfa-olefins), in *Synthetic Lubricants and High-Performance Functional Fluids*, 2nd ed., Marcel Dekker Inc., New York, 1999. With permission.

The effect of mineral and synthetic PAO base stocks on the oxidative thickening of engine oils is shown in Table 9.27. After oxidation at 160°C for 5 days and in the presence of Fe and Cu metals used as catalysts, the viscosity of the mineral-based engine oil increased from 95 to 146.3 cSt at 40°C by 54%. After oxidation under the same conditions, the viscosity of PAO-based engine oil increased from 94 to 96.8 cSt at 40°C by only 3%. The use of synthetic PAO base stock leads to a drastically lower oxidative thickening of engine oils.

The effect of mineral and synthetic PAO base stocks on the oxidative thickening of engine oils and the engine wear is shown in Table 9.28. The oxidation of mineral 15W-40 engine oil leads to a high oxidative thickening and engine wear. The oxidation of synthetic 15W-40 engine oil containing 100% PAO base stock, leads to a low

TABLE 9.28
Effect of Mineral and PAO Base Stocks on Engine Wear

	15W-40 Engine Oils	
	Mineral Oil	PAO (100%) Oil
Petter WI engine test (after 108 h)		
Viscosity increase at 40°C (%)	108	20
Bearing weight loss (mg)	14.1	14.5

Source: Rudnick, L.R. and Shubkin, R.L., Poly(alfa-olefins), in *Synthetic Lubricants and High-Performance Functional Fluids*, 2nd ed., Marcel Dekker Inc., New York, 1999. With permission.

oxidative thickening but an increase in the engine wear is observed. The synthetic PAO base stock has limited ability to dissolve some polar additives and tends to shrink seals (Rudnick and Shubkin, 1999). PAO base stocks were reported to have limited ability to dissolve polar additives, such as rust inhibitors, and have a difficult time in preventing rust formation (Pillon, 2007).

The oxidative thickening of part-synthetic 10W-40 engine oils and the engine wear is shown in Table 9.29. The part-synthetic 10W-40 engine oil containing 25% of PAO was reported to have the oxidative thickening of 54% and the bearing weight loss of 9.7 mg. With an increase in the PAO content from 25% to 50%, the part-synthetic 10W-40 engine oil was reported to have a lower oxidative thickening of 45% but an increase in the bearing weight loss of 11.5 mg was reported.

Some qualities of engine oils cannot be met with the use of mineral base stocks and the use of synthetic base stocks is required. It is widely accepted that the use of PAO/ester blends offer a number of performance advantages over the use of conventional petroleum-based oils, such as improved cold starting, better fuel and oil economy, improved engine cleanliness, wear protection, and viscosity retention during the service (Brown et al., 1997).

The effect of the calcium sulfonate detergent on the wear and the appearance of the engine are shown in Table 9.30. While dispersants are used to disperse the oil-insoluble by-products, such as sludge and soot particles, detergents are used to clean engine surfaces and neutralize the acidic by-products of combustion (Lee and Harris, 2003). The use of the calcium sulfonate detergents #1–2, produced by the sulfonation of naphthenic oil, was reported to lead to a variation in the wear and the appearance of the engine. The literature reports that the concentration of oxidation by-products in engine oils would vary depending on the oil composition, the temperature, and the engine design (Rasberger, 1997).

The variation in the temperature of the internal surfaces in a V-8 engine is shown in Table 9.31. The literature reports that heavy sludge can plug oil filters, leading to oil starvation and thus to catastrophic wear especially during cold-temperature start-ups. Contaminants are bonded by polar attraction to dispersant molecules,

---

TABLE 9.29
Oxidative Thickening of Part-Synthetic Engine Oils and Engine Wear

	Part-Synthetic 10W-40 Engine Oils	
	PAO Content (25%)	PAO Content (50%)
Petter WI engine test (after 108 h)		
Viscosity increase at 40°C (%)	54	45
Bearing weight loss (mg)	9.7	11.5

Source: Rudnick, L.R. and Shubkin, R.L., Poly(alfa-olefins), in *Synthetic Lubricants and High-Performance Functional Fluids*, 2nd ed., Marcel Dekker Inc., New York, 1999. With permission.

TABLE 9.30
Effect of Ca Sulfonate Detergent on Wear and Appearance of Engine

SAE 30 Engine Oil, Use of Detergent (3%)	Copper-Lead Bearing, Weight Loss (mg)	Appearance of Engine
Ca sulfonate #1	1000	Clean
Ca sulfonate #2	1510	Lacquer on piston skirt

Source: Colyer, C.C. and Gergel, W.C., Detergents and dispersants, in *Chemistry and Technology of Lubricants*, 2nd ed., Mortier, R.M. and Orszulik, S.T., Eds., Blackie Academic & Professional, London, U.K., 1997. With permission.

TABLE 9.31
Variation in Temperatures of Internal Surfaces in a V-8 Engine

Area of Engine	Temperature Range (°C)
Exhaust valve head	650–730
Exhaust valve stem	635–675
Combustion chamber gases	2300–2500
Combustion chamber wall	204–260
Piston crown	204–426
Piston rings	149–315
Piston (wrist) pin	120–230
Piston skirt	93–204
Top cylinder wall	93–371
Bottom cylinder wall	Up to 149
Main bearings	Up to 177
Connecting rod bearings	93–204

Source: Reproduced from Bardasz, E.A. and Lamb, G.D., Additives for crankcase lubricant applications, in *Lubricant Additives Chemistry and Applications*, Rudnick, L.R., Ed., Marcel Dekker, Inc., New York, 2003. With permission.

prevented from agglomerating and kept in suspension due to the solubility of the dispersant. The literature reports that the most common group of dispersants are polyisobutenylsuccinimides. They are known to be effective in reducing the black sludge problem and the oil thickening caused by high soot levels in diesel engine oils (Bardasz and Lamb, 2003). Different engine tests are used to measure the oxidation resistance of engine oils and the engine wear at different temperatures.

TABLE 9.32
Engine Testing Requirements of Engine Oils

Engine Test Requirements	5W-30	10W-30	15W-40
ASTM D 5844 seq. IID, engine rusting	Pass	Pass	Pass
ASTM D 5533 seq. IIIE, oil thickening and wear	Pass	Pass	Pass
ASTM D 5302 seq. VE, sludge and wear	Pass	Pass	Pass
ASTM D 5119 L-38, bearing corrosion	Pass	Pass	Pass

Source: American Petroleum Institute, *Engine Oil Licensing and Certification System,* API 1509, 15th ed., American Petroleum Institute, Washington, DC, April 2002.

The engine testing requirements of engine oils are shown in Table 9.32. The Sequence VE engine test was reported to be used to measure the sludge, the varnish, and the wear at a relatively low temperature of 68.3°C. The literature reports that the Sequence VE test is used to measure the lubricant ability to prevent low temperature sliding wear in the valve train of a gasoline engine. In an internal combustion engine, a variety of many different factors can cause the oil deterioration, the sediment formation, and the engine wear. The patent literature reports on the RHC process and the effect of RHC base stocks having a lower volatility, on the oxidation resistance of engine oils.

The effect of the mineral and the RHC base stocks on the oxidation resistance of 5W-30 engine oils under Seq. VE test conditions is shown in Table 9.33. At a low temperature of 68.3°C, the oxidation of 5W-30 engine oil containing mineral base stocks resulted in the formation of sludge and varnish, and an average cam wear of 83.6 μm. Under the same engine test conditions, the oxidation of 5W-30 engine oil containing RHC base stocks and the same additive package also resulted in the formation of sludge and varnish and an average cam wear of only 18 μm. The patent literature reports that the use of RHC base stocks decreased the engine wear when tested at lower temperature of 68.3°C. The Sequence IIIE engine test was reported to be used to measure the oxidation resistance of engine oils, such as oil thickening, deposits, and wear, at a higher temperature of 149°C.

The effect of the mineral and the RHC base stocks on the oxidation resistance of 5W-30 engine oils under the Seq. IIIE test conditions is shown in Table 9.34. At a higher temperature of 149°C, the viscosity of mineral 5W-30 engine oil was reported to increase by 182% after 64 h and further increased to max 375% after 71.2 h. With an increase in the oxidative thickening of mineral 5W30 engine oil, the sludge, the varnish, and the deposit formed, and an average wear of 15.4 μm was reported. Under the same engine test conditions, the viscosity of 5W-30 engine oil containing RHC

TABLE 9.33
Effect of Base Stocks on Oxidation Resistance of 5W-30 Oils in Seq. VE Test

5W-30 Engine Oils Additive Package	Mineral Oil Same	RHC Oil Same	Performance Limits
Sequence VE engine test (68.3°C)			
Average engine sludge (merits)	9.14	9.49	Min 9.0
Rocker cover sludge (merits)	8.28	9.04	Min 7.0
Piston skirt varnish (merits)	7.02	6.90	Min 6.5
Average engine varnish (merits)	5.43	6.25	Min 5.0
Oil screen clogging (%)	3	0	Max 20
Hot stuck rings	None	None	None
Average cam wear (μm)	83.6	18	Max 130
Max cam wear (μm)	231	27	Max 380

Source: Cody, I.A. et al., Raffinate hydroconversion process, CA Patent 2,429,500, 2002.

TABLE 9.34
Effect of Base Stocks on Oxidation Resistance of 5W-30 Oils in Seq. IIIE Test

5W-30 Engine Oils Additive Package	Mineral Oil Same	RHC Oil Same	Performance Limits
Sequence IIIE engine test (149°C)			
% Viscosity increase at 64 h	182	63	Max 375
Hours to 375% viscosity increase	71.2	78.9	Min 64
Average engine sludge (merits)	9.57	9.51	Min 9.2
Average piston skirt varnish (merits)	9.31	9.17	Min 8.9
Oil ring land deposits (merits)	3.02	3.96	Min 3.5
Stuck lifters	None	None	None
Scuffed/worn cam or lifters	None	None	None
Average cam and lifter wear (μm)	15.4	9	30
Max cam and lifter wear (μm)	74	20	64
Oil consumption (L)	3.85	2.55	Report

Source: Cody, I.A. et al., Raffinate hydroconversion process, CA Patent 2,429,500, 2002.

base stocks was reported to increase by only 63% after 64 h and further increased to max 375% after 78.9 h. With an increase in the oxidative thickening of RHC-based 5W-30 engine oil, the sludge, the varnish, and the deposit formed, and an average wear of 9 μm was reported. At a higher temperature the use of RHC base stocks in 5W-30 engine oils also resulted in a decrease in the engine wear. Additives such as

TABLE 9.35

Effect of Different Chemistry ZnDTP on Engine Wear

ZnDTP		Average (μm)	
% Zn	Alcohol Type	Seq. VD Wear	Seq. IIID Wear
0.13	C_6 secondary	18	
0.13	C_6/C_8 mixed	48	
0.13	C_8 primary		175
0.13	C_6/C_8 mixed		25

Source: Bardasz, E.A. and Lamb, G.D., Additives for crankcase lubricant applications, in *Lubricant Additives Chemistry and Applications*, Rudnick, L.R., Ed., Marcel Dekker, Inc., New York, 2003. With permission.

zinc dialkyldithiophosphates (ZDDP) were initially used to reduce bearing corrosion. The ZDDP type compounds have excellent antiwear and antioxidant properties and are commonly used at high temperatures to prevent the engine wear (Lee and Harris, 2003).

The effect of different chemistry ZnDTP additives on the engine wear under different engine test conditions is shown in Table 9.35. The literature reports that at a lower temperature and under the sequence VD engine test conditions, the ZDDP additive containing C_6 secondary alcohol was more effective in preventing the engine wear. At a higher temperature and under the sequence IIID engine test conditions, the ZDDP additive containing C_6 secondary alcohol was also reported more effective than ZnDTP molecule containing C_8 primary alcohol in preventing the engine wear (Bardasz and Lamb, 2003). Some additives were reported to have an effect on the sediment formation at a high temperature of 230°C (Konovich and Dracheva, 1989). The literature reports that on combustion, detergents containing metals, leave an ashy residue. These deposits can interfere with the oil circulation and build up behind piston rings to cause ring sticking and rapid ring wear. Once formed, such deposits are hard to remove except by mechanical cleaning (Pirro and Wessol, 2001). In some cases, this may be detrimental in that the ash can contribute to combustion chamber deposits. In other cases, it may be beneficial in that ash provides wear-resistant coatings for the surfaces such as valve faces and seats (Pirro and Wessol, 2001). Ashless additives were reported to be used in some engine oils.

REFERENCES

American Petroleum Institute, *Engine Oil Licensing and Certification System*, API 1509, 15th ed., American Petroleum Institute, Washington, DC, April 2002.

Bardasz, E.A. and Lamb, G.D., Additives for crankcase lubricant applications, in *Lubricant Additives Chemistry and Applications*, Rudnick L.R., Ed., Marcel Dekker, Inc., New York, 2003.

Bondi, A., *Physical Chemistry of Lubricating Oils*, Reinhold Publishing Corporation, New York, 1951.

Braun, J. and Omeis, J., *Additives, in Lubricants and Lubrications*, Mang, T. and Dresel, W., Eds., Wiley-VCH GmbH, Weinheim, 2001.

Brown, M. et al., Synthetic base fluids, in *Chemistry and Technology of Lubricants*, 2nd ed., Mortier, R.M. and Orszulik, S.T., Eds., Blackie Academic & Professional, London, U.K., 1997.

Cody, I.A. et al., Raffinate hydroconversion process, CA Patent 2,429,500, 2002.

Colyer, C.C. and Gergel, W.C., Detergents and dispersants, in *Chemistry and Technology of Lubricants*, 2nd ed., Mortier, R.M. and Orszulik, S.T., Eds., Blackie Academic & Professional, London, U.K., 1997.

Crawford, J., Psaila, A., and Orszulik, S.T., Miscellaneous additives and vegetable oils, in *Chemistry and Technology of Lubricants*, 2nd ed., Blackie Academic & Professional, London, U.K., 1997.

Crosthwait, H.K., May, C.J., and Deane, B.C., Method for optimizing the fuel economy of lubricant base stock, CA Patent 2,390,229, 2001.

Gao, J.Z., A partly synthetic multigrade crankase lubricant, CA Patent 2,263,562, 1999.

Gao, J. et al., Antioxidant additive composition for lubricating oils, CA Patent 2,397,885, 2001.

Henderson, H.E., Chemically modified mineral oils, in *Synthetics, Mineral Oils, and Bio-Based Lubricants Chemistry and Technology*, Rudnick, L.R., Ed., CRC Press, Boca Raton, FL, 2005a.

Henderson, H.E., Gas to liquids, in *Synthetics, Mineral Oils, and Bio-Based Lubricants Chemistry and Technology*, Rudnick, L.R., Ed., CRC Press, Boca Raton, FL, 2005b.

Instrumentation Scientifique de Laboratoire, Product News Letter 4602, 2002.

Koenitzer, B.A. and May, C.J., Reducing low temperature scanning Brookfield gel index value in engine oils, CA Patent 2,385,419, 2001.

Konovich, L.G. and Dracheva, S.I., *Russian Journal of Applied Chemistry*, (4), 22 (1989).

Kubo, K. and Lillywhite, J.R., *Toraiborojisuto*, 39(1), 57 (1994).

Lee, F.L. and Harris, J.W., Long-term trends in industrial lubricant additives, in *Lubricant Additives: Chemistry and Applications*, Rudnick, L.R., Ed., Marcel Dekker, Inc., New York, 2003.

MacAlpine, G.A., Halls, P.M.E., and Asselin, A.E., Wax isomerate having a reduced pour point, CA Patent 2054096, 1991.

Migdal, C.A., Antioxidants, in *Lubricant Additives Chemistry and Applications*, Rudnick, L.R., Ed., Marcel Dekker, Inc., New York, 2003.

Mills, A.J., Lindsay, C.M., and Atkinson, D.J., The formulation of automotive lubricants, in *Chemistry and Technology of Lubricants*, 2nd ed., Blackie Academic & Professional, London, U.K., 1997.

Phillipps, R.A., Highly refined mineral oils, in *Synthetic Lubricants and High-Performance Functional Fluids*, 2nd ed., Rudnick L.R. and Shubkin R.L., Eds., Marcel Dekker Inc., New York, 1999.

Pillon, L.Z., *Petroleum Science and Technology*, 20(1,2), 101 (2002).

Pillon, L.Z., *Interfacial Properties of Petroleum Products*, CRC Press, Boca Raton, FL, 2007.

Pillon, L.Z. and Asselin, A.E., Wax isomerate having a reduced pour point, U.S. Patent 5,229,029, 1993.

Pillon, L.Z. and Asselin, A.E., Antifoaming agents for lubricating oils, U.S. Patent 5,766,513, 1998.

Pillon, L.Z., Vernon, P.D.F., and Asselin, A.E., Method for reducing foaming of lubricating oils, U.S. Patent 6,090,758, 2000.

Pirro, D.M. and Wessol, A.A., *Lubrication Fundamentals*, 2nd ed., Marcel Dekker Inc., New York, 2001.

Prince, R.J., Base oils from petroleum, in *Chemistry and Technology of Lubricants*, 2nd ed., Mortier R.M. and Orszulik S.T., Eds., Blackie Academic & Professional, London, U.K., 1997.

Rasberger, M., Oxidative degradation and stabilisation of mineral oil based lubricants, in *Chemistry and Technology of Lubricants*, 2nd ed., Mortier, R.M. and Orszulik, S.T., Eds., Blackie Academic & Professional, London, U.K., 1997.

Ross, S., Foaminess and capillarity in apolar solutions, in *Interfacial Phenomena in Apolar Media*, Eicke, H.F. and Parfitt, G.D., Eds., Marcel Dekker Inc., New York, 1987.

Rudnick, L.R. and Shubkin, R.L., Poly(alfa-olefins), in *Synthetic Lubricants and High-Performance Functional Fluids*, 2nd ed., Marcel Dekker Inc., New York, 1999.

Stambaugh, R.L., Viscosity index improvers and thickeners, in *Chemistry and Technology of Lubricants*, 2nd ed., Mortier R.M. and Orszulik S.T., Eds., Blackie Academic & Professional, London, U.K., 1997.

Takakura, Y., *Idemitsu Giho*, 42(4), 375 (1999).

Wang, N. and Zhu, Y., *Shiyou Lianzhi Yu Huagong*, 31(3), 9 (2000).

10 Additive Interactions

10.1 RUST INHIBITORS SYNERGISM

The primary function of lubricants is to protect the metal surface and to reduce friction. The rusting of the metal surfaces results in the formation of rust particles, which might contribute to the wear and the sludge formation. The literature also reports that the rusting of the metal surfaces could cause rapid seal and packing wear resulting in the increased leakage and the system contamination (Pirro and Wessol, 2001). The typical acidic type rust inhibitors, such as partially esterified alkylsuc-cinic acid (EASA), do not degrade the demulsibility properties of solvent-refined base stocks but were found effective only in lower viscosity mineral oils. The basic type rust inhibitors, such as succinic anhydride amine (SAA), were found effective in low and high viscosity mineral base stocks but degrade their demulsibility proper-ties. The patent literature reports that a synergism between two different rust inhibi-tors allows obtaining the rust protection at a lower concentration of the mixture than can be obtained when using a single rust inhibitor (Pillon, Reid, and Asselin, 1995). The rust-preventive properties of lubricating oils are tested at 60°C following the ASTM D 665 method for rust-preventing characteristics of inhibited mineral oils and the evaluation pass/fail is usually used. The literature reports on the variation in the severity of rusting but the oil passes the test only if no rusting is observed, not even one rust spot (Pillon, 2007).

The performance of different rust inhibitors and the synergistic EASA/SAA blend, 1:1 weight ratio in different viscosity mineral base stocks is shown in Table 10.1. The use of hydroprocessing affects the composition and the solvency properties of base stocks leading to a decrease in the solubility of some polar addi-tives such as rust inhibitors. The use of a typical acidic rust inhibitor, such as par-tially EASA, is not effective in preventing the rust formation in SWI 6 base stock due to poor solubility. The use of basic type rust inhibitors, such as SAA, is effective in preventing the rust formation in SWI 6 base stock but some increase in the emulsion stability is also observed.

The effect of different rust inhibitors and their blends on the rust-preventive prop-erties of SWI 6 base stock is shown in Table 10.2. At the treat rate of 0.03 wt%, the partially EASA is not effective in preventing the rust formation in SWI 6 base stock due to poor solubility. At the treat rate of 0.03 wt%, the SAA is also not effec-tive because its treat rate is too low. However, the rust inhibitor EASA/SAA blends, containing the weight ratio of about 1:1, were found effective in preventing the rust formation at a low total treat rate of 0.03 wt%. The use of synthetic base stocks, such as polyalfaolefins (PAO) 6, having poor solvency properties also affects the solubility and the performance of polar additives such as rust inhibitors. The use of

TABLE 10.1
Performance of Different Rust Inhibitors and EASA/SAA Blend in Mineral Oils

Paraffinic Base Stocks	Inhibitor (wt%)				
	0.03	0.04	0.05	0.1	0.15
ASTM D 665 B test					
150N (EASA)	Fail	Pass			
150N (SAA)	Fail	Fail	Fail	Pass	
150N (EASA/SAA-1:1)	Pass				
600N (EASA)	Fail	Fail	Pass		
600N (SAA)	Fail	Fail	Fail	Pass	
600N (EASA/SAA-1:1)	Pass				
1400N (EASA)	Fail	Fail	Fail	Fail	Fail
1400N (SAA)	Fail	Fail	Fail	Fail	Pass
1400N (EASA/SAA-1:1)	Fail	Fail	Fail	Pass	

Source: Pillon, L.Z. et al., Lubricating oil for inhibiting rust formation, U.S. Patent 5,397,487, 1995.

TABLE 10.2
Effect of Different Rust Inhibitors and Their Blends on Rust Prevention in SWI 6 Oil

SWI 6 Base Stock EASA:SAA Ratio	EASA:SAA Content (Total in wt%)	ASTM D 665B Rust Test
100:0	0.03	Fail
95:5	0.03	Fail
90:10	0.03	Fail
80:20	0.03	Fail
70:30	0.03	Fail
60:40	0.03	Pass
50:50	0.03	Pass
40:60	0.03	Pass
30:70	0.03	Pass
20:80	0.03	Pass
10:90	0.03	Pass
0:100	0.03	Fail

Source: Pillon, L.Z. et al., Lubricating oil for inhibiting rust formation, U.S. Patent 5,397,487, 1995.

TABLE 10.3

Performance of Synergistic EASA/SAA Blend in Synthetic PAO 6 and White Oil

Saturated Base Stocks	Inhibitor (wt%)				
	0.03	0.04	0.05	0.1	0.15
ASTM D 665 B test					
SWI 6 (EASA)	Fail	Fail	Fail	Fail	Fail
SWI 6 (SAA)	Fail	Fail	Fail	Pass	
SWI 6 (EASA/SAA-1:1)	Pass				
PAO 6 (EASA)	Fail	Fail	Fail	Fail	Fail
PAO 6 (SAA)	Fail	Fail	Fail	Pass	
PAO 6 (EASA/SAA-1:1)	Pass				
White oil (EASA)	Fail	Fail	Fail	Pass	
White oil (SAA)	Fail	Fail	Fail	Pass	
White oil (EASA/SAA-1:1)	Fail	Pass			

Source: Pillon, L.Z. et al., Lubricating oil for inhibiting rust formation, U.S. Patent 5,397,487, 1995.

the partially EASA is also not effective in preventing the rust formation in PAO 6 base stock due to poor solubility. The use of basic type rust inhibitors, such as SAA, is effective in preventing the rust formation in PAO 6 base stock at a significantly higher treat rate.

The effect of different rust inhibitors and the synergistic EASA/SAA blend 1:1 weight ratio in synthetic PAO 6 base stock and white oil is shown in Table 10.3.

At a very high treat rate of 0.15 wt%, the partially EASA is not effective in preventing the rust formation in SWI 6 base stock due to poor solubility. At a high treat rate of 0.1 wt%, the SAA is effective while the synergistic EASA/SAA blend containing the weight ratio of about 1:1 was found to be effective in preventing the rust formation at a low total treat rate of 0.03 wt%. At a very high treat rate of 0.15 wt%, the partially EASA is also not effective in preventing the rust formation in synthetic PAO 6 base stock due to poor solubility. At a high treat rate of 0.1 wt%, the SAA is effective while the synergistic EASA/SAA blend, containing the weight ratio of about 1:1, was found effective in preventing the rust formation at a low total treat rate of 0.03 wt%. White oil contains practically 100% saturate content, which includes a high content of the naphthenic hydrocarbons. At a treat rate of 0.1 wt%, the partially EASA is effective in preventing the rust formation in white oil due to an increase in its solvency properties. At the same treat rate of 0.1 wt%, the SAA is also effective while the synergistic EASA/SAA blend, containing the weight ratio of about 1:1, was found effective in preventing the rust formation at a lower total treat rate of 0.04 wt%. The use of the synergistic EASA/SAA blend, about 1:1 weight ratio, was found effective in preventing the rust formation in low and high viscosity mineral oils and highly saturated base stocks, including white oil and synthetic PAO 6 fluid.

10.2 "THICKENING" EFFECT OF VI IMPROVERS

The circulating oils are required to have good oxidation stability, resistance to foaming, ability to quickly release entrained air, separate water, and have good rust-preventive properties. The circulating oils are blended in low and high viscosity grades, ranging from low viscosity ISO VG 22 oils to high viscosity ISO VG 680 oils or even higher. The low viscosity circulating oils are used as turbine oils. Some high viscosity circulating oils are used to lubricate and protect precision industrial machinery having the most exact lubrication. High viscosity circulating oils are used as paper machine oils. They need to have an excellent oxidation stability to provide a long life without the formation of the harmful sludge. The literature reports that the modern paper machines operate at higher speeds and higher dryer bearing temperatures and use less make-up oil than older machines. Oil change-outs are infrequent and the oil must have the oxidation stability to withstand years of service. Industrial lubricants use paraffinic base stocks to assure their good oxidation stability. Most refiners produce different viscosity grade paraffinic base stocks and a bright stock from which they blend their range of different viscosity lubricating oils.

The properties of different viscosity grade hydrofinished paraffinic base stocks and 150 bright stock are shown in Table 10.4. A low viscosity hydrofinished 100N paraffinic base stock, having a VI of 101 and a dewaxed pour point of $-12°C$, was reported to contain 0.07 wt% of S and has a flash point of 192°C. A higher viscosity hydrofinished 150N paraffinic base stock, having a VI of 96, was reported to contain 0.06 wt% of S. Another higher viscosity hydrofinished 600N paraffinic base stock, having a VI of 89, was reported to contain 0.11 wt% of S. A high viscosity hydrofinished 1400N paraffinic base stock, having also a VI of 89, was reported to contain 0.16 wt% of S. Very high viscosity hydrofinished 150 bright stock, having a VI of 95 and a dewaxed pour point of $-18°C$, was reported to contain 0.26 wt% of S and has a flash point of 302°C. With an increase in the viscosity of the hydrofinished paraffinic base stocks, including the bright stock, an increase in the S content and flash point is observed. Lower viscosity turbine oils and ISO VG 100 circulating oils

TABLE 10.4
Properties of Different Viscosity Paraffinic Base Stocks and 150 Bright Stock

	Viscosity Grade				
	100N	150N	600N	1400N	150 Bright Stock
KV at 40°C (cSt)	20.4	29.7	111.4	301.7	438.0
KV at 100°C (cSt)	4.1	5.1	11.6	22.0	29.5
VI	101	96	89	89	95
Pour point (°C)	−13	—	—	—	−18
Sulfur (wt%)	0.07	0.06	0.11	0.18	0.26

Sources: Pillon, L.Z. et al., Lubricating oil for inhibiting rust formation, U.S. Patent 5,397,487, 1995; Lubrizol, Wickliffe, OH, http://www.lubrizol.com (accessed 2001).

can be formulated by blending paraffinic base stocks with 1 vol% of the additives that include antioxidants, metal deactivators, pour point depressants, rust inhibitors, demulsifiers, and antifoaming agents.

When formulating higher viscosity circulating oils, VI improvers are added to increase their VI needed to resist large changes in viscosity with variations in temperature. The VI improvers are long chain, high molecular weight polymers that function by causing the relative viscosity of an oil to increase more at a high temperature than at a low temperature. The literature reports that at low temperatures, the polymer adopts a coiled position and its effect on the oil viscosity is minimized. At high temperatures, the polymer tends to straighten out and produces a thickening effect (Pirro and Wessol, 2001). The literature reports that the VI improvers stabilize foams and emulsions (Crawford, Psaila, and Orszulik, 1997).

The effects of the Si 12500 antifoaming agent and the VI improver on the foaming tendency and the air release time of paraffinic 150N base stock are shown in Table 10.5. The addition of 3 ppm of Si 12500 prevents the foaming tendency of 150N base stock at 24°C but an increase in the air release time from 2 to 6 min at 50°C, is observed. The addition of the VI improver increases the foaming tendency of 150N base stock at 24°C but no significant increase in the air release time is observed. The addition of the VI improver and 3 ppm of Si 12500 prevents the foaming tendency of paraffinic 150N base stock at 24°C but a drastic increase in the air release time to 9 min at 50°C is observed. This is the "thickening effect" of the VI improver, which leads to a higher foaming tendency and an increase in the air entrainment due to an increase in the viscosity of the oil at higher temperatures. Most refiners settle for three to four different viscosity grade paraffinic base stocks from which they blend their range of lubricating oils that require good oxidation stability (Prince, 1997). Some industrial oils that require good oxidation stability use liquid paraffins.

The properties of liquid paraffins are shown in Table 10.6. Highly refined liquid paraffins are produced from paraffinic or naphthenic feedstocks and are colorless, odorless, and tasteless viscous liquids. Liquid paraffin, having a boiling point

TABLE 10.5
Effect of Si 12500 and VI Improver on Foaming and Air Release Time of 150N Oil

Paraffinic 150N Base Stock	Neat 150N	Neat 150N	150N/VII	150N/VII
Si 12500 (ppm)	0	3	0	3
ASTM D 892 foam test				
Seq. I at 24°C (mL)	300/0	0/0	500/0	0/0
Seq. II at 93.5°C (mL)	20/0	10/0	30/0	10/0
Seq. III at 24°C (mL)	300/0	5/0	500/0	0/0
ASTM D 3427 (50°C)				
Air release time (min)	2	6	3	9

Source: Pillon, L.Z. and Asselin, A.E., Antifoaming agents for lubricating oils, U.S. Patent 5,766,513, 1998.

TABLE 10.6
Properties of Liquid Paraffins

Liquid Paraffins	Paraffin	Light Paraffin	Heavy Paraffin
Boiling point (°C)	>300		
KV at 40°C (cSt)	21.49	Max 30	Min 64
KV at 100°C (cSt)	4.42		
VI	117		
Sulfur (ppm)	<1		

Sources: Huang, W. et al., *Lubr. Sci.*, 18, 77, 2006; Panama Petrochem Ltd., Mumbai, India, http://www.panamapetro.com (accessed 2002).

TABLE 10.7
Effect of Temperature and VI Improver on Surface Activity of Si 12500

	Si 12500		
	Neat 150N	150N/VII Blend	Neat SWI 6
Spreading (*S*) at 24°C (mN/m)	8	8	7
Spreading (*S*) at 93.5°C (mN/m)	6	2	2

Source: Pillon, L.Z. and Asselin, A.E., Antifoaming agents for lubricating oils, U.S. Patent 5,766,513, 1998.

above 300°C, was reported to have a high VI of 117 and practically no S content (Huang et al., 2006). The viscosity of liquid paraffins can vary. The light liquid paraffin has a kinematic viscosity of a max 30 cSt at 37.8°C while the heavy liquid paraffin has a kinematic viscosity of a min 64 cSt at 37.8°C (Panama Petrochem, 2002). Some lubricating oils that require good oxidation stability use paraffinic base stocks, VI improvers, and liquid paraffins.

The effect of the temperature and the VI improver on the surface activity of Si 12500 antifoaming agent in mineral 150N and SWI 6 base stocks is shown in Table 10.7. With an increase in the temperature from 24°C to 93.5°C, the spreading coefficient (*S*) of Si 12500 in neat paraffinic 150N base stock decreases from 8 to 6 mN/m. With an increase in the temperature, the spreading coefficient (*S*) of Si 12500 in paraffinic 150N base stock, containing the VI improver, drastically decreases from 8 mN/m at 24°C to only 2 mN/m at 93.5°C. With an increase in the temperature, the spreading coefficient (*S*) of Si 12500 in neat SWI 6 base stock, containing over 99 wt% of the saturate content that comprises mostly of

TABLE 10.8

Foaming Tendency of High Viscosity Circulating Oils

Mineral Circulating Oils	ISO VG 320 Oil	ISO VG 400 Oil	ISO VG 460 Oil
Si 12500 (10 wt%), vol%	0.0005	0.0005	0.0005
ASTM D 892			
Foaming tendency at 24°C (mL)	<50	<50	<50
Foaming tendency at 93.5°C (mL)	300–400	500–600	700

iso-paraffins, also drastically decreases from 7 mN/m at 24°C to only 2 mN/m at 93.5°C.

To formulate very high viscosity circulating oils, having good oxidation stability, high viscosity paraffinic base stocks, VI improvers, and heavy liquid paraffins are used. Other additives include the synergistic blend of two antioxidants, metal deactivator, pour point depressant, rust inhibitor, demulsifier, and a silicone antifoaming agent. Silicones are the most surface-active antifoaming agents and are used in lubricating oils that need to prevent foaming at low and high temperatures. To prevent air entrainment, only 0.5 ppm of Si 12500, active ingredient is used in high viscosity circulating oils. The ASTM D 1401 demulsibility test recommends increasing the temperature to 82°C when testing oils more viscous than 90 cSt at 40°C. An increase in the temperature decreases the viscosity of the oil and helps to separate the emulsion.

The effect of the temperature on the foaming tendency of high viscosity mineral ISO VG 320, 400, and 460 circulating oils is shown in Table 10.8. The mineral ISO VG 320 circulating oil, containing a high viscosity paraffinic base stock and the VI improver, was found to have a low foaming tendency at 24°C but a high foaming tendency of 300–400 mL at 93.5°C. The higher viscosity mineral ISO VG 400 oil, containing a high viscosity paraffinic base stock and a higher content of VI improver, was found to have a low foaming tendency at 24°C but a higher foaming tendency of 500–600 mL at 93.5°C. The high viscosity ISO VG 460 oil, containing a high viscosity paraffinic base stock, a VI improver, and a heavy liquid paraffin, was also found to have a low foaming tendency at 24°C but a very high foaming tendency of 700 mL at 93.5°C. With an increase in the temperature, the surface activity of silicone antifoaming agents in terms of spreading on the oil surface will decrease and a high temperature foaming caused by the thickening effect of the VI improver is observed. The literature reports that the values for foaming tendency should be maintained below 50 mL and values for foaming tendency above 100 mL are considered unacceptable for many industrial lubricating oils. The industry standards require a max foaming tendency of 100 mL at 93.5°C (Butler and Henderson, 1990). While the use of non-silicone antifoaming agents such as polyacrylate is limited to low temperatures, the silicone overtreatment needs to be avoided to prevent severe air entrainment. To prevent a high temperature foaming, the lubricant needs to be reformulated.

10.3 FORMULATION OF INDUSTRIAL GEAR OILS

Most industrial operations use equipment driven by the different types of gears, such as rock drills, crushers, hoists, mobile equipment, grinders, mixers, and speed increasers or speed reducers. The gear oil performance requirements vary depending on the amount of sliding and rolling action and the load-carrying capacity (Lee and Harris, 2003). In the case of warm gears, their action is almost all sliding, and an oil containing friction modifier, such as acidless tallow, may be sufficient. Lightly loaded spur gears require an oil with only rust and oxidation inhibitors while heavily loaded hypoid gears require an oil with high levels of extreme-pressure (EP) additives (Lee and Harris, 2003). The EP type industrial gear oils were reported to use shear stable VI improvers, antioxidants, metal deactivators, soluble EP additives, rust inhibitors, pour point depressants, demulsifiers, and antifoaming agents (Lee and Harris, 2003). The literature reports that for large, slow-moving gears that cannot be enclosed in oil-tight casings, an adhesive, and a high viscosity material is required to stick to the gears. Such open gear lubricants are generally applied by spraying or brushing. Such lubricants generally contain high viscosity base stock, tackifier, corrosion inhibitor, oil soluble and solid EP additives. In cold climates, a synthetic base stock may be required to provide a high viscosity base fluid that can be sprayed at low temperatures (Lee and Harris, 2003). The literature reports on what one needs to know when selecting gear oils to maximize machinery reliability under normal and abnormal conditions (Potteiger, 2006).

The performance of the mineral and the synthetic industrial gear oils is shown in Table 10.9. The mineral industrial gear oils were reported to have a VI of 80–100 and the pour points varying from as low as $-40°C$, indicating the presence of naphthenic base stocks, to as high as $-10°C$, indicating the presence of solvent-refined paraffinic

TABLE 10.9
Performance of Mineral and Synthetic Industrial Gear Oils

Industrial Gear Oils	Mineral Oils	Synthetic Oils	Synthetic Oils
Base Stocks	Naphthenic/Paraffinic	PAO	Polyglycol
VI	80–100	130–160	150–270
Pour point (°C)	−40 to −10	−50 to −30	−56 to −23
Lubricity	Good	Good	Very good
Compatibility with elastomers	Good	Good	Insufficient to good
Flash point (°C)	<250	>200	150–300
Thermal stability	Moderate	Good	Good
Oxidation resistance	Moderate	Good	Good

Source: Reproduced from Lauer, D.A., Industrial gear lubricants, in *Synthetics, Mineral Oils, and Bio-Based Lubricants Chemistry and Technology*, Rudnick, L.R., Ed., CRC Press, Boca Raton, FL, 2005. With permission.

base stocks. The synthetic industrial gear oils, containing PAO fluids, were reported to have a high VI of 130–160 and very low pour points varying from as low as –50°C to –30°C. Another type of synthetic industrial gear oils, containing polyglycol fluids, was reported to have a very high VI of 150–270 and also very low pour points varying from as low as –56°C to –23°C. The industrial gear oils are formulated to provide rust protection to ferrous metal surfaces and are noncorrosive to steel, copper, brass, bronze, and other common bearing materials. The corrosion of lubricating oils is usually tested following the ASTM D 130 test by immersing the freshly polished strip for 3 h at 100°C (Bock, 2001).

The chemistry of the additives used in industrial gear oils and their performance testing is shown in Table 10.10. The literature reports on the use of a statistical analysis to study the effect of neutral base stocks and bright stocks on the interfacial properties of turbine oils, hydraulic oils, and industrial gear oils. The properties, such as demulsibility and weld load, were found to have a similar correlation but a different correlation between the composition of base stocks and the foaming tendency of lubricating oils was observed (Abou El Naga, Bendary, and Salem, 1990). The industrial gear oils need to have resistance to foaming and good demulsibility properties. The literature reports that the skill of the formulation of lubricating oils, having resistance to foaming and good demulsibility properties is a "black art." Finding an effective antifoaming agent and demulsifier is often done by "trial and error" (Crawford, Psaila, and Orszulik, 1997). The typical EP mineral industrial gear oils contain a VI improver, an EP additive package, a pour point depressant, a demulsifier, and a silicone antifoaming agent. The EP type mineral industrial gear oils, for use at very low temperatures, are formulated by the addition of low viscosity naphthenic oils, having very low natural pour points.

The effect of the demulsifier on the silicone performance in EP type mineral ISO VG 32 industrial gear oils containing low viscosity naphthenic oils is shown in Table 10.11. The EP type mineral ISO VG 32 industrial gear oil #1 containing a low viscosity naphthenic oil, the EP additive package, the pour point depressant, the antifoaming agent but without a demulsifier, was found to have a low foaming tendency at 24°C and 93.5°C but poor demulsibility properties. After the addition of the demulsifier, the EP type mineral ISO VG 32 industrial gear oil #2 was found to have a high foaming tendency of 200 mL at 24°C but required less time to separate the emulsion. Low viscosity mineral ISO VG 32 oils, containing naphthenic oils have a low ST and very good solvency properties that leads to a decrease in the surface activity of antifoaming agents. The presence of a highly surface-active demulsifier at the oil/water and at the oil/air interface can prevent antifoaming agents from adsorbing on the oil surface and foaming will be observed. The surface-active additive, while designed to adsorb at the specific interface, is usually surface-active toward other interfaces. In many cases, there is a need for an antifoaming agent that will operate effectively at low and high temperatures, without degrading the air separation properties of lubricating oils.

The EP type mineral ISO VG industrial gear oils were found to have good rust-preventive properties and passed the ASTM D 665B rust test. Following the ASTM D 130 corrosion test, the EP type mineral industrial gear oils were rated 1b showing a slight tarnish but no corrosion.

TABLE 10.10

Chemistry of Additives and the Performance Testing of Industrial Gear Oils

Additive Function	Additive Chemistry	Performance Testing
Shear stable VI improvers	Poly(alkylmethacrylates)	ASTM D 5621
	Olefin copolymers	
	Styrene diene copolymers	
Antioxidants	Diaryl amines	
	Hindered phenols	ASTM D 6186
	Phenothiazine	ASTM D 2893
	Metal dialkyldithiocarbamates	
	Ashless dialkyldithiocarbamates	
	Metal dialkyldithiophosphates	
Metal deactivators	Thiadiazoles	ASTM D 130
	Benzotriazoles	
	Tolutriazole derivative	
	2-Mercaptobenzothiazoles	
Rust inhibitors	Alkylsuccinic acid derivatives	ASTM D 665
	Metal sulfonates	
	Ethoxylated phenols	
Corrosion inhibitors	Fatty amines	ASTM D 130
	Imidazoline derivatives	
	Ammonium sulfonates	
	Salts of phosphate esters and amines	
Antiwear additives	Alkylphosphoric acid ester	ASTM D 4172
	Alkylphosphoric salts	
	Dialkyldithiophosphates	
	Metal dialkyldithiocarbamates	
	Dithiophosphate esters	
Soluble EP additives	Sulfurized esters	ASTM D 5182
	Sulfurized olefins	ASTM D 2782
	Dialkyldithiophosphate esters	ASTM D 2783
	Lead naphthenate	
Solid EP additives	Molybdenum disulfide	
	Graphite	
Friction modifiers	Acidless tallow	ASTM D 4857
	Long chain esters and amides	
Pour point depressant	Poly(alkyl methacrylates)	ASTM D 97
		ASTM D 5949
Demulsifiers	Polyalkoxylated phenols	ASTM D 1401
	Polyalkoxylated polyols	ASTM D 2711
	Polyalkoxylated polyamines	

TABLE 10.10 (continued)
Chemistry of Additives and the Performance Testing of Industrial Gear Oils

Additive Function	Additive Chemistry	Performance Testing
Foam inhibitors	Polydimethylsiloxanes	ASTM D 892
	Polyacrylates	
Tackifier	Polyisobutylenes	
Antimisting additive	Polyisobutylenes	ASTM D 3705

Source: Reproduced from Lee, F.L. and Harris, J.W., Long-term trends in industrial lubricant additives, in *Lubricant Additives Chemistry and Applications*, Rudnick, L.R., Ed., Marcel Dekker, Inc., New York, 2003. With permission.

TABLE 10.11
Effect of Demulsifier on Silicone Performance in EP Mineral ISO VG 32 Oils

	EP ISO VG 32 Industrial Gear Oils	
	Naphthenic Oil #1	Naphthenic Oil #2
VI improver	No	No
Demulsifier	No	Yes
Si 12500 (10 wt%)	Yes	Yes
ASTM D 892		
Foaming tendency at 24°C (mL)	<50	200
Foaming tendency at 93.5°C (mL)	<25	<25

The recent literature reports on the variation in the foaming tendency of mineral industrial gear oils. The low viscosity oils, such as ISO VG 10 to ISO VG 32, were reported to have the highest foaming tendency at cold start-up temperatures ranging from −10°C to 10°C. Conversely, the high viscosity oils such as ISO VG 460 to ISO VG 1,000 were reported to have the greatest foaming tendency at higher operating temperatures ranging from 30°C to 60°C (Fitch, 2007). Higher viscosity mineral industrial gear oils contain VI improvers and foam at high temperatures. The tendency of the lubricating oils to foam can be a serious problem in systems such as high volume pumping, splash lubrication, and high-speed gearing. The EP type mineral gear oils, designed to be used in outdoor gear drives, have a high VI to give them multigrade characteristics for a year-round applications. To prevent a high temperature foaming, different chemistry VI improvers, or different viscosity silicone antifoaming agents need to be selected. The food-grade gear lubricants require the use of food-grade base oils and food-grade approved additives.

The properties of the food-grade gear lubricants containing mineral white oil and synthetic fluids are shown in Table 10.12. The mineral food-grade ISO VG 320 gear

TABLE 10.12

Properties of Mineral and Synthetic Food-Grade Gear Lubricants

Food Grade Gear Lubricants	ISO VG 320 Gear Oil Mineral White Oil	ISO VG 320 Gear Oil Synthetic PAO	Gear Oil Synthetic PAG
KV at 40°C (cSt)	327	295	285
KV at 100°C (cSt)	24.8	28	40.7
VI	98	127	198
Pour point (°C)	−12	−35	−26
Flash point (°C)	220	230	221
FZG gear test (pass)	12	12	12

Source: Reproduced from Raab, M.J., Synthetic-based food-grade lubricants and greases, in *Synthetics, Mineral Oils, and Bio-Based Lubricants Chemistry and Technology*, Rudnick, L.R., Ed., CRC Press, Boca Raton, FL, 2005. With permission.

oil, containing white oil, was reported to have a VI of 98 and a pour point of −12°C. The literature reports that the synthetic fluids such as PAO and polyalkylglycols (PAG) are used in the food-grade gear lubricants that require good oxidation stability (Raab, 2005). The same viscosity synthetic food-grade ISO VG 320 gear oil, containing PAO fluid, was reported to have a higher VI of 127 and a very low pour point of −35°C. Another synthetic food-grade gear oil, containing PAG fluid, was reported to have a very high VI of 198 and a low pour point of −26°C. The biodegradable gear lubricants require the use of biodegradable oils and biodegradable approved additives. The vegetable oils from soy, corn, sunflower, and canola are used to formulate the outdoor biodegradable gear oils and they were reported to contain high oleic acid content (United Bio Lube, 2008).

The properties of different viscosity outdoor biodegradable gear oils containing vegetable oils are shown in Table 10.13. With an increase in the viscosity of the outdoor biodegradable gear oils from ISO VG 32 to ISO VG 460, their pour point increases from −25°C to −5°C. The outdoor biodegradable ISO VG 32, 100 and ISO VG 460 gear oils have a TAN of 1.1 mg KOH/g and no foaming tendency was reported. The oils required only 10 min to separate the emulsions and were found to have good rust-preventive properties. With an increase in their viscosity, their flash point was reported to increase from 230°C to 275°C. The ASTM D 2272 method is used to compare the oxidation stability of steam turbine oils by the rotating pressure vessel at 150°C. The EP type biodegradable gear oils were reported to have an RBOT life of 250 min. The variation in petroleum-derived base stocks from naphthenic to paraffinic oils, and the use of different synthetic fluids and vegetable oils affect the low temperature fluidity and oxidation stability of different gear oils. Additionally, the use of additives such as demulsifiers and VI improvers can further affect their foaming tendency and indicates the presence of additive interactions.

TABLE 10.13

Properties of Outdoor Biodegradable Gear Oils Containing Vegetable Oils

	EP Biodegradable Gear Oils		
	ISO VG 32	**ISO VG 100**	**ISO VG 460**
KV at 40°C (cSt)	31.6	97.9	442.5
KV at 100°C (cSt)	7.0	20.0	62.8
VI	193	229	216
Pour point (°C)	−25	−16	−5
TAN (mg KOH/g)	1.1	1.1	1.1
Foaming tendency	0	0	0
Demulsibility	40/40/0	40/40/0	40/40/0
Emulsion separation time	10	10	10
Rust prevention	Pass	Pass	Pass
Corrosion	1b	1b	1b
Four-ball wear (mm)	0.35	0.35	0.35
FZG test	11	11	11
Flash point (°C)	230	257	275
RBOT life (min)	250	250	250

Source: United Bio Lube, Mountain View, CA, http://www.biofoodgradegearoils .com (accessed 2008).

10.4 OXIDATION RESISTANCE AND SURFACE ACTIVITIES OF FOOD-GRADE LUBRICANTS

Almost all commercial lubricants contain additives to enhance their performance and the selection of the base stocks and the additives involves the consideration of the equipment to be lubricated. The food-compatible lubricants for the food processing machinery, such as can seamer equipment, conveyor belts, grinders, and mixers, require the use of the food-grade base oils and the food-grade additives. The patent literature reports that the modern machines, operating at higher temperatures with a higher speed, require the food-grade lubricating oils having a higher resistance to deterioration. The food-grade base stocks include the petroleum-derived white oils, the synthetic hydrocarbons, such as PAO, polyisobutenes, polyalkylene glycol, dimethylpolysiloxane, and the natural oils. The polyisobutylenes, polybutenes, and polyethylenes are used as VI improvers and can also be used as the thickeners. The synthetic-based lubricants and the synthetic-based greases were reported to be used as "problem solvers" in the food processing industry that require high temperature applications (Raab and Hamid, 2003).

The oxidative stability of a food-grade lubricating oil can be measured following the ASTM D 2272 test. The majority of lubricants require good demulsibility properties and use demulsifiers; however, depending on their applications, some lubricants are required to form emulsions and need to contain emulsifiers. The ASTM D 1401 test can be used to test the demulsibility or the emulsibility of lubricating oils when mixed with distilled water. Some bench testing procedures are modified to better reflect the actual operating conditions. The patent literature reports on the modified ASTM D 1401 test, which used 16:64 carbonated beverage:oil at 82°C to better reflect the performance requirements of the equipment to be lubricated (Dewalt, Butler, and Kent, 2001). The rust-preventive properties of lubricating oils are usually tested following the ASTM D 665 method. The antiwear properties of lubricants are usually tested following the ASTM D 4172 test. Some commercial food-grade lubricants, useful in slower machines operating at lower temperatures, were found not to perform properly when used in the newer high-speed machines at higher temperatures (Dewalt, Butler, and Kent, 2001).

The oxidation resistance and the surface activity of some commercial food-grade lubricants are shown in Table 10.14. The commercial food-grade lubricant #1 having a short RBOT life of 52 min, did not form a stable emulsion indicating the absence of emulsifiers. It was found to have good rust-preventive properties and produced a wear of 0.36 mm. The commercial food-grade lubricant #2 having an increased RBOT of 205 min, did not form a stable emulsion; it was found to have good rust-preventive properties and produced a wear of 0.34 mm. The commercial food-grade lubricant #3 having a higher RBOT life of 292 min, did not form a stable emulsion; it was found to have good rust-preventive properties but produced an increased wear of 0.47 mm. Another commercial food-grade lubricant #4 having a high RBOT life of 495 min, also did not form a stable emulsion but it was found to have poor rust-preventive properties and produced a wear of 0.42 mm. Highly polar rust inhibitors are not effective in preventing the rust formation in nonpolar synthetic base stocks due to their poor solubility (Pillon, 2007). The specifications set by the can seamer equipment

TABLE 10.14

Oxidation Resistance and Surface Activity of Commercial Food-Grade Lubricants

	Food-Grade Lubricants			
	Oil #1	Oil #2	Oil #3	Oil #4
KV at 40°C (cSt)	146	150	93	171
VI	97	97	122	98
ASTM D 2272 (150°C)				
RBOT life (min)	52	205	292	495
Modified ASTM D 1401 (82°C)				
Emulsion stability	None	None	None	None
ASTM D 665B rust test (60°C)	Pass	Pass	Pass	Fail
ASTM D 4172 wear test (75°C)				
Four-ball wear (mm)	0.36	0.34	0.47	0.42

Source: Dewalt, R.D. et al., Non-sludging, high temperature resistant food compatible lubricant for food processing machinery, CA Patent 2,351,447, 2001.

TABLE 10.15

Oxidation Resistance and Surface Activity of Different
Food-Grade Lubricants

	Food-Grade Lubricants			
	Oil #5	Oil #6	Oil #7	Oil #8
KV at 40°C (cSt)	150	97	150	233
VI	97	106	97	93
ASTM D 2272 (150°C)				
RBOT life (min)	182	173	48	80
Modified ASTM D 1401 (82°C)				
Emulsion stability	Fluid	Thick	Thick	Stable
ASTM D 665B rust (60°C)	Pass	Pass	Pass	Fail
ASTM D 4172 wear (75°C)				
Four-ball wear (mm)	0.32	0.41	0.4	0.7

Source: Dewalt, R.D. et al., Non-sludging, high temperature resistant food compatible lubricant for food processing machinery, CA Patent 2,351,447,
2001.

manufacturer required that the oils have a high oxidation resistance and form fluid
emulsions at 82°C as a way of removing aqueous contaminants from the lubrication
system by the flow of the lubricating oil (Dewalt, Butler, and Kent, 2001).

The oxidation resistance and the surface activity of different commercial food-
grade lubricants are shown in Table 10.15. The commercial food-grade lubricant #5
having an RBOT life of 182 min, was reported to form a fluid emulsion indicating
some presence of an emulsifier. It was found to have good rust-preventive properties
and produced a wear of 0.32 mm. The commercial food-grade lubricant #6 having
a similar RBOT life of 173 min, was found to form a thick emulsion indicating a
higher content of emulsifiers. It was found to have good rust-preventive properties
and produced a wear of 0.41 mm. Another commercial food-grade lubricant #7 having a short RBOT life of 48 min, was also found to form a thick emulsion indicating
a higher content of emulsifiers. It was found to have good rust-preventive properties
and produced a wear of 0.4 mm. The commercial food-grade lubricant #8 having an
RBOT life of 80 min, was found to form a stable emulsion. It failed the rust test and
produced a high wear of 0.7 mm. Rust inhibitors protect the metal by adsorbing on
the metal surface at low temperatures. Antiwear agents protect the metal surface at
high temperatures. The patent literature reports on the technology of formulating
food-grade lubricating oils meeting the specific performance requirements used by
the can seamer equipment manufacturer. The food-grade lubricant can be formulated by blending white oil with polyisobutylene, used as a thickener, and the food-
grade additives, including the sorbitan mono-oleate emulsifier and the polyglycerol
oleate coupling agent needed to form a fluid emulsion. Coupling agents have a similar chemistry to emulsifiers but they also form associations with contaminants, such
as fats (Dewalt, Butler, and Kent, 2001).

TABLE 10.16
Effect of Emulsifier and Coupling Agent on Oxidation Resistance of Lubricants

	Food-Grade Lubricating Oils		
	Formulation #1	Formulation #2	Formulation #3
White oil (vol%)	90.3	90.2	86.3
Thickener (vol%)	9	9	9
Hindered phenol antioxidant (%)	0.5	0.5	0.5
Antiwear additive (%)	0.2	0.2	0.2
Emulsifier (%)	0	0.02	2
Coupling agent (%)	0	0.1	2
Diluted silicone (vol%)	0.002	0.002	0.002
KV at 40°C (cSt)	150	150	150
VI	97	97	97
ASTM D 2272 (150°C)			
RBOT life (min)	205	182	48
Modified ASTM D 1401 (82°C)			
Emulsion stability	None	Fluid	Thick
ASTM D 665B rust (60°C)	Pass	Pass	Pass
ASTM D 4172 wear (75°C)			
Four-ball wear (mm)	0.34	0.4	0.4

Source: Dewalt, R.D. et al., Non-sludging, high temperature resistant food compatible lubricant for food processing machinery, CA Patent 2,351,447, 2001.

The effect of the emulsifier and the coupling agent on the oxidation resistance and the surface activity of food-grade lubricating oils are shown in Table 10.16. The food-grade formulation #1 containing white oil, the polyisobutylene thickener, 0.5 wt% of hindered phenol antioxidant, antiwear additive, and the silicone anti-foaming agent, was found to have an RBOT life of 205 min. It did not form a fluid emulsion indicating a need for an emulsifier but was reported to have good rust-preventive properties and produced a wear of 0.34 mm. The food-grade formulation #2 containing an additionally 0.02 wt% of the emulsifier and 0.1 wt% of a coupling agent, was found to have a shorter RBOT life of 182 min. It formed a fluid emulsion and was reported to have good rust-preventive properties but produced an increased wear of 0.4 mm. The food-grade formulation #3 containing a high content of emul-sifier and a high content of 2 wt% of the coupling agent, was found to have a short RBOT life of only 48 min. It formed a thick emulsion but it was reported to have good rust-preventive properties and produced a wear of 0.4 mm. After the addi-tion of the emulsifier and the coupling agent, the oxidation resistance of food-grade lubricating oils decreases.

The effect of different antioxidants and the metal deactivator on the RBOT life of the food-grade lubricating oils containing emulsifier and coupling agent is shown

TABLE 10.17
Effect of Antioxidants and Metal Deactivator on RBOT
Life of Lubricants

	Food-Grade Lubricating Oils		
	Formulation #2	Formulation #4	Formulation #5
White oil (vol%)	90.2	90.2	90.1
Thickener (vol%)	9	9	9
Hindered phenol (%)	0.5	0	0.5
S-Hindered phenol (%)	0	0.5	0
Tolutriazole derivative (%)	0	0	0.08
Antiwear additive (%)	0.2	0.2	0.2
Emulsifier (%)	0.02	0.02	0.02
Coupling agent (%)	0.1	0.1	0.1
Diluted silicone (%)	0.002	0.002	0.002
KV at 40°C (cSt)	150	150	150
VI	97	97	97
ASTM D 2272 (150°C)			
RBOT life (min)	182	195	263

Source: Dewalt, R.D. et al., Non-sludging, high temperature resistant food compatible lubricant for food processing machinery, CA Patent 2,351,447, 2001.

in Table 10.17. The antioxidants are usually used at the treat rate of 0.5 wt% in lubricant products. The food-grade formulation #2 containing 0.5 wt% of the standard hindered phenolic antioxidant, an emulsifier, and a coupling agent was found to have an RBOT life of 182 min. The use of 0.5 wt% of the sulfur-containing hindered phenolic antioxidant was found to increase the RBOT life of food-grade formulation #4 to only 195 min. The use of 0.5 wt% of hindered phenolic antioxidant and the addition of 0.08 wt% of the metal deactivator, *N,N*-dioctylaminomethyl 1,2,4-benzotriazole, was found to further increase the RBOT life of food-grade formulation #3 to 263 min. The food-grade lubricants require the use of the food-grade base stocks and the food-grade additives and their performance will vary depending on their composition. While hundreds of additives are available, in many cases, the use of one additive might improve one aspect of performance while degrading another aspect of the lubricant performance. White oils have excellent water separation properties and require the use of emulsifiers and coupling agents to form fluid emulsions. The addition of an emulsifier and a coupling agent decreased the oxidation resistance and increased wear on the metal surface. The list of additives available from various suppliers is extensive and the use of bench tests can be used to screen them in a relatively short time (Raab and Hamid, 2003). The use of different crude oils affects the paraffinic and the naphthenic content of white oils, which might lead to a variation in their solvency properties and affect the solubility and the performance of some additives.

10.5 ADDITIVE ANTAGONISM AND INCOMPATIBILITY

The additive interactions can take place with or without chemical reactions. The lubricant solvency and the poor solubility of an additive can cause a poor storage stability that might lead to a sediment formation. The compatibility issue is related to the chemical reactions between different additives and can also lead to a poor stability of lubricating oils and cause filter plugging. The literature reports that the additive interactions lead to a poor stability of some engine oils and a sediment formation. The interaction among additives of the diesel engine oils such as detergent, dispersant, oxidation inhibitor, and corrosion inhibitor was reported (Jiang, Guo, and Dong, 1998). The interactions between dispersant, detergent, antiwear additive, oxidation inhibitor, and the VI improver, studied using FTIR and NMR, were reported to be affected by the blending order of additives (Kapur et al., 1999). The literature reports on the use of the ultracentrifuging and the FTIR to study the reversibility of the adsorption of the poly(isobutene)succinimide dispersant onto a calcium alkylarylsulfonate particle. The dispersant adsorption onto the calcium detergent particle was shown to occur through the interaction between polar succinimide/amine groups and the sulfonate molecule (Papke, 1992). The additive compatibility and the filterability issue of lubricant products are affected by the chemistry of selected additives. The additives are used in lubricants to impart special performance features to the finished oil. Some additives affect the physical or the chemical properties of the finished oil or are added for the cosmetic purpose (Shugarman, 2001).

The classification of additives by their physical/chemical function in the finished lubricants is shown in Table 10.18. The literature reports that when selecting an additive such as an antioxidant, its performance and other factors such as volatility, color, odor, physical form, toxicity, solubility, and the compatibility with other additives must be considered (Migdal, 2003). The chemistry of the rust inhibitors needs to be carefully selected to prevent degradation in the demulsibility properties of lubricating oils.

TABLE 10.18
Classification of Additives by Physical/
Chemical Function

Additive Physical Function	Additive Chemical Function
Viscosity modifier	Rust inhibitor
Thickener	Oxidation inhibitor
Pour point depressant	Corrosion inhibitor
Antifoaming agent	Metal deactivator
Demulsifier	Metal passivator
Emulsifier	Detergent
Tackiness agent	Antiwear agent
Dispersant	EP additive
	Friction modifier

Source: Shugarman, A.L., *Machinery Lubrication Magazine,* September 2001.

The spindle oils are low viscosity lubricants, which need to have excellent oxidation stability, good rust- and corrosion-preventive properties, and a low fluid friction. They are recommended for bearing lubrication of high-speed textile spindles, machine tool spindles, timing gears, centrifugal separators, positive displacement blowers, and hydraulic systems of certain high-precision tools. Spindle oils can be formulated to meet the viscosity requirements of nearly all spindle applications found in the textile industry and contain antioxidants, antiwear additives, rust inhibitors, and antifoaming agents (Apar Industries, 2003). The addition of acidic type rust inhibitors increased the emulsion stability of antiwear type spindle oil. The use of basic type rust inhibitors was effective in preventing rust formation and a decrease in the emulsion stability was observed. The literature reports that the lubricants that require good demulsibility should not be mixed with the lubricants containing dispersants or detergents. A small amount of oil with good emulsion characteristics will destroy the water-shedding properties of a highly demulsible lubricant (Shugarman, 2001).

The classification of some lubricants by their demulsibility or their emulsibility characteristics is shown in Table 10.19. The mixing of different lubricants may lead to degradation in the oil performance. In mineral turbine oils, the use of the acidic EASA rust inhibitor was found to have no effect on the emulsion stability while the use of the basic SAA rust inhibitor caused a drastic increase in the emulsion stability. In antiwear mineral spindle oils containing different chemistry additives, the use of the acidic EASA rust inhibitor was found to increase the emulsion stability while the use of the basic SAA rust inhibitor decreased the emulsion stability. The lubricants used in the marine engines must protect the engine parts from rust and corrosion. Rust is produced when ferrous engine parts are exposed to water, which could be produced during combustion or come from outside in the form of fresh water or saline water. The patent literature reports that calcium sulfurized alkyl phenolate has been effective as the oxidation and the corrosion inhibitors (Russo et al., 1996). Rust and corrosion inhibitors are added to lubricant products in the amount of 0.5–2.5 wt% and should be totally miscible with lubricating oils and also compatible with other additives, if not, a haze or a solution turbidity is observed (Russo et al., 1996).

TABLE 10.19
Classification of Some Lubricants by Demulsibility/Emulsibility Characteristics

Lubricant Demulsibility Characteristics	Lubricant Emulsibility Characteristics
Turbine oil	Arbor oil
Circulating oil	Bar and chain oil
Hydraulic oil	Concrete form oil
Industrial gear oil	Cutting oil
Paper machine oil	Way oil

Source: Shugarman, A.L., *Machinery Lubrication Magazine*, September 2001.

TABLE 10.20

Effect of Rust Inhibitors on the Turbidity and the Rust-Preventive Properties of Lubricants

	Lubricant Composition			
	Oil #1	Oil #2	Oil #3	Oil #4
Paraffinic oil #1 (vol%)	39.3	39.3	39.3	39.3
Paraffinic oil #2 (vol%)	55.2	55.2	55.2	55.8
Additive package (wt%)	3.6	3.6	3.6	3.6
Additive A (wt%)	0.65	0.65	0.65	0.65
Additive B (wt%)	0.3	0.3	0.3	0.3
Additive C (ppm)	150	150	150	150
Additive D (wt%)	—	—	—	0.35
Hazetron turbidity	9	5	7	55
ASTM D 665B rust test	Fail	Pass	Pass	Pass

Source: Russo, J.M. et al., Polymeric thioheterocyclic rust and corrosion inhibiting marine diesel engine lubricant additives, U.S. Patent 5,516,442, 1996.

The effect of the rust inhibitors on the turbidity and the rust-preventive properties of the marine lubricant oils containing different additive packages are shown in Table 10.20. The marine lubricant oil #1 containing paraffinic base stocks, additive package, and the rust and corrosion inhibitors was reported to have a hazetron turbidity of 9 and failed the rust test. The marine lubricant oils #2–3 containing paraffinic base stocks, different additive packages, and the rust and corrosion inhibitors were reported to have a hazetron turbidity of 5–7 and passed the rust test. Another marine lubricant oil #4 containing paraffinic base stocks, additive package, rust and corrosion inhibitors, and an additional additive D, was reported to have a high hazetron turbidity of 55 and passed the rust test. Many different chemistry additives are used to protect the metal surface. At low temperatures, the rust inhibitors are used to protect the metal surface from rusting. The lubricating oils containing the acidic rust inhibitors are incompatible with the oils containing the basic rust inhibitors. Under high temperatures and oxidation conditions, corrosion inhibitors, metal deactivators, and metal passivators are used to prevent the corrosion of the metal surface. The antiwear additives are also used in some lubricant products to protect the metal surface at higher temperatures.

The classification of some lubricants by their acid/base chemistry of additives is shown in Table 10.21. Mixing lubricating oils having the acidic and the basic additive systems can lead to sediment formation. When the oils are mixed in the presence of water, grease-like insolubles are formed that can clog filters (Shugarman, 2001). The literature reports that adding turbine oil to antiwear hydraulic oil in a hydraulic pump could lead to a disaster. Deposits may form that could increase the wear and plug filters (Shugarman, 2001). Lubricants formulated with antioxidant

TABLE 10.21

Classification of Some Lubricants by Acid/Base Chemistry of Additives

Acidic Chemistry Type Additives	Basic Chemistry Type Additives
Turbine oil	Zinc-containing antiwear hydraulic oil
Circulating oil	Hydraulic/tractor fluid
Non-zinc antiwear hydraulic oil	Zinc-containing paper machine oil
Industrial gear oil	ATF
Non-zinc paper machine oil	Passenger car engine oil

Source: Shugarman, A.L., *Machinery Lubrication Magazine*, September 2001.

and non-zinc antiwear additives, such as railroad engine oils; and ashless or low-ash gas engine oils will cause engine damage if they are contaminated with additives containing zinc (Shugarman, 2001). Lubricants are formulated to be neutral to seals or cause them to swell slightly. Some incompatible lubricant mixtures may affect rubber seals. The EP gear oils are known to deteriorate silicone seals. Engine oils formulated with certain types of dispersants were reported to attack the fluorocarbon seals (Shugarman, 2001). The mixing of mineral engine oil with synthetic oil will not damage the engine but the oil performance will not be the same. The additive antagonism can lead to a decrease in the lubricant performance while the additive incompatibility can cause filter plugging.

REFERENCES

Abou El Naga, H.H., Bendary, S.A., and Salem, A.E.M., *Lubrication Science*, 3(1), 3 (1990).

Apar Industries, http://www.aparind.com (accessed 2003).

Bock, W., Hydraulic oils, in *Lubricants and Lubrications*, Mang, T. and Dresel, W., Eds., Wiley-VCH GmbH, Weinheim, 2001.

Butler, K.D. and Henderson, H.E., Adsorbent processing to reduce base stock foaming, U.S. Patent 4,600,502, 1990.

Crawford, J., Psaila, A., and Orszulik, S.T., Miscellaneous additives and vegetable oils, in *Chemistry and Technology of Lubricants*, 2nd ed., Mortier, R.M. and Orszulik, S.T., Eds., Blackie Academic & Professional, London, U.K., 1997.

Dewalt, R.D., Butler, K.D., and Kent, C.J.S., Non-sludging, high temperature resistant food compatible lubricant for food processing machinery, CA Patent 2,351,447, 2001.

Fitch, J., *Machinery Lubrication Magazine*, September 2007, http://www.machinerylubrication.com (accessed 2008).

Huang, W. et al., *Lubrication Science*, 18, 77 (2006).

Jiang, M., Guo, X., and Dong J., *Science in China* (in Chinese), 13(6), 47 (1998).

Kapur, G.S. et al., *Tribology Transactions*, 42(4), 807 (1999).

Lauer, D.A., Industrial gear lubricants, in *Synthetics, Mineral Oils, and Bio-Based Lubricants Chemistry and Technology*, Rudnick, L.R., Ed., CRC Press, Boca Raton, FL, 2005.

Lee, F.L. and Harris, J.W., Long-term trends in industrial lubricant additives, in *Lubricant Additives Chemistry and Applications*, Rudnick, L.R., Ed., Marcel Dekker, Inc., New York, 2003.

Lubrizol, http://www.lubrizol.com (accessed 2001).

Migdal, C.A., Antioxidants, in *Lubricant Additives Chemistry and Applications*, Rudnick, L.R., Ed., Marcel Dekker, Inc., New York, 2003.

Panama Petrochem, http://www.panamapetro.com (accessed 2002).

Papke, B.L., *ACS Symposium Series, Mixed Surfactant System*, 501, 377 (1992).

Pillon, L.Z., *Interfacial Properties of Petroleum Products*, CRC Press, Boca Raton, FL, 2007.

Pillon, L.Z. and Asselin, A.E., Antifoaming agents for lubricating oils, U.S. Patent 5,766,513, 1998.

Pillon, L.Z., Reid, L.E., and Asselin, A.E., Lubricating oil for inhibiting rust formation, U.S. Patent 5,397,487, 1995.

Pirro, D.M. and Wessol, A.A., *Lubrication Fundamentals*, 2nd ed., Marcel Dekker Inc., New York, 2001.

Potteiger, J., *Machinery Lubrication Magazine*, September 2006, http://www.machinerylubrication.com (accessed 2008).

Prince, R.J., Base oils from petroleum, in *Chemistry and Technology of Lubricants*, 2nd ed., Mortier, R.M. and Orszulik, S.T., Eds., Blackie Academic & Professional, London, U.K., 1997.

Raab, M.J., Synthetic-based food-grade lubricants and greases, in *Synthetics, Mineral Oils, and Bio-Based Lubricants Chemistry and Technology*, Rudnick, L.R., Ed., CRC Press, Boca Raton, FL, 2005

Raab, M.J. and Hamid, S., Additives for food-grade lubricant applications, in *Lubricant Additives Chemistry and Applications*, Rudnick, L.R., Ed., Marcel Dekker, Inc., New York, 2003.

Russo, J.M. et al., Polymeric thioheterocyclic rust and corrosion inhibiting marine diesel engine lubricant additives, U.S. Patent 5,516,442, 1996.

Shugarman, A.L., *Machinery Lubrication Magazine*, September 2001, http://www.machinery-lubrication.com (accessed 2008).

United Bio Lube, http://www.biofoodgradegearoils.com (accessed 2008).

11 Scope and Limits of Lubricant Testing

11.1 FRESH OIL TESTING OF INDUSTRIAL OILS

The literature reports that to shorten the time and lower the cost of testing, the laboratory bench tests are to be used during the development of new lubricant products. Some of these bench tests have been used for years because they have shown the ability to predict how lubricating oils will perform in actual service (Pirro and Wessol, 2001). A review of the analyses and testing of lubricating oils was published by Denis, Briant, and Hipeaux (2000). The literature also reports that there is a growing demand to improve the performance of industrial oils such as turbine, hydraulic, transmission, and compressor oils (Dawson and Kerkemeyer, 1999). The thermal breakdown or the oxidation of turbine and hydraulic oils can result in the formation of insoluble solids and acids and render the oil unfit for further use. Turbine oils can be used in hydraulic systems requiring lubricants having outstanding properties (Apar Industries, 2003). Turbine oils are required to have good oxidation stability and good surface-active properties, such as resistance to foaming, good air and water separation properties, and good rust-inhibiting and corrosion-preventing properties. The use of antioxidants cannot prevent oil oxidation, and their use can only extend the lubricant life by reducing the degradation caused by oxygen and heat. Severe agitation can lead to foaming, and the literature reports that the formation of foam could accelerate the oxidation of the oil, leading to poor lubrication (Pirro and Wessol, 2001). In many instances, such as in the gears of a steam turbine, water can become mixed with the lubricating oil and rusting of ferrous metal parts can occur. Rust and corrosion are the two phenomena that increase the wear and might lead to the breakdown of the equipment.

The list of standard test methods used for testing fresh turbine oils is shown in Table 11.1. The American Society of Testing Materials (ASTM) D 2272 (RBOT) oxidation test is used to test the oxidation stability of turbine oils. Many different oxidation tests were developed and are used to test the oxidation stability of other industrial oils such as hydraulic oils. There is one foam test, the ASTM D 892 method, which measures the effect of the temperature on the foaming tendency of lubricating oils. The ASTM D 1401 method for water separability of petroleum oils and synthetic fluids is often used. Another test, the IP 19 test, is used to determine a demulsibility index for new uninhibited oils. The rust-preventive properties of lubricating oils are usually tested following the ASTM D 665 method. Other methods for testing the rust-preventive properties of lubricants include the ASTM D 1748 (humidity cabinet) and the ASTM D 3603 (horizontal disc) methods (Denis, Briant, and Hipeaux, 2000). The ASTM D 130 method is used to measure the copper corrosion. According to the

TABLE 11.1

Standard Test Methods for Testing Fresh Turbine Oils

Properties	Test Method
Kinematic viscosity at 40°C (cSt)	ASTM D 445
Kinematic viscosity at 100°C (cSt)	ASTM D 445
VI	ASTM D 2270
Pour point (°C)	ASTM D 97
TAN (mg KOH/g)	ASTM D 664
Flash point (°C)	ASTM D 92
Oxidation stability	ASTM D 2272 (RBOT)
Foaming characteristics	ASTM D 892
Air release time	ASTM D 3427
Demulsibility	ASTM D 1401
Rust protection	ASTM D 665B
Copper corrosion	ASTM D 130

Source: Reprinted from El-Ashry, El-S.H. et al., *Lubr. Sci.*, 18(2), 109, 2006. With permission.

ASTM D 130 corrosion test, the strip showing a slight tarnish is rated 1a,b; moderate tarnish is rated 2a–e; dark tarnish is rated 3a,b; and corrosion is rated 4a–c.

The early literature reports that the foaming of lubricating oils is affected by the operating conditions, such as temperature and pressure oil circulation rate, and by the presence of some additives, such as VI improvers and antiwear additives (Kichkin, 1966). Different viscosity mineral base stocks and synthetic fluids are used to formulate industrial oils. Synthetic fluids such as polyalkylglycols (PAG) are used in the food-grade gear lubricants that require good oxidation stability (Raab, 2005). The surface activity of antifoaming agents needs to be measured to prevent the foaming of lubricants. The present laboratory methods for fluid surface tension (ST) measurements are Wilhelmy plate, du Nouy ring, sessile drop, pendant drop, capillary rise, drop weight, and the maximum bubble pressure methods. According to the literature, all these methods should produce the same results; however, difficulties in the mathematical treatments of the phenomena can lead to discrepancies (SensaDyne Instrument Division, 2000).

The effect of the temperature on the ST of mineral oil, poly(ethylene glycol), and polydimethylsiloxane (PDMS) having a specific molecular weight (MW) is shown in Table 11.2. With an increase in the temperature from 25°C to 100°C, the ST of mineral oils decreases from 31–32 to 25–26 mN/m. The poly(ethylene glycol), having an MW of 1000, was reported to have a very high ST of 45 mN/m at 25°C, which gradually decreases to 38 mN/m at 100°C. With an increase in the temperature, the ST of PDMS having an MW of 32,000 was also reported to gradually decrease from 22 to 17 mN/m. While engine oils represent the largest application of VI improvers, they are also used in automatic transmission fluids, power steering fluids, shock absorber

TABLE 11.2
Effect of Temperature on STs of Mineral Oil, PEG, and PDMS

Temperature (°C)	Mineral Oil	Poly(Ethylene Glycol) (MW 1,000)	PDMS (MW 32,000)
		ST (mN/m)	
25	31–32	45	22
50	—	42	19
100	25–26	38	17

Sources: Bondi, A., *Physical Chemistry of Lubricating Oils*, Reinhold Publishing Corporation, New York, 1951. With permission; Dee G.T. and Sauer B.B., *TRIP*, 5(7), July 1997. With permission.

TABLE 11.3
Extrapolated STs of Different Polymers

Polymers	MW	Extrapolated ST (mN/m)
Polypropylene	3000	28.3
Poly(propylene oxide)	4000	30.4
PIB	2700	33.6
Poly(methyl methacrylate)	3000	41.1

Source: Reprinted from Owen, M.J., *Ind. Eng. Chem. Prod. Res. Dev.*, 19, 97, 1980. With permission.

fluids, turbine engine oils, some circulating oils, hydraulic oils, and industrial gear oils. The literature reports that the ST of polymers is very difficult to measure and it is extrapolated to 20°C from the studies of their melts (Owen, 1980). The extrapolated STs of polymers vary depending on their chemistry and their MW.

The STs of different polymers, extrapolated to 20°C from the studies of their melts, is shown in Table 11.3. The extrapolated ST of polyisobutylene (PIB) having an MW of 2700 was reported to be 33.6 mN/m while the extrapolated ST of poly(methyl methacrylate) having a similar MW of 3000 was reported to be 41.1 mN/m. The use of PIB, which has a lower ST, might lead to a decrease in the surface activity of antifoaming agents and foaming might be observed. There are limits to the amount of silicone used in circulating, hydraulic, and industrial gear oils that are required to have the ability to separate entrained air. The literature reports that the PIB and the polyalkymethacrylates (PAMA) chemistry VI improvers are used in gear oils (Braun and Omeis, 2001). The literature also reports that for non-engine oil applications, the chemistry of VI improvers is optimized for the specific base stock used. The use of naphthenic and synthetic base stocks require changes in the polymer solubility and the elimination of the high alkyl monomer when the wax interaction is not an issue

TABLE 11.4

Effect of High Temperatures on STs of Different
***n*-Paraffins**

Temperature (°C)	*n*-Paraffins		
	n-$C_{26}H_{54}$	*n*-$C_{32}H_{66}$	*n*-$C_{60}H_{122}$
	ST (mN/m)		
75	26.1	27.0	26.5
100	24.1	25.2	23.1
150	20.0	21.6	22

Source: Bondi, A., *Physical Chemistry of Lubricating Oils*,
Reinhold Publishing Corporation, New York, 1951.
With permission.

(Stambaugh, 1997). The early literature reports on the effect of the temperature on
the ST of *n*-paraffins having a different MW (Bondi, 1951).

The effect of the temperature on the STs of *n*-paraffins, having a different MW,
is shown in Table 11.4. With an increase in the temperatures from 75°C to 100°C,
the ST of *n*-C_{26}, *n*-C_{32} and *n*-C_{60} paraffins decreased from 26–27 to 23–25 mN/m, a
decrease of 2–3 mN/m. With a further increase in the temperatures to 150°C, their
STs further decreased to 20–22 mN/m, a decrease of an additional 1–4 mN/m. To
assure the surface activity of antifoaming agents and their efficiency to prevent
foaming, the effect of the liquid paraffins on the STs of lubricating oils needs to
be measured. The literature reports that beyond the chain length of C_{14}, hydrocarbons become waxes or solids. The paraffin wax was reported to have a critical ST
of 23 mN/m measured at 20°C (Owen, 1980). Circulating high viscosity minerals
use liquid paraffins and VI improvers, hence a severe high temperature foaming is
observed. To prevent air entrainment, antifoaming agents need to be used at their
minimum treat rates. In some cases, the lubricant products might need to be reformulated to prevent foaming without an increase in the effective treat rate of the antifoaming agents. The literature reports on the demulsibility of commercial industrial
lube oils for film lubrication bearings (FLB) in rolling mills and the variation in their
oxidation stability (Grigoreva et al., 2003).

The demulsibility of fresh minerals ISO VG 46 and ISO VG 220 industrial oils for
FLB in rolling mills and their oxidation stabilities are shown in Table 11.5. With an
increase in the viscosity of fresh commercial rolling mill FLB oils from ISO VG 46
to 220, their color darkens, which might be related to a darker color of higher viscosity mineral base stocks. The emulsion separation time of low viscosity mineral oils is
tested at 54°C and the fresh commercial ISO VG 46 was reported to require 20 min
to separate the emulsions. The emulsion separation time of high viscosity mineral
oils was tested at 82°C and the fresh commercial ISO VG 220 oils were reported
to require 20 min to separate the emulsions. However, after oxidation at 120°C for
75 h, their total acid number (TAN) was reported to increase from 0.1 mg KOH/g,

TABLE 11.5

Demulsibility and Oxidation Stabilities of Mineral ISO VG 220 Industrial FLB Oils

	Mineral Industrial Rolling Mill FLB Oils		
	Commercial ISO VG 46	Commercial ISO VG 220	New ISO VG 220
KV at 40°C (cSt)	41.5–50.5	198–242	220
VI	100	95	95
Color (CST unit)	3	4.5	4.5
Demulsibility			
O/W/E (mL)	40/38/2	40/38/2	42/38/0
Separation time (min)	20	20	20
BMP friction wear (mm)	0.37	0.37	0.35
Oxidation at 120°C for 75 h			
Viscosity increase (%)	5	5	3
TAN increase (mg KOH/g)	0.1	0.15	0.3

Source: Grigoreva, N.I. et al., *Chem. Technol. Fuels Oils*, 39(3), 117, 2003.

for lower viscosity ISO VG 46 oil, to 0.15–0.3 mg KOH/g, for higher viscosity ISO VG 220 oils. While new ISO VG 220 oil has improved the demulsibility properties, it produces a higher TAN when oxidized.

The demulsibility of fresh mineral ISO VG 320 and ISO VG 460 industrial oils for FLB in rolling mills and their oxidation stabilities are shown in Table 11.6. With an increase in the viscosity of fresh commercial rolling mill FLB oils from ISO VG 320 to 460, their color darkens, which might also be related to a darker color of high viscosity mineral base stocks. Higher viscosity mineral oils require more time to separate the emulsions, and the fresh commercial ISO VG 320 oil was reported to require 25 min while ISO VG 460 oils were reported to require a longer time of 30–40 min. However, after oxidation at 120°C for 75 h, their TAN was reported to increase from 0.2 mg KOH/g (for lower viscosity ISO VG 320 oils) to 0.25–0.5 mg KOH/g (for higher viscosity ISO VG 460 oils). While the new ISO VG 460 oil has improved properties, it produces a higher TAN when oxidized. The literature reports on the rolling oil emulsions that contained fats, fatty acids, and fatty esters and showed poor performance in actual applications. Their emulsion characteristics were found to be affected by the chemistry of the fatty acids (Joseph et al., 2000). The literature reports that selecting lubricants based on meeting a set of laboratory bench tests will not always guarantee meeting the requirements of an application. Testing under actual service conditions is the best way to evaluate the performance of a new product before it is placed on the market (Pirro and Wessol, 2001).

The effects of the operation time on the TAN and the solid content of used oil from a reciprocating compressor are shown in Table 11.7. The compressor oil was reported to contain hydrotreated base stock and two thermally stable phenolic and sterically hindered phosphate antioxidants and was used in the continuous operation in excess of 8000 h (Rasberger, 1997). After the operation time of only 34 h, the

TABLE 11.6

Demulsibility and Oxidation Stability of Mineral ISO VG 460 Industrial FLB Oils

	Mineral Industrial Rolling Mill FLB Oils		
	Commercial ISO VG 320	Commercial ISO VG 460	New ISO VG 460
KV 40°C (cSt)	288–352	400–520	460
VI	95	95	95
Color (CST unit)	5	6	6
Demulsibility			
O/W/E (mL)	40/38/2	40/38/2	42/38/0
Separation time (min)	25	30	40
BMP friction wear (mm)	0.37	0.37	0.34
Oxidation at 120°C for 75 h			
Viscosity increase (%)	5	5	3
TAN increase (mg KOH/g)	0.2	0.25	0.5

Source: Grigoreva, N.I. et al., *Chem. Technol. Fuels Oils*, 39(3), 117, 2003.

TABLE 11.7

Effect of Operation Time on TAN and Solids of Used Compressor Oil

	Compressor Operation Time	
	34 h	8600 h
Viscosity at 40°C (cSt)	72	73
TAN (mg KOH/g)	0.7	0.8
Solids (%)	0.01	0.04

Source: Rasberger, M., Oxidative degradation and stabilisation of mineral oil–based lubricants, in *Chemistry and Technology of Lubricants*, 2nd ed., Mortier, R.M. and Orszulik, S.T., Eds., Blackie Academic & Professional, London, U.K., 1997. With permission.

compressor oil was reported to have a kinematic viscosity of 72 cSt at 40°C, a TAN of 0.7 mg KOH/g, and a solid content of 0.01%. After the operation time of 8600 h, the compressor oil was reported to have a kinematic viscosity of 73 cSt at 40°C, a TAN of 0.8 mg KOH/g, and a solid content of 0.04%. While the hydroprocessed base stocks and the antioxidants are used to prevent the oxidation of lubricating oils, their use cannot totally prevent the formation of the acid by-products and the sludge. The literature reports that although the formulated lubricants may have many desirable properties when they are new, oxidation can lead to a dramatic loss of performance in service leading to viscosity change, sludge, corrosion, and a loss of electrical resistivity (Shanks et al., 2006). The degradation of the lubricants by the oxidative process will vary depending on the severity of the application and the operating conditions.

11.2 PROPERTIES OF USED TURBINE OILS

The correlation between the lab bench test results and the field performance is a major issue when testing lubricating oils. Most laboratory test procedures were developed for evaluating new (fresh) oils and it is difficult to duplicate operating conditions and contamination for equipment that operates for years (Pirro and Wessol, 2001). The aging of the aviation turbine oils in the presence of air, simulated by their oxidation at 24°C and 94°C, was reported to increase their foaming tendency and their air release time (Antipina, Chesnokov, and Radchenko, 1989). The literature reports that the oil viscosity and the concentration of the oil-degraded compounds affect the demulsibility properties of industrial lubricating oils (Watanabe, Fukushima, and Nose, 1976). The early literature reports that under the conditions simulating those in a steam turbine, some rust and corrosion inhibitors were found to decompose forming corrosive products (Losikov, Khalif, and Aleksandrova, 1948). The typical contaminants found in the circulating oil system were reported to include air, water, rust particles, pipe scale, wear particles, dust, and dirt. The literature reports that the rust inhibitors are adsorbed on the metal surface and the oil can be depleted of the rust inhibitor in time (Pirro and Wessol, 2001).

The different contaminants found in the circulating oil system are shown in Table 11.8. The turbine oils deteriorate during their use and a significant degradation in their color was reported. The ASTM D 4378 Standard Practise Test for In-Service Monitoring of Mineral Turbine oils for Steam and Gas Turbines is used to detect the presence of the contaminants and the oxidation by-products. The additional testing of used turbine oils includes the FTIR analysis and the IFT measurement. The literature reports on the use of x-ray and FTIR to study the effect of the oxidation by-products on the wear of stainless steel (Chandrasekaran, Batchelor, and Loh, 2000). The ASTM D 971 test is frequently applied to service-aged lubricating oils to measure their oil/water interfacial tension (IFT) as an indication of the degree of deterioration. A high IFT value of fresh mineral lubricating oil indicates the absence of most undesirable polar contaminants and oxidation by-products.

The ASTM D 4378 standard analysis of used turbine oils is shown in Table 11.9. The literature reports that the use of different base stocks and different chemistry antioxidants can lead to a variation in the oxidation resistance of turbine oils in terms of preventing an increase in the TAN and the sludge formation. While the ASTM D 2272 oxidation test (RBOT life) is used to test the oxidation stability of fresh and used turbine oils, the ASTM D 943 oxidation test (TOST life) is used when the oxidized oil reaches a TAN of 2 mg KOH/g. A shorter TOST life indicates an increase in the acidic oxidation by-products and leads to more sludging but it

TABLE 11.8

Different Contaminants Found in Circulating Oil System

Typical Contaminants	Other Contaminants
Air	Assembly lubes
Water	Cleaning solvents
Rust particles	Drawing compounds
Pipe scale	Gasket sealants
Wear particles	Grease from pump bearings
Dust	Lint from cleaning rags
Dirt	Metal chips
	Paint flakes
	Weld spatter
	Wrong oil

Source: Reproduced from Pirro, D.M. and Wessol, A.A., *Lubrication Fundamentals*, 2nd ed., Marcel Dekker Inc., New York, 2001. With permission.

TABLE 11.9

ASTM D 4378 Standard Analysis of Used Turbine Oils

ASTM D 4378 Test	Additional Tests
Appearance	Ultracentrifuge
Viscosity	FTIR
Water content	Filter patch
TAN	IFT
Particle count	
Trace metals	
Foaming	
Demulsibility	
Rust protection	
Oxidation stability (RBOT life)	

Source: Stein, W.H. and Bowden, R.W., *Mach. Lubr.*, July 2004, http://www.machinerylubrication. com (accessed 2008).

might also affect the IFT of turbine oils. The recent literature reports on the foaming of the oxidized turbine oils when their TAN increased from 0.3 to 0.4 mg KOH/g (Duncanson, 2003).

The properties and the composition of fresh and used steam mineral turbine oils are shown in Table 11.10. While no significant increase in the viscosity of used

TABLE 11.10

Composition and Properties of Fresh and Used Steam Mineral Turbine Oils

	Steam Turbine Oils	
	Fresh Mineral Oil	Used Mineral Oil
KV at 40°C (cSt)	48.4	48.9
KV at 100°C (cSt)	6.9	6.9
VI	96	96
Sulfur (wt%)	0.11	0.12
Nitrogen (ppm)	35	49
ASTM D 2272 (150°C)		
RBOT life (min)	479	154
ASTM D 1401 (54°C)		
Oil/water/emulsion (mL)	41/40/0	3/24/53
Emulsion separation time (min)	10	30
ASTM D 665B (60°C)	Pass	Fail

Source: Butler, K.D., Process to produce lube oil base stock by low severity hydrotreating of used industrial circulating oils, CA Patent 2,170,138, 1996.

turbine oil was observed, following the ASTM D 2272 test method for the oxidation stability of steam turbine oils, the RBOT life of used mineral steam turbine oil drastically decreased from 479 to 154 min. Following the ASTM D 1401 test at 54°C, the fresh turbine oil required only 10 min to separate the emulsion while the used steam turbine oil formed a stable emulsion that would not separate after 30 min on standing. Following the ASTM D 665B rust test, the fresh turbine oil was reported to have good rust-preventive properties and passed the test, while the used steam turbine oil was found to have poor rust-preventive properties and failed the test. The recent literature reports on the reclamation of used lubricants, which refers to cleaning and reconditioning by the removal of water and solid particles. Some segregated industrial oils require more reprocessing, such as filtration, the removal of water and vacuum distillation, and can be returned to their original service (Stein and Bowden, 2004). The patent literature reports that the used steam mineral turbine oils, containing low levels of additives, can be re-refined by mild hydrotreating (Butler, 1996).

The contamination limits for used lubricating oils suitable for mild hydrotreatment are shown in Table 11.11. Before the hydrotreatment process, used oils are filtered to remove the sludge that can plug or foul the catalyst. To protect the catalyst, the used turbine oil needs to have a maximum of 2 vol% of sediment, a maximum TAN of 5, and a maximum of 200 ppm of chlorine. The limit on the chlorine content is to avoid the production of hydrogen chloride causing corrosion (Butler, 1996). To protect the catalyst, the used turbine oil also needs to meet the limits for specific metals such as a maximum of 250 ppm of iron and a maximum of 75 ppm of copper.

TABLE 11.11

Contamination Limits for Hydrotreatment of Used Lubricating Oils

Used Lubricating Oils	Maximum Limit	Preferred Limit
Sediment (vol%)	2	1
TAN (mg KOH/g)	5	2.5
Chlorine (ppm)	200	40
Metals (ppm)	150	70
Iron (ppm)	250	100
Copper (ppm)	75	40

Source: Butler, K.D., Process to produce lube oil base stock by low severity hydrotreating of used industrial circulating oils, CA Patent 2,170,138, 1996.

The refortification refers to adding additives to used lubricant that was cleaned. It was reported that in most cases, reclamation and refortification are used together (Stein and Bowden, 2004). The hydrotreated used turbine oils can be readditized to produce lower quality product or blended with base stocks or other re-refined products (Butler, 1996). Different hydroprocessing conditions can affect the removal of contaminants and the quality of the re-refined oils.

The effect of the hydrotreating temperatures and the readditization on the properties of re-refined steam mineral turbine oils is shown in Table 11.12. The use of a mild hydrotreatment at 250°C and readditization produced re-refined steam mineral turbine oil having an increased RBOT life of 393 min, a decreased emulsion separation time of 10 min and good rust-preventive properties. With an increase in the

TABLE 11.12

Effect of Hydrotreating and Readditization on Properties of Mineral Turbine Oils

Readditized Steam Mineral Turbine Oils	Hydrotreatment At 250°C	Hydrotreatment At 275°C	Hydrotreatment At 300°C
ASTM D 2272 (150°C)			
RBOT life (min)	393	407	276
ASTM D 1401 (54°C)			
Oil/water/emulsion (mL)	41/40/0	41/40/0	41/40/0
Separation time (min)	10	5	10
ASTM D 665B (60°C)	Pass	Pass	Pass

Source: Butler, K.D., Process to produce lube oil base stock by low severity hydrotreating of used industrial circulating oils, CA Patent 2,170,138, 1996.

hydrotreating temperature from 250°C to 275°C, the RBOT life was found to further increase to 407 min, the emulsion separation time was found to further decrease to 5 min, and good rust-preventive properties were observed. However, with a further increase in the hydrotreating temperature to 300°C, the RBOT life was reported to drastically decrease to 276 min, the emulsion separation time was reported to increase to 10 min, while good rust-preventive properties were observed.

Many industrial oils, such as turbine, transformer, and hydraulic oils can be readily collected and recycled. The recycling involves using a spent lubricant in a different, less critical application, or re-refining. The use of a mild hydrotreatment was reported to be the most effective in improving the properties of re-refined mineral turbine oils. Additives are added to re-refined oils to produce lubricants. Some lubricants are "lost" to the environment, such as two-stroke oils, greases, and oils used on railways, chainsaw bar lubricants, rubber oils used in tires, and many white oils.

11.3 FRESH OIL TESTING OF ENGINE OILS

Automotive lubricant performance trends are toward lower viscosity base stocks, having a higher VI, and a lower volatility to meet the growing demand for high quality engine oils having improved low temperature performance, fuel economy, and oxidation stability. The literature reports that up to 1992, there were three organizations that set specifications and classified oils for passenger cars, light trucks, and commercial vehicles. These organizations were listed as the American Petroleum Institute (API, 2002) responsible for administering the licensing and certification of engine oils, the Society of Automotive Engineers (SAE) that actually defines the need for new oil specifications in conjunction with the ASTM which sets the performance parameters and targets for each specification (AMSOIL, 2008). Many of the ASTM testing procedures are based on the same technical principles and measure the same properties.

The list of ASTM methods used to test engine oils is shown in Table 11.13. In 1992, the American Automobile Manufacturers Association (AAMA) and the Japanese Automobile Standards Organization (JASO) formed the International Lubricant Standardization and Approval Committee (ILSAC) that sets minimum performance standards for gasoline powered passenger car and noncommercial light trucks. The ILSAC and the tripartite system (SAE, ASTM, and API) were reported to come together to form the Engine Oil Licensing and Certification System (EOLCS), which is administered by the API (AMSOIL, 2008). The additional testing procedures are listed by ILSAC, USA government, some car manufacturers and other associations, such as Coordinating European Council (CEC).

The list of many other methods used to test engine oils is shown in Table 11.14. The literature reports that automotive gasoline engine lubricants require an improved oxidation stability to meet the increased service interval requirements (Phillipps, 1999). The poor oxidation stability of engine oils can lead to the formation of acids that are corrosive and insoluble sludge which in turn can lead to the formation of a hard layer in the grooves of the piston rings and affect the engine's performance (Crawford, Psaila, and Orszulik, 1997). Many petroleum companies were reported to have a favored list of tests used for the lubricant development that were internally

TABLE 11.13

ASTM Methods Used for Testing Engine Oils

Test	Procedure
ASTM D 92	Flash and fire points by Cleveland open cup
ASTM D 93	Flash point by Pensky-Martens closed cup tester
ASTM D 445	Kinematic viscosity of transparent and opaque liquids
ASTM D 892	Foaming characteristics of lubricating oils
ASTM D 1552	Sulfur in petroleum products
ASTM D 2007	Characteristic groups by the clay–gel absorption chromatography
ASTM D 2270	Calculating viscosity index from kinematic viscosity at 40°C and 100°C
ASTM D 2622	Sulfur in petroleum products
ASTM D 2887	Boiling range distribution of petroleum fractions
ASTM D 3120	Trace quantities of sulfur in light liquid petroleum hydrocarbons
ASTM D 3244	Utilization of test data to determine conformance with specifications
ASTM D 4057	Manual sampling of petroleum and petroleum products
ASTM D 4294	Sulfur in petroleum and petroleum products
ASTM D 4485	Performance of engine oils
ASTM D 4628	Analysis of barium, calcium, magnesium, and zinc in unused oils
ASTM D 4629	Trace nitrogen in liquid petroleum hydrocarbons
ASTM D 4683	Measuring viscosity at high shear rate and high temperature
ASTM D 4684	Determination of yield stress and apparent viscosity at low temperature
ASTM D 4741	Measuring viscosity at high temperature and high shear rate
ASTM D 4927	Analysis of lubricant and additive components, Ba, Ca, P, S, and Zn
ASTM D 4951	Determination of additive elements in lubricating oils
ASTM D 5119	Evaluation of automotive engine oils in CRC L-38 engine
ASTM D 5133	Low temperature, low shear rate, viscosity/temperature dependance
ASTM D 5185	Determination of additive elements, wear metals and contaminants
ASTM D 5293	Apparent viscosity of engine oils between −5°C and −30°C (CCS)
ASTM D 5302	Evaluation of engine oils for inhibition of deposit formation and wear
ASTM D 5480	Motor oil volatility by gas chromatography
ASTM D 5481	Measuring apparent viscosity at high-temperature and high-shear rate
ASTM D 5533	Evaluation of automotive engine oils in the seq. IIIE test
ASTM D 5800	Evaporation loss of lubricating oils by the NOACK method
ASTM D 5844	Evaluation of engine oils for inhibition of rusting (seq. IID)
ASTM D 5862	Evaluation of engine oils in 6V92TA diesel engine
ASTM D 6082	High temperature foaming characteristics of lubricating oils
ASTM D 6202	Fuel economy of passenger cars and light-duty trucks (seq. VIA)
ASTM D 6335	Determination of high temperature deposits by thermo-oxidation
ASTM D 6417	Estimation of engine oil volatility by capillary GC
ASTM D 6557	Evaluation of rust preventative characteristics of engine oils
ASTM D 6593	Evaluation of engine oils for inhibition of deposit formation

Source: American Petroleum Institute, *Engine Oil Licensing and Certification System*, API 1509, 15th ed., American Petroleum Institute, Washington, DC, 2002.

TABLE 11.14

Other Methods Used for Testing Engine Oils

Test	Procedure
ILSAC GF-1	Minimum performance standards for passenger car engine oils
ILSAC GF-2	Minimum performance standards for passenger car engine oils
ILSAC GF-3	Minimum performance standards for passenger car engine oils
DOD CID A-A-52039A	Lubricating oil, automotive engine, API service SG
DOD MIL-L-2104	Lubricating oil, internal combustion engine, technical service
GSA	Federal test 791B, method 3470 homogeneity and miscibility
SAE J183	Engine oil performance and engine oil classification
SAE J300	Engine oil viscosity classification
SAE J357	Physical and chemical properties of engine oils
SAE J1423	Energy-conserving engine oils
GM 9099P	Engine oil filterability test (EOFT)
GM 9099P	EOFT (modified)
CEC L-36-A-90	High temperature/high shear viscosity
CEC L-40-A-93	Evaporation loss of lubricating oils
JPI 5S-41–93	Evaporation loss

Source: American Petroleum Institute, *Engine Oil Licensing and Certification System*, API 1509, 15th ed., American Petroleum Institute, Washington, DC, 2002.

developed (Raab and Hamid, 2003). The literature also reports on the selected testing procedures used by the petroleum additive suppliers (Lubrizol, 2001). The commercial additive packages used in engine oils contain antioxidants, antiwear additives, dispersants, detergents, pour point depressants, and antifoaming agents. Some engine oil applications require additional additives such as rust inhibitors, corrosion inhibitors, seal swelling additive, biocide, demulsifiers, and friction modifiers (Lee and Harris, 2003). The literature reports on many different performance upgrades needed for the gasoline engines (Henderson, 2005).

The performance upgrades needed for gasoline engines are shown in Table 11.15. Despite the use of the petroleum-derived base stocks with a lower volatility and with the continuous improvement in the chemistry of the additive packages, there is a problem with the oxidation resistance of engine oils and the engine wear. Many applications involving the use of lubricating oils are carried out at increased temperatures and the literature reports that it is difficult to differentiate between their thermal degradation and their volatility. Under high temperature conditions, a portion of oil can evaporate, which might affect the oil viscosity. The evaporation loss may also contribute to oil consumption in an engine (Rudnick and Shubkin, 1999). According to the literature, the one test that is used to avoid misinterpreting the volatility for the thermal instability is the panel coker thermal stability test. In this test, an aluminum panel heated to 310°C is splashed by the test oil for 6 min and baked for 1.5 min. At the end of the test, the panels are rated for cleanliness and a completely clean panel has a rating of 10 (Rudnick and Shubkin, 1999).

TABLE 11.15

Performance Upgrades Needed for Gasoline Engines

Year 2001	Year 2002	Year 2005
Fuel economy	Fuel economy retention	Emission compatibility (low P)
Oil volatility	Oxidation control	Fuel economy retention
Foaming	Volatility	Oxidation control
Catalyst compatibility	Foaming	High temperature wear control
High temperature deposits	Ball rust	Low temperature pumpability
	Oil seal compatibility	

Source: Henderson, H.E., Chemically modified mineral oils, in *Synthetics, Mineral Oils, and Bio-Based Lubricants Chemistry and Technology*, Rudnick, L.R., Ed., CRC Press, Boca Raton, FL, 2005. With permission.

TABLE 11.16

Performance of Mineral and Synthetic Oils in Panel Coker Thermal Stability Test

Base Fluids	KV at 100°C (cSt)	Cleanliness
Polyol ester	4	9.5
PAO	4	8
Mineral oil	4	0
Alkylated aromatic	5	2
Dibasic ester	5.4	8

Source: Reproduced from Rudnick, L.R. and Shubkin, R.L., Poly(alfa-olefins), in *Synthetic Lubricants and High-Performance Functional Fluids*, 2nd ed., Marcel Dekker Inc., New York, 1999. With permission.

The performance of mineral base stock and different synthetic fluids in the panel coker thermal stability test is shown in Table 11.16. The mineral oil having a kinematic viscosity of 4 cSt at 100°C, was rated 0 indicating the presence of many molecules having poor thermal stability and leading to deposit formation. The same viscosity synthetic polyol ester was reported to be the most clean, having a high rating of 9.5, followed by the synthetic polyalfaolefin (PAO) having a lower rating of 8. Higher viscosity alkylated aromatic fluid having a kinematic viscosity of 5 cSt at 100°C, was rated 2 while dibasic ester having a kinematic viscosity of 5.4 cSt at 100°C, was rated 8. The literature reports that the oxidation of lubricating oils in service is slow, and, to shorten the testing time, many different bench-testing procedures are used. The tests involving thin oil films were reported to produce different results than the tests using bulk oils (Rudnick and Shubkin, 1999). The literature reports on the properties and the composition of the mono-grade SAE 40 engine oil

TABLE 11.17
Properties and Composition of Mono-grade SAE 40 Engine Oil

Oil Properties	Base Stock	SAE 40 Engine Oil
KV at 100°C (cSt)	13.5	14.8
KV at 40°C (cSt)	146.6	154
VI	84	97
TAN (mg KOH/g)	0.23	0.39
Paraffins (wt%)		39
Naphthenes (wt%)		28
Aromatics (wt%)		38
Zinc (wt%)		0.11
Calcium (wt%)		0.03
Magnesium (wt%)		0.07
TBN (mg KOH/g)		5.3

Source: Egharevba, F. and Maduako, A.U.C., *Ind. Eng. Chem. Res.*, 41, 3473, 2002.

containing the poly(methyl acrylate) VI improver, zinc dithiophosphates, a sulfonate type dispersant, and a detergent (Egharevba and Maduako, 2002).

The properties and the composition of mono-grade SAE 40 engine oil are shown in Table 11.17. After the addition of the additives, the VI of base stock increased from 84 to 97, and the TAN increased from 0.23 to 0.39 mg KOH/g. The mono-grade SAE 40 engine oil was reported to contain 0.11 wt% of Zn, 0.03 wt% of Ca, 0.07 wt% of Mg, and have a TBN of 5.3 mg KOH/g. The detergents are described in terms of the total base number (TBN) and all detergents are overbased to some extent (Pirro and Wessol, 2001). The calcium-based detergent was reported to be preferred for the specialized diesel engine application and the magnesium-based detergent was preferred for the rust control in gasoline car engines. The sodium carbonate detergent was sometimes used, in addition to calcium and magnesium, to provide additional benefits in rust and oxidation control (Marsh, 1987). The literature reports that the barium carbonate additives were extensively used until the 1960s and since then their use has declined (Papke, 1988). The literature also reports on the properties and the composition of the multigrade 20W-50 engine oil containing the poly(methyl acrylate) VI improver, zinc dithiophosphates, sulfonate type dispersant, and a detergent (Egharevba and Maduako, 2002).

The properties and the composition of multigrade SAE 20W-50 engine oil are shown in Table 11.18. After the addition of the additives, the VI of base stock increased from 97 to 130, and the TAN increased from 0.22 to 0.44 mg KOH/g. The multigrade SAE 20W-50 engine oil was reported to contain 0.13 wt% of Zn, 0.04 wt% of Ca, 0.09 wt% of Mg, and has a TBN of 6.3 mg KOH/g. During the operation of combustion engines, lubricating oils are subjected to heat and mechanical shearing and various combustion products such as moisture, acidic gases, and carbon particles

TABLE 11.18
Properties and Composition of Multigrade SAE 20W-50 Engine Oil

Oil Properties	Base Stock	SAE 20W-50 Oil
KV at 100°C (cSt)	14	17
KV at 40°C (cSt)	127	131
VI	97	130
TAN (mg KOH/g)	0.22	0.44
Paraffins (wt%)		63
Naphthenes (wt%)		24
Aromatics (wt%)		10
Zinc (wt%)		0.13
Calcium (wt%)		0.04
Magnesium (wt%)		0.09
TBN (mg KOH/g)		6.3

Source: Egharevba, F. and Maduako, A.U.C., *Ind. Eng. Chem. Res.*, 41, 3473, 2002.

are formed. The literature reports that some highly alkaline and metallic type antioxidants can lead to an increase in the TAN and the TBN of engine oils (Pirro and Wessol, 2001). The dispersants are added to engine oils to keep the insoluble contaminants dispersed and to assure their stability (Colyer and Gergel, 1997). The engine oil detergents are used to neutralize acids formed during the fuel combustion. Different chemistry detergents were developed to prevent the engine wear. The additives, such as zinc phosphates, thiophosphates, phosphonates, and thiophosphonates used as antioxidants, can also serve as detergents (Bardasz and Lamb, 2003).

The variation in the chemistry of different detergents is shown in Table 11.19. Although many metals have been incorporated into detergents, only three metal

TABLE 11.19
Variation in Chemistry of Different Detergents

	Detergent Types		
	Sulfonates	**Phenates**	**Salicylates**
Chemistry	Sulfonic acid	Alkylphenol	Carboxylic acid
Metal cation	Ca, Mg, Na	Ca, Mg	Ca, Mg
TBN	0–500	50–400	50–400

Source: Colyer, C.C. and Gergel, W.C., Detergents and dispersants, in *Chemistry and Technology of Lubricants*, 2nd ed., Mortier, R.M. and Orszulik, S.T., Eds., Blackie Academic & Professional, London, U.K., 1997. With permission.

cations, that is, calcium, magnesium, and sodium are commonly used (Bardasz and Lamb, 2003). The calcium-based detergent was reported to consist of a calcium carbonate core surrounded by an alkyl-aryl sulfonate or a sulfurized alkyl-phenol shell (Crawford, Psaila, and Orszulik, 1997). The literature reports that in engine oils, the water is provided as a by-product of the combustion and the oil must resist the emulsion formation (Crawford, Psaila, and Orszulik, 1997). The polar oxidation by-products and the soot were reported to increase the stability of foams and emulsions. The patent literature reports on the in-house bench oxidation test used to compare the oxidative thickening of engine oils containing different additive packages. The oils were subjected to air sparging in the presence of a soluble iron catalyst at 165°C and the change in the viscosity was measured (Cody et al., 2002). The volatility of engine oils, which has been shown to affect the oil consumption, is determined almost entirely by the base stock. The raffinate hydroconversion (RHC) process was reported to produce base stocks having a lower volatility that decreased the engine oil consumption (Cody et al., 2002).

The effect of different additive packages on the oxidative thickening of 5W-30 engine oils containing mineral and RHC base stocks is shown in Table 11.20. The mineral 5W-30 engine oil containing a typical additive package was reported to have an increased viscosity by 375% after 82.5 h. The use of another additive package containing additionally a demulsifier and a friction modifier increased the oxidative thickening of mineral 5W-30 engine oil by 375% after only 57.5 h indicating a decrease in the oxidation resistance. The 5W-30 oil containing RHC base stock and a typical additive package, was reported to have an increased viscosity by 375% after 83.3 h indicating no significant difference. The use of another additive package containing additionally a demulsifier and a friction modifier increased the oxidative thickening of 5W-30 engine oil containing RHC base stock, by 375% after 72 h also indicating a decrease in the oxidation resistance. The oxidative thickening can increase the stability of foams. The organic acids may form as a result of oxidation

TABLE 11.20
Effect of Different Additive Packages on Oxidative Thickening of 5W-30 Oils

5W-30 Engine Oils	Oil #1	Oil #3	Oil #2	Oil #4
Base Stocks	Mineral	Mineral	RHC	RHC
Additive packages	Typical	Extra	Typical	Extra
Additional additives		Demulsifier		Demulsifier
		Friction modifier		Friction modifier
Oxidation at 165°C (Fe)				
375% viscosity increase				
Time (h)	82.5	57.5	83.3	72

Source: Cody, I.A. et al., Raffinate hydroconversion process, CA Patent 2,429,500, 2002.

and stabilize the emulsions. The literature reports that foams and emulsions can impair the proper lubrication of an engine by the starvation of the lubricant or by the blockage of oil ducts (Crawford, Psaila, and Orszulik, 1997). The conditions that influence the formation of rust (iron surface) and corrosion are the entrainment of atmospheric oxygen, moisture from the combustion of fuel, and temperature changes (Crawford, Psaila, and Orszulik, 1997). Rust and corrosion are two phenomena which increase wear.

11.4 WATER CONTAMINATION

Engine oils can become contaminated with water during their storage or their use. The early literature reports that the dispersion of seawater in marine-lubricating systems and the condensation of water in the crankcases of internal combustion engines operating intermittently at temperatures below normal running can lead to water contamination (Hughes, 1969). The literature reports that detergent type additives adsorb at the oil/water interface and stabilize the emulsions (Denis, Briant, and Hipeaux, 2000). The more recent literature reports that the addition of a high concentration of dispersants can provide foam control in automotive oils while detergents can act as pro-foamants (Duncanson, 2008). Engine oils contain antifoaming agents and the silicone antifoaming agents are nonpolar, nonionic, and insoluble in water. The silicones are hydrophobic to the extent that they were reported to be used as water repellents (Pape, 1981). Under high temperature and high mechanical shear conditions, dispersants, detergents, and some VI improvers, were reported to increase the stability of foams and emulsions (Crawford, Psaila, and Orszulik, 1997). To prevent foaming, the silicone antifoaming agents need to have a high efficiency to spread on the oil surface.

The spreading coefficient (S) of Si 12500 antifoaming agent on the oil surface of different petroleum-derived base stocks at 24°C is shown in Table 11.21. At 24°C, Si 12500 has a spreading coefficient of 8.2 mN/m in hydrofined 150N mineral base stock and a lower spreading coefficient of 6.6 mN/m in hydroprocessed SWI 6 base stock. Despite a relatively low ST of petroleum-derived base stocks in the range of 30–35 mN/m, the silicones have a relatively high efficiency to spread on the oil surface due to a low oil/silicone IFT that is below 5 mN/m. The

TABLE 11.21

Spreading Coefficient of Si 12500 on Oil Surface of Different Base Stocks

Base Stocks	Neat Oil, ST (mN/m) at 24°C	Si 12500, ST (mN/m) at 24°C	Oil/Si 12500, IFT (mN/m) at 24°C	Oil/Si 12500, (S) (mN/m) at 24°C
Mineral 150N	33.5	24.0	1.5	8.2
SWI 6	32.7	24.0	2.2	6.6

Source: Pillon, L.Z. and Asselin, A.E., Antifoaming agents for lubricating oils, U.S. Patent 5,766,513, 1998.

spreading coefficient of the silicone antifoaming agents can decrease or increase if there are other surface-active molecules present and capable of changing the ST or the IFT of the oil. Water has a high ST of 72.8 mN/m measured at 20°C and the silicone antifoaming agents have a high spreading coefficient to spread on the water surface. The early literature reports on the spreading coefficient of different low viscosity siloxane polymers on the water surface at 20°C (Bondi, 1951).

The spreading (S) coefficient of different viscosity siloxane polymers on the water surface at 20°C is shown in Table 11.22. With an increase in the viscosity of siloxane polymer from 5 to 35 cSt at 25°C, its ST was reported to increase from 19 to 19.9 mN/m and the water/siloxane IFT increased from 42.2 to 43.1 mN/m leading to a decrease in the spreading coefficient from 11.6 to 9.8 mN/m at 20°C. The literature reports on the octamethyltrisiloxane polymer, having a low ST of 17.0 mN/m and a water/polymer IFT of 42.5 mN/m, which was found to have a high spreading coefficient of 12.5 mN/m on the water surface at 20°C (Owen, 1980). The literature reports on the variation in the surface activity of the PDMS antifoaming agent, having an ST of 20.6 mN/m, in water containing different chemistry surfactants (Bergeron et al., 1997).

The effect of different chemistry surfactants on the spreading coefficient of the PDMS antifoaming agent on the water surface is shown in Table 11.23. The PDMS antifoaming agent was found to have a high efficiency to spread on the surface of water of 13.1 mN/m at 20°C that gradually decreased in the presence of different surfactants. The selected surfactants were nonionic penta(ethylene glycol)-mono-n-decyl ether ($C_{10}E_5$), cationic alkyltrimethylammonium bromides (CnTAB), anionic aerosol-OT (AOT), and zwitterionic fluorinated betaine (Zonyl FSK). After the addition of different surfactants, the ST of water decreased from 72.8 to 16.0 mN/m, and a decrease in the water solution/silicone IFT from 39.1 to 3.5 mN/m, was also observed leading to a decrease in the spreading coefficient of the PDMS antifoaming agent from a positive 9.7 mN/m to a negative 14 mN/m (Bergeron et al., 1997). To prevent foaming, the silicone antifoaming agents need to have a high efficiency to

---

TABLE 11.22
Spreading Coefficient of Different Viscosity Siloxane Polymers on Water Surface

	Siloxane Viscosity	
	5 cSt at 25°C	35 cSt at 25°C
Water ST at 20°C (mN/m)	72.8	72.8
Siloxane ST at 20°C (mN/m)	19.0	19.9
Water/siloxane IFT at 20°C (mN/m)	42.2	43.1
Siloxane/water (S) at 20°C (mN/m)	11.6	9.8

Source: Bondi, A., *Physical Chemistry of Lubricating Oils*, Reinhold Publishing Corporation, New York, 1951. With permission.

TABLE 11.23

Effect of Different Surfactants on Spreading (S) of PDMS on Water Surface

Water (Surfactant)	Solution, ST (mN/m) at 20°C	Solution/PDMS, IFT (mN/m) at 20°C	Solution/PDMS, (S) (mN/m) at 20°C
Neat water	72.8	39.1	13.1
Water (C_9TAB)	41.0	10.7	9.7
Water (C_{10}TAB)	39.8	10.4	8.8
Water (C_{12}TAB)	38.8	9.8	8.4
Water ($C_{10}E_5$)	31.5	3.5	7.4
Water (C_{14}TAB)	37.3	9.4	7.3
Water (C_{16}TAB)	37.7	9.8	7.3
Water (AOT)	28.0	4.7	2.7
Water (FSK)	16.0	10.1	−14

Source: Bergeron, V. et al., *Colloid Surf.*, 122(1–3), 103, 1997. With permission.

spread on the water surface. In some cases, depending on the chemistry of surfactant, a very low spreading or negative spreading of the PDMS on the water surface was calculated indicating that the antifoaming agent would not be effective in the prevention of foaming.

The composition of bearing steel and magnesium alloy is shown in Table 11.24. The presence of iron and copper in an internal combustion engine was reported to catalyze the process of oxidation indicating a need for oils having a high resistance to oxidation

TABLE 11.24

Composition of Bearing Steel and Magnesium Alloy

	Bearing Steel	Magnesium Alloy
	Composition (wt%)	
Fe	>96–97	0.01
Mg		>90–91
C	1–1.1	
Al		8.5
Cr	1.3–1.6	
Zn		0.55
Mn	0.25–0.45	0.20
Si	0.15–0.35	0.01
Cu		0.01

Source: Reprinted from Huang, W. et al., *Lubr. Sci.*, 18, 77, 2006. With permission.

in the presence of metals (Migdal, 2003). The literature also reports on the use of two different chemistry ZDDP additives, amine type antioxidant, phenol type antioxidant, and copper salts as an effective antioxidant system for engine oils. The copper salt of succinic anhydride derivatives was reported to be an effective antioxidant, rust inhibitor, and corrosion inhibitor. The copper complexed with oxazoline dispersant was also effective in improving the oxidation stability of engine oils (Migdal, 2003). The solid and metal surfaces are difficult to study. Only mercury (Hg) is a liquid metal and was reported to have an ST of 435 mN/m measured at 20°C (Pillon, 2007).

The ST of different metals is shown in Table 11.25. The ST of metals, such as iron and copper, is above 1000 mN/m, when melted above the temperature of 1000°C (Pillon, 2007). Different molecules found in petroleum-derived base stocks and the surface-active additives can spread on the metal surface and protect it from water and rusting. The detergents, dispersants, and some VI improvers are surfactants. The VI improvers are selected to incorporate pour point depressancy or dispersancy in addition to an increase in the VI. Polymers used as dispersant VI improvers contain functional polar groups such as amines, alcohols, and amides (Stambaugh, 1997).

The spreading coefficient of some organic molecules and water on the surface of mercury is shown in Table 11.26. The benzene molecule having an ST of 28.9 mN/m and an IFT of 357 mN/m, was found to have a high spreading coefficient (S) of 98.1 mN/m at 20°C on the mercury surface. Another organic molecule such as ethyl ether having an ST of 17 mN/m, was found to have a lower spreading coefficient (S) of 88 mN/m at 20°C due to a higher IFT of 379 mN/m. The molecule of water having a high ST of 72.8 mN/m and a high IFT of 375 mN/m, was found to have a drastically lower spreading coefficient (S) of 36.2 mN/m at 20°C. The presence of other surface-active molecules capable of changing the ST or the IFT can change the surface activity of organic liquids such as benzene on the metal surface. An increase in the temperature leads to a decrease in the ST of liquids including petroleum-derived base stocks and water.

TABLE 11.25
STs of Different Metals

Metals	Gas Phase	ST (mN/m)	Temperature (°C)
Mercury	Air	435	20
Mercury	Vacuum	484	20
Iron	Vacuum	1700	1535 (m.p.)
Copper	Vacuum	1300	1083
Copper	Vacuum	1145	1150
Aluminum	Vacuum	845	800
Nickel	Vacuum	1756	1455 (m.p.)
Vanadium	Vacuum	1950	1710 (m.p.)
Molybdenum	Vacuum	2250	2620 (m.p.)

Source: Lide, D.R., Ed., *CRC Handbook of Chemistry and Physics*, 86th ed., CRC Press, Boca Raton, FL, 2005. With permission.

TABLE 11.26

Spreading Coefficient of Organic Molecules and Water on Mercury Surface

Liquid Compounds	Liquid, ST (mN/m) at 20°C	Liquid/Hg, IFT (mN/m) at 20°C	Liquid/Hg, (S) (mN/m) at 20°C
Benzene	28.9	357	98.1
Ethyl ether	17.0	379	88.0
Water	72.8	375	36.2

Source: Lide, D.R., Ed., *CRC Handbook of Chemistry and Physics*, 86th ed., CRC Press, Boca Raton, FL, 2005. With permission.

TABLE 11.27

Effect of Temperature on Density and ST of Water

Temperature (°C)	Water Density (g/cm³)	Water ST (mN/m)
0	0.99984	75.6
10	0.99970	74.2
20	0.99821	72.8
30	0.99565	71.2
40	0.99222	69.6
50	0.98803	67.9
60	0.98320	66.2
70	0.97778	64.5
80	0.97182	62.7
90	0.96535	60.8
100	0.95840	58.9

Source: Lide, D.R., Ed., *CRC Handbook of Chemistry and Physics*, 86th ed., CRC Press, Boca Raton, FL, 2005. With permission.

The effect of the temperature on the density and the ST of water is shown in Table 11.27. An increase in the water content of engine oils was reported to cause deterioration of dispersing properties and to lead to sedimentation of contaminants and additives (Artemev, Boikov, and Lebedeva, 1991). An increase in the temperature will further decrease the surface activity of the silicone antifoaming agent in terms of the prevention of foaming. With an increase in the temperature and under oxidation conditions, the oxidation by-products are formed. An increase in the temperature and a decrease in the ST of water will increase its surface activity on the metal surface leading to an increase in rust. The organic acids may form as a result of oxidation and the heavy metal soaps are the results of the reaction between acids and metals

(Pirro and Wessol, 2001). The literature reports that used automotive lubricants, such as mono- and multigrade crankcase oils, gear oils, and transmission fluids, are contaminated with degradation products from the in-service use and other contaminants such as water (combustion by-product or rain), fuels (residual components of gasoline and diesel fuel), and solids (additives, soot, rust, wear metals, and dirt) (Betton, 1997). Waste-lubricating oils are collected and disposed of either by burning as fuel or incineration as waste, or are reclaimed. The automotive engine oil "consumption" can be seen as a black coating on the road (Betton, 1997). The used industrial oils need to be collected and kept segregated from engine oils, which contain high levels of additives, metals, oxidation by-products, and wear contamination.

REFERENCES

American Petroleum Institute, *Engine Oil Licensing and Certification System*, API 1509, 15th ed., American Petroleum Institute, Washington, DC, April 2002.

AMSOIL, http://www.performanceoiltechnology.com (accessed 2008).

Antipina, N.A., Chesnokov, A.A., and Radchenko, L.A., *Russian Journal of Applied Chemistry*, (3), 14 (1989).

Apar Industries, http://www.aparind.com (accessed 2003).

Artemev, V.A., Boikov, D.V., and Lebedeva, V.L., *Khimija i Tekhnologiya Topliv i Masel* (in Russian), 4, 19 (1991).

Bardasz, E.A. and Lamb, G.D., Additives for crankcase lubricant applications, in *Lubricant Additives Chemistry and Applications*, Rudnick L.R., Ed., Marcel Dekker, Inc., New York, 2003.

Bergeron, V. et al., *Colloids and Surfaces, A: Physicochemical and Engineering Aspects*, 122(1–3), 103 (1997).

Betton, C.I., Lubricants and their environmental impact, in *Chemistry and Technology of Lubricants*, 2nd ed., Mortier, R.M. and Orszulik, S.T., Eds., Blackie Academic & Professional, London, U.K., 1997.

Bondi, A., *Physical Chemistry of Lubricating Oils*, Reinhold Publishing Corporation, New York, 1951.

Braun, J. and Omeis, J., Additives, in *Lubricants and Lubrications*, Mang, T. and Dresel, W., Eds., Wiley-VCH GmbH, Weinheim, 2001.

Butler, K.D., Process to produce lube oil base stock by low severity hydrotreating of used industrial circulating oils, CA Patent 2,170,138, 1996.

Chandrasekaran, M., Batchelor, A.W., and Loh, N.L., *Wear*, 243, 68 (2000).

Cody, I.A. et al., Raffinate hydroconversion process, CA Patent 2,429,500, 2002.

Colyer, C.C. and Gergel, W.C., Detergents and dispersants, in *Chemistry and Technology of Lubricants*, 2nd ed., Mortier, R.M. and Orszulik, S.T., Eds., Blackie Academic & Professional, London, U.K., 1997.

Crawford, J., Psaila, A., and Orszulik, S.T., Miscellaneous additives and vegetable oils, in *Chemistry and Technology of Lubricants*, 2nd ed., Blackie Academic & Professional, London, U.K., 1997.

Dawson, R.B. and Kerkemeyer, M.E., Industrial trends, in *Synthetic Lubricants and High-Performance Functional Fluids*, 2nd ed., Rudnick, L.R. and Shubkin, R.L., Eds., Marcel Dekker Inc., New York, 1999.

Dee, G.T. and Sauer, B.B., *TRIP*, 5(7), 230, 1997.

Denis, J., Briant, J., and Hipeaux, J.C., *Lubricant Properties Analysis and Testing*, Editions Technip, Paris, 2000.

Duncanson, M., *Journal of the Society of Tribologists and Lubrication Engineers*, May, 9 (2003).

Duncanson, M., *Oil Analysis and Lubrication Learning Center*, http://www.noria.com (accessed 2008).

Egharevba, F. and Maduako, A.U.C., *Industrial and Engineering Chemistry Research*, 41, 3473 (2002).

El-Ashry, El-S.H. et al., *Lubrication Science*, 18(2), 109 (2006).

Grigoreva, N.I. et al., *Chemistry and Technology of Fuels and Oils*, 39(3), 117 (2003).

Henderson, H.E., Chemically modified mineral oils, in *Synthetics, Mineral Oils, and Bio-Based Lubricants Chemistry and Technology*, Rudnick, L.R., Ed., CRC Press, Boca Raton, 2005.

Huang, W. et al., *Lubrication Science*, 18, 77 (2006).

Hughes, R.I., *Corrosion Science*, 9, 535, 1969.

Joseph, P.V. et al., *2nd International Symposium on Fuel and Lubricants,* New Delhi, India, 2000.

Kichkin, G.I., *Chemistry and Technology of Fuels and Oils*, 4, 49 (1966).

Lee, F.L. and Harris, J.W., Long-term trends in industrial lubricant additives, in *Lubricant Additives Chemistry and Applications*, Rudnick, L.R., Ed., Marcel Dekker, Inc., New York, 2003.

Lide, D.R., Ed., *CRC Handbook of Chemistry and Physics*, 86th ed., CRC Press, Boca Raton, 2005.

Losikov, V.V., Khalif, A.L., and Aleksandrova, L.A., *Engineering and Mining Journal* (in Russian), 26(8), 47 (1948).

Lubrizol, http://www.lubrizol.com (accessed 2001).

Marsh J.F., *Chemistry and Industry*, July, 20 (1987).

Migdal, C.A., Antioxidants, in *Lubricant Additives Chemistry and Applications*, Rudnick, L.R., Ed., Marcel Dekker, Inc., New York, 2003.

Owen, M.J., *Industrial and Engineering Chemistry Product Research and Development*, 19, 97 (1980).

Pape, P.G., Silicones: Unique chemicals for petroleum processing, SPE 10089, *56th Annual Fall Technical Conference and Exhibition of SPE*, San Antonio, 1981.

Papke, B.L., *Tribology Transactions*, 31(4), 420 (1988).

Phillipps, R.A., Highly refined mineral oils, in *Synthetic Lubricants and High-Performance Functional Fluids*, 2nd ed., Rudnick L.R. and Shubkin R.L., Eds., Marcel Dekker Inc., New York, 1999.

Pillon, L.Z., *Interfacial Properties of Petroleum Products*, CRC Press, Boca Raton, 2007.

Pillon, L.Z. and Asselin, A.E., Antifoaming agents for lubricating oils, U.S. Patent 5,766,513, 1998.

Pirro, D.M. and Wessol, A.A., *Lubrication Fundamentals*, 2nd ed., Marcel Dekker Inc., New York, 2001.

Raab, M.J., Synthetic-based food-grade lubricants and greases, in *Synthetics, Mineral Oils, and Bio-Based Lubricants Chemistry and Technology*, Rudnick, L.R., Ed., CRC Press, Boca Raton, 2005.

Raab, M.J. and Hamid, S., Additives for food-grade lubricant applications, in *Lubricant Additives Chemistry and Applications*, Rudnick, L.R., Ed., Marcel Dekker, Inc., New York, 2003.

Rasberger, M., Oxidative degradation and stabilisation of mineral oil based lubricants, in *Chemistry and Technology of Lubricants*, 2nd ed., Mortier, R.M. and Orszulik, S.T., Eds., Blackie Academic & Professional, London, U.K., 1997.

Rudnick, L.R. and Shubkin, R.L., Poly(alfa-Olefins), in *Synthetic Lubricants and High-Performance Functional Fluids*, 2nd ed., Rudnick, L.R. and Shubkin, R.L., Eds., Marcel Dekker Inc., New York, 1999.

SensaDyne Instrument Division, *Comparison of Surface Tension Measurements Methods*, Chem-Dyne Research Corporation, Mesa, Arizona, 2000.

Shanks, D. et al., *Lubrication Science*, 18, 87 (2006).

Stambaugh, R.L., Viscosity index improvers and thickeners, in *Chemistry and Technology of Lubricants*, 2nd ed., Mortier R.M. and Orszulik S.T., Eds., Blackie Academic & Professional, London, U.K., 1997.

Stein, W.H. and Bowden, R.W., *Machinery Lubrication*, July 2004, http://www.machinerylubrication.com (accessed 2008).

Watanabe, J., Fukushima, T., and Nose, Y., *Proceedings of the JSLE–ASLE International Lubrication Conference*, Amsterdam, 1976, p. 641.

12 Lubricant Storage

12.1 LUBRICANT LIFE

The lubricant industry formulates the products to meet the performance requirements of different applications, which can range from severe low temperature fluidity specifications to high temperature oxidation resistances. The literature reports that the petroleum products might be re-qualified for an additional storage period after the satisfactory quality testing (Pirro and Wessol, 2001). The lubricating oils can deteriorate in the storage as a result of exposure to low or high temperatures. The literature reports that the major problem associated with the storage outdoors during cold weather is with dispensing of products not intended for low temperature service. The literature reports that at moderate temperatures, the long-term storage has little effect on most premium lubricating oils, hydraulic fluids, process oils, and waxes (Pirro and Wessol, 2001). Below-freezing temperatures will not affect the quality of most fuels, solvents, naphthas, engine oils, and greases. The prolonged storage at higher temperatures might cause the darkening of oils due to oxidation (Pirro and Wessol, 2001). Different containers used for the packing and the storing of the turbine oils were reported to affect their demulsibility properties. The clear glass and the white plastic containers were found to lead to a fast aging caused by the photooxidation (De Araujo Monte, 1998). The literature reports that any new engine oil formulation should be tested for the additive solubility under both cold and hot storage conditions (Rudnick and Shubkin, 1999).

The typical storage life of petroleum products is shown in Table 12.1. The commercial lubricants contain additives to enhance their performance in amounts ranging from less than 1 wt% used in some industrial oils, such as turbine oils, to 25 wt% or more used in engine oils. Petroleum additives are generally more polar than the lube oil base stocks and can precipitate during storage. Many lubricating oils can become contaminated with water during their storage and the proper chemistry rust inhibitors are needed to prevent rust formation without degradation in their demulsibility properties. The early literature reports that the contamination with water, gasoline, or gasket-sealing compounds may affect the foaming characteristics of turbine oil (Tourret and White, 1952). The recent literature reports on the variation in the foaming tendency of turbine oils contaminated with water (Duncanson, 2003). The polyacrylate (PA) polymer, used as a non-silicone type antifoaming agent in turbine oils, has a hydrocarbon (HC) backbone and polar ester side chain groups. The presence of the ester side chains makes the PA polymer surface active toward the oil/air interface and the oil/water interface. When selecting the additive, their chemistry and their physical properties need to be considered. Additives must also be capable of being handled in the conventional blending equipment and of being stable in storage.

The chemistry and the physical properties of some additives used in lubricants are shown in Table 12.2. The polyisobutylene polymer is used as a VI improver and

TABLE 12.1

Typical Storage Life of Petroleum Products

Petroleum Products	Maximum Storage
Gasoline	6 months
Diesel fuels	6 months
Custom blended lubricating oils	3 months
Emulsion type fire resistant fluids	6 months
Wax emulsions	6 months
Greases	6–36 months

Source: Reproduced from Pirro, D.M. and Wessol, A.A., *Lubrication Fundamentals*, 2nd ed., Marcel Dekker, New York, 2001. With permission.

a thickener in many lubricant products. It was reported to be a clear to light yellow viscous liquid, having a viscosity of 35,000–140,000 cSt at 40°C, and a pour point varying from as low as −40°C to 15°C. The fatty acids and their esters are used as rust inhibitors in food grade lubricants containing mineral white oil or synthetic polyalfaolefin (PAO) base stocks (Raab and Hamid, 2003). Highly polar compounds, such as glycerol mono-oleate, glycerol di-oleate, glycerol mono stearate, lecithin, oleic acid, lard oils, and other oleic acid esters, are also used as the lubricity agents in some lubricants. The antiwear and extreme pressure (EP) agents are organic compounds containing P and S, such as amine phosphates and triphenylphosphorothionate (TPPT), which form metal salts. The non-silicone type antifoaming agents, such as dispersion of PA, are used in low viscosity circulating oils, such as turbine oils, which need to have resistance to foaming at 24°C and meet the air release time specifications at 50°C (Pillon, 2007). While the most surface active antifoaming agents are polydimethylsiloxane polymers, the PA polymers do not degrade the air release time of the lubricating oils to the same extent as the silicones do and replace silicone antifoaming agents in some lubricant applications. The PA antifoaming agent is commercially available as a 40 wt% solution in the petroleum diluent.

The effective treat rates and the storage stability at room temperature (RT) of Si 12500 and PA (40%) antifoaming agents in SWI 6 base stock are shown in Table 12.3. The silicone antifoaming agents prevent foaming by entering the oil/air interface and spreading on the oil surface, which can cause a severe air entrainment. The use of 3 ppm of Si 12500 is effective in preventing the foaming tendency at 24°C of SWI 6 base stock; however, a significant increase in the air release time from 1.1 to 2.8 min at 50°C is also observed. The SWI 6 base stock, containing Si 12500, was found to have an excellent storage stability at RT since no increase in the foaming tendency after 35 days was observed. The PA antifoaming agents prevent foaming by entering the oil/air interface and destabilizing it with a little or no spreading taking place, which leads to less air entrainment. The PA (40 wt%) antifoaming agent is less surface active and higher treat rates are required to prevent foaming. The SWI 6 base stock requires a high treat rate of 80 ppm of PA (40%), active ingredient, to

TABLE 12.2
Chemistry and Physical Properties of Some Additives Used in Lubricants

Additive Chemistry	Physical Properties
Polyisobutylene polymer	Clear, light yellow viscous liquid
Viscosity at 40°C (cSt)	35,000–140,000
Pour point (°C)	−40 to 15
Fire point (°C)	Min 230
Diphenylamine antioxidant	Clear, yellow to brown viscous liquid
Viscosity at 40°C (cSt)	276
Flash point (°C)	185
Solubility	Mineral oil
Triazole metal deactivator	Brown liquid
Viscosity at 40°C (cSt)	83
Flash point (°C)	>150
Solubility	Soluble in mineral and synthetic oils
Oleic acid (fatty acid)	Colorless to pale red
Melting point (°C)	16.3
Flash point (°C)	180
Glycerol mono-oleate (fatty acid ester)	Clear liquid
Flash point (°C)	242
TAN (mg KOH/g)	6
Amino acid sarcosine (derivative of fatty acid)	Dark amber liquid
Viscosity at 40°C (cSt)	660
Free fatty acid (%)	6
Solubility	Soluble in mineral oil and synthetic PAO
TPPT antiwear and EP additive	Gray or yellow powder
Melting point (°C)	51–54
Phosphorus content (%)	8–10
TAN (mg KOH/g)	0.03
Polydimethylsiloxane polymer	Crystal clear liquid
Viscosity at 40°C (cSt)	50–100,000
Pour point (°C)	−50 to −70
Flash point (°C)	>318

Source: Raab, M.J. and Hamid, S., Additives for food-grade lubricant applications, in *Lubricant Additives Chemistry and Applications*, Rudnick, L.R., Ed., Marcel Dekker, New York, 2003. With permission.

prevent foaming tendency at 24°C. Despite the use of a high treat rate of PA (40%), no significant increase in the air release time is observed. However, after only 1–2 days of storage at RT, SWI 6 base stock, containing PA (40%), was found to foam. A poor solvency of SWI 6 base stock and a low surface activity of PA (40 wt%) antifoaming agent on its oil surface leads to a decrease in its storage stability in terms of the foam inhibition.

TABLE 12.3

Effective Treat Rates and Storage Stability of Antifoaming Agents in SWI 6 Oil

SWI 6 Base Stock, ST of 32.5 mN/m	Effective Treat Rate (ppm) (Active)	Air Release Time at 50°C (min)	Increased Foaming Storage at RT (Days)
Si 12500	3	2.8	>35
PA (40 wt%)	80	1.1	1–2

Source: Pillon, L.Z. et al., Method for reducing foaming of lubricating oils, U.S. Patent 6,090,758, 2000.

12.2 COLLOIDAL STABILITY

A great many everyday materials ranging from paints, inks, to engine oils contain colloidal particles as concentrated or dilute dispersions. The literature reports that the colloidal dispersions, which are clear and transparent have particles that are less than 100 nm. The particles having the size above 400 nm are opaque and can be viewed under a microscope (Inoue and Watanabe, 1982). The neutron scattering technique can be used to measure the size of the colloidal dispersions at low concentrations. The x-ray scattering technique can be used to measure the colloidal dispersions at high concentrations (Ottewill et al., 1992). When the dispersed particles are so large that they exceed the limit of colloidal dimensions, and when the density of the particles relative to that of the suspension medium is sufficiently great, they settle out under the force of gravitation. A high additive content and the use of detergents can lead to the colloidal instability of engine oils. Initially, the haze formation is observed. The literature reports that the use of large amounts of overbased detergents, such as calcium sulfonates and phonates, can lead to the colloidal instability. To increase the additive stability, the inorganic carbonates are solubilized in the HC solvents by an adsorbed layer of the sulfonate surfactant. In partial synthetic blends, even a change in the ratio of synthetic/mineral base stocks was found to affect the solubility of additives (Rudnick and Shubkin, 1999).

The HC composition and the pour points of different petroleum-derived base stocks and synthetic PAO 6 fluids, having a similar viscosity, are shown in Table 12.4. The solvent-refined 150N base stock, having a VI of 95 and a dewaxed pour point of −12°C, was reported to contain 49.9 wt% of naphthenes and 22.5 wt% of aromatic HCs, which are known to have excellent solvency properties. Similar viscosity HC base stock, having a VI of 99 and a dewaxed pour point of −12°C, was reported to contain 63.1 wt% of naphthenes and 3.5 wt% of aromatic HCs, which might show some variation in the solvency properties. Another similar viscosity VHVI base stock, having a high VI of 133 and a lower dewaxed pour point of −15°C, was reported to contain 43.7 wt% of naphthenes and only 0.8 wt% of aromatic HCs, which will show a significant decrease in the solvency properties. A similar viscosity synthetic PAO 6 base stock, having a high VI of 135 and a very low

TABLE 12.4
Composition and Pour Points of Different Base Stocks and Synthetic PAOs

	Processing			
	Solvent Refined	HC	VHVI	PAO 6
KV at 40°C (cSt)	30.1	29.6	32.5	31.3
KV at 100°C (cSt)	5.1	5.1	6.0	5.9
VI	95	99	133	135
Pour point (°C)	−12	−12	−15	−60
Paraffins (wt%)	27.6	33.4	55.5	100
Naphthenes (wt%)	49.9	63.1	43.7	0
Aromatics (wt%)	22.5	3.5	0.8	0
Sulfur (ppm)	5800	300	<10	0
Nitrogen (ppm)	12	4	<1	0

Source: Henderson, H.E., Chemically modified mineral oils, in *Synthetics, Mineral Oils, and Bio-Based Lubricants Chemistry and Technology*, Rudnick, L.R., Ed., CRC Press, Boca Raton, FL, 2005. With permission.

TABLE 12.5
Effect of Different Additive Packages on Foaming and Colloidal Stability

	Mineral Engine Oils	
	Oil #1	Oil #2
Additive package	Package A	Package B
Detergent content	Higher	Lower
ASTM D 892 foam test	Fresh oil	Fresh oil
Foaming tendency at 24°C (mL)	0	0
Foaming tendency at 93.5°C (mL)	<50	50
Colloidal static stability (after 10 days at RT)	Haze	Clear

natural pour point of −60°C, contains 100% of paraffinic type HCs and is known to have very poor solvency properties. The literature reports that the use of some automotive gear oil additive packages, containing polyisobutene (PIB) dispersant and an antifoaming agent, was leading to a variation in their foaming tendency and their stability.

The effect of different additive packages on the foaming tendency and the colloidal stability of some mineral engine oils is shown in Table 12.5. The fresh mineral engine oil #1, containing the VI improver and the additive package A, was found to have a total resistance to foaming at 24°C and a foaming tendency below 50 mL at 93.5°C. However, after 10 days of the storage stability testing at RT and under static conditions, a haze was formed indicating the additive precipitation. Another fresh mineral engine oil #2, containing the same VI improver and a different additive package B, was also found to have a total resistance to foaming at 24°C but an increased foaming tendency of 50 mL at 93.5°C. After 10 days of the storage stability testing at RT and under static conditions, the oil remained clear and no haze formation was observed. The variation in the foaming tendency of some fresh mineral engine oils might be related to some differences in the chemistry or the treat rate of the antifoaming agents. However, the variation in the colloidal stability is related to some differences in the chemistry or the treat rate of detergents. With a decrease in the detergent content, an increase in the colloidal stability of mineral engine oils is observed. Traditionally, the marine lubricants were reported to contain naphthenic oils, which have very good solvency properties. The modern marine lubricants contain paraffinic base stocks, which have a higher VI, a lower volatility, and an improved oxidation resistance (Carter, 1997).

The pour points and the total base number (TBN) values of the marine diesel engine lubricants are shown in Table 12.6. The additives used in the marine lubricants include antioxidants, dispersants, alkaline detergents, corrosion inhibitors, antiwears and EP agents, and pour point depressants. The generally available commercial detergents are diluted about 50% in mineral oil and are used in engine oils in amounts ranging from 0.5% to 30%. While gasoline engine oils may contain up to 5%, heavy duty diesel truck oils use up to 8% detergents. Marine engine oils have a TBN as high as 100 and may contain up to 30% of a detergent additive (Colyer and Gergel, 1997). Deposit formation, mainly of calcium carbonate, can lead to the performance problems of engine oils (Colyer and Gergel, 1997). In unstable marine engine oils, the colloidal phase separation can lead to the formation of deposits causing the blockages of feed lines, filters, and lubricating quills (Carter, 1997). The colloidal dispersions can be clear solutions. With time, the particle agglomeration might take place and the solutions become opaque indicating the sediment formation.

TABLE 12.6
Pour Points and TBN Values of Marine Diesel Engine Lubricants

Marine Diesel Engine Lubricants	System Oil	Cylinder Oil	Trunk Piston Engine Oil
KV at 100°C (cSt)	11.5	19	14
KV at 40°C (cSt)	103	218	138
Pour point (°C)	−18	−12	−18
TBN (mg KOH/g)	5	70–100	20–40

Source: Carter, B.H., Marine lubricants, in *Chemistry and Technology of Lubricants*, 2nd ed., Mortier, R.M. and Orszulik, S.T., Eds., Blackie Academic & Professional, London, U.K., 1997. With permission.

It takes days or weeks to visually detect the process of the additive agglomeration leading to haze formation. To detect the process of particle agglomeration, different techniques are used. The marine quality additives with good colloidal stability are selected through extended static and accelerated centrifuge testing.

12.3 LOW-TEMPERATURE STABILITY

The literature reports that some specialty industrial products, such as transformer oils, refrigeration oils, and some greases, require very low pour points (Shell, 2001). Some industrial oils are designed specifically for the lubrication of chains and guide bars on modern high speed chain sows under different weather conditions. The major problem associated with the storage outdoors, during cold weather, is with products that have high pour points (Pirro and Wessol, 2001). The pneumatic tool oils, formulated with selected base stocks, special tackiness, antiwear and EP additives, are used to lubricate pneumatic equipments, such as rock drills, jack hammers, chippers, and other tools, drills, and grinders (Apar Industries, 2003). Some pneumatic tool oils are recommended for ambient temperatures up to 25°C. Other oils are recommended to be used at winter ambient temperatures as low as −25°C. In some cases, to meet the severe winter temperature requirements, a maximum pour point of −45°C is required.

The general purpose machinery oils are recommended for the lubrication of all types of industrial machinery using once through lubrication systems, such as machine tools and textile machinery. They are used in different machinery parts, such as bottle oiler and drip feed systems. The heavy viscosity grade machinery oils are used to lubricate small open gears under light load conditions where the oil is applied intermittently (Apar Industries, 2003). The literature reports that some machinery oils are high viscosity straight mineral oils containing stable paraffinic oils that are resistant to oxidation and contain no additives (Kajdas, 1997). Some high viscosity machinery oils are dark colored oils for "once through" lubrication where high oil consumption justifies the use of an economical short life product. With an increase in temperature, the viscosity decreases and some high temperature applications require high viscosity oils having good oxidation stability. The use of high viscosity paraffinic base stocks, having a high dewaxed pour points, is required to meet the viscosity and the oxidation stability requirements of some high viscosity machinery oils.

The literature reports that the cylinder oils can be used to lubricate paper and textile mill calendar bearings and sugar mill roller bearings. They are also suitable for lubricating warm gears (Apar Industries, 2003). The cylinder oils are manufactured to have good film strength and thermal degradation characteristics, and can be produced from bright stocks or liquid paraffins and fatty oils. The literature reports that the use of fatty oils improves their water displacement characteristics and adhesion to cylinder walls (Kajdas, 1997). Some cylinder oils are used in applications involving wet steam or saturated steam, and they are manufactured to provide excellent lubrication in steam engines under a variety of steam conditions. The cylinder oils are used for lubricating cylinders, valves, and other moving parts of steam engines. The use of a high content of fatty oil additives leads to an increase in the depressed pour points of cylinder oils and can affect their storage stability if exposed to low temperatures.

The effect of petroleum-derived base stocks and additives on the pour points of different industrial oils is shown in Table 12.7. The solvent-refined paraffinic base stocks have relatively high dewaxed pour points and require the use of pour point depressants. Some naphthenic oils, having very low natural pour points, are blended with paraffinic base stocks in products that require low pour points. The use of low viscosity naphthenic oils, having very low natural pour points, is the most effective way to produce industrial oils that require very low pour points. Some petroleum products can deteriorate in storage as a result of the water contamination and the exposure to low temperatures. The freezing of the wax-emulsions will cause the separation of wax and the water phase that cannot be restored to its original condition. The literature reports that the products containing a significant amount of water should not be stored below 4°C (Pirro and Wessol, 2001). Most engine oils have low pour points and are not affected by the prolonged storage time under the conditions of low temperature. The literature reports on the typical properties and the pour points of the conventional aircraft piston engine oils and the aircraft turbine engine oils (Lansdown, 1997).

The variation in the pour points of the aircraft engine oils is shown in Table 12.8. The polymeric components of lubricating oils have a tendency to separate out under

TABLE 12.7
Effect of Base Stocks and Additives on Pour Points of Industrial Oils

Industrial Oils	EP Pneumatic		ISO VG 680	
Content	Tool Oil #1	Tool Oil #2	Machinery Oil	Cylinder Oil
Base stocks	Paraffinic	Naphthenic	Paraffinic	Paraffinic
Additives	Same	Same	Different	Different
Pour point depressant	Yes	No	Yes	Yes
Pour point (°C)	−36	−51	−12	3

TABLE 12.8
Variation in the Pour Points of Aircraft Engine Oils

	Piston Engine		Turbine Engine	
Aircraft Engine Oils	Oil #1	Oil #2	Oil #1	Oil #2
KV at 40°C (cSt)	270	140	34	14
KV at 100°C (cSt)	23	20	7.5	3
Pour point (°C)	−20	−30	−60	−65

Source: Lansdown, A.R., Aviation lubricants, in *Chemistry and Technology of Lubricants*, 2nd ed., Mortier, R.M. and Orszulik, S.T., Eds., Blackie Academic & Professional, London, U.K., 1997. With permission.

slow cooling conditions, for example, in drums and tanks stored in unheated areas (Crawford et al., 1997). The literature reports that the use of certain chemistry pour point depressants caused the formation of a haze indicating the additive precipitation under the conditions of the long-term storage and a low temperature. The interaction between the wax and some VI improvers was reported to produce sediment and to affect the low temperature properties of engine oils (Rhodes, 1993). The synthetic base stocks are used in engine oils that require very low pour points. The literature reports that the VI improvers and the detergents often pose compatibility problems in hydrocracked or synthetic base stocks because their structure has been optimized for use in the mineral base stocks (Denis, Briant, and Hipeaux, 2000). The use of different petroleum-derived base stocks affects the depressed pour points of engine oils, and some oils can deteriorate in storage as a result of the exposure to low temperatures.

12.4   HOT-TEMPERATURE STABILITY

The solvent-dewaxed paraffinic base stocks are usually hydrofined to remove any residual solvents and to improve their color, filterability, demulsibility, and the oxidation stability that is tested as the hot temperature storage stability. The hot temperature storage stability indicates the number of days for oil to produce a detectable change, other than color, when stored in the dark at 70°C and in the presence of air. The time until formation of haze, flocculation, or deposit is measured and the hot temperature storage time of less than 60 days was reported to be unacceptable (Dagom et al., 1996). The filterability of oil is a measure of the filter-blocking tendency and is expressed in terms of the time needed to filter a certain volume of oil through a specified filter. The test is known as the CETOP filterability method. The solvent-refined mineral base stocks have good solvency properties. The hydroprocessed base stocks have lower aromatic and heteroatom contents, which leads to an improvement in not only their oxidation stability but also leads to a decrease in their solvency properties. The patent literature reports on the effect of the activated carbon treatment at different temperatures on the properties and the hot temperature storage stability of the solvent-refined and the hydroprocessed base stocks (Dagom et al., 1996).

The effect of the activated carbon (C) treatment at 70°C on the properties of the solvent-dewaxed base stock, having good hot temperature stability, is shown in Table 12.9. The solvent-dewaxed mineral base stock, having a good hot temperature storage stability of over 60 days, was reported to have a VI of 106 and contain 49.8 wt% of aromatics, 0.7 wt% of S, and 20 ppm of N. It required 19 min to filter the 1000 mL of oil, and it required 20 min to separate the emulsion. After the activated C treatment at 70°C, the VI increased to 113 and the aromatic content increased to 51.1 wt% with no significant change in the S and N contents. The time to filter 1000 mL of oil decreased to 16 min, and the time to separate the emulsion decreased to 15 min indicating an improvement in the filterability and a decrease in the surface active contaminants, having emulsifying properties.

The effect of the activated C treatment, at different temperatures, on the demulsibility and the hot temperature storage stability of the solvent-dewaxed mineral

TABLE 12.9
Effect of Activated C at 70°C on the Properties of Mineral Base Stocks

Solvent-Dewaxed Base Stock	Before Treatment	After C Treatment at 70°C
KV at 40°C (cSt)	28.0	27.7
KV at 100°C (cSt)	5.0	5.1
VI	106	113
Aromatics (%)	49.8	51.1
Sulfur (wt%)	0.7	0.7
Nitrogen (ppm)	21	20
Stability at 70°C (days)	>60	>60
Filterability (1000 mL) (min)	19	16
ASTM D 1401 (54°C)		
O/W/E (mL)	40/37/3	40/40/0
Separation time (min)	20	15

Source: Dagom, D.N. et al., Process for improving lubricating base oil quality, U.S. Patent 5,554,307, 1996.

TABLE 12.10
Effect of Activated C Treatments on Demulsibility and Hot Temperature Stability

Solvent-Dewaxed Mineral Oil	Before Treatment	After C Treatment at 70°C	After C Treatment at 250°C
KV at 40°C (cSt)	71.4	71.1	71.7
KV at 100°C (cSt)	9.0	9.0	9.0
VI	99	99	99
ASTM D 1401 (54°C)			
O/W/E (mL)	40/37/3	40/37/3	40/40/0
Separation time (min)	15	50	5
Stability at 70°C (days)	17	18	>60

Source: Dagom, D.N. et al., Process for improving lubricating base oil quality, U.S. Patent 5,554,307, 1996.

base stock is shown in Table 12.10. The solvent-dewaxed mineral base stock, having a hot temperature storage stability of only 17 days at 70°C, was reported to have a VI of 99, and it required 15 min to separate the emulsion. After the activated carbon treatment at 70°C, the emulsion separation time actually increased to 50 min indicating an increase in the surface active contaminants, and no improvement in the hot temperature storage stability was observed. After the activated C treatment

TABLE 12.11

Effect of Activated C Treatment at 70°C on the Properties of Hydroprocessed Oil

Hydroprocessed Base Stock	Before Treatment	After C Treatment at 70°C
KV at 40°C (cSt)	71.4	71.4
KV at 100°C (cSt)	9.1	9.0
VI	101	100
Aromatics (wt%)	63.5	62.5
Sulfur (ppm)	106	100
Nitrogen (ppm)	2	4
Air release time at 50°C (min)	8	8
ASTM D 1401 (54°C)		
O/W/E (mL)	40/37/3	40/40/0
Separation time (min)	15	5
Filterability (1000 mL) (min)	80	45
Stability at 70°C (days)	31	>60

Source: Dagom, D.N. et al., Process for improving lubricating base oil quality, U.S. Patent 5,554,307, 1996.

at high temperature of 250°C, the emulsion separation time decreased to 5 min, and a drastic improvement in the hot temperature storage stability to over 60 days was reported.

The effect of activated C treatment at 70°C on the composition, the properties, and the hot temperature storage stability of the hydroprocessed base stock is shown in Table 12.11. The hydroprocessed base stock, having the hot temperature storage stability of 31 days at 70°C, was reported to have a VI of 101 and contain 63.5 wt% of the aromatics, 106 ppm of the S content, and 2 ppm of the N content. It required 80 min to filter the 1000 mL of oil, and it was reported to have the air release time of 8 min at 50°C and it required 15 min at 54°C to separate the emulsion. After the activated C treatment at 70°C, no significant change in the VI, the aromatic content, and the S and N contents was observed. The time to filter 1000 mL of oil decreased to 45 min and the time to separate the emulsion decreased to 5 min indicating a small decrease in the surface active contaminants. After the activated C treatment at 70°C, the hot temperature storage stability at 70°C of hydroprocessed base stock also increased to over 60 days indicating a decrease in the O-containing contaminants having the surface active properties towards the oil/water interface.

The effect of the activated C treatment at 130°C on the interfacial property and the hot temperature storage stability of the hydroprocessed base stock is shown in Table 12.12. The hydroprocessed base stock, having a poor hot temperature storage stability of only 6 days at 70°C, was reported to have a VI of 101 and contain 51.2 wt% of the aromatics, 148 ppm of the S content, and 4 ppm of the N content. It was reported to have a high foaming tendency of 290 mL and have the air release

TABLE 12.12

Effect of Activated C Treatment at 130°C on Interfacial Properties and Stability

Hydroprocessed Base Stock	Before Treatment	After C Treatment at 130°C
KV at 40°C (cSt)	71.9	71.8
KV at 100°C (cSt)	9.1	9.1
VI	101	101
Aromatics (wt%)	51.2	51.2
Sulfur (ppm)	148	142
Nitrogen (ppm)	4	3
Foaming tendency (mL)	290	30
Air release time at 50°C (min)	9	6
ASTM D 1401 (54°C)		
O/W/E (mL)	40/33/7	40/40/0
Separation time (min)	60	9
Filterability (1000 mL) (min)	>60	45
Stability at 70°C (days)	6	>60

Source: Dagom, D.N. et al., Process for improving lubricating base oil quality, U.S. Patent 5,554,307, 1996.

time of 9 min at 50°C. It required over 60 min to filter the 1000 mL of oil and 60 min at 54°C to separate the emulsion. After the activated C treatment at 130°C, no change in the VI and the aromatic content but a small decrease in the S and N contents was reported. The foaming tendency decreased to 30 mL and the air release time also decreased to 6 min at 50°C. The time to filter 1000 mL of oil decreased to 45 min and the time to separate the emulsion also decreased to 9 min at 54°C. The hot temperature storage stability at 70°C drastically increased to over 60 days. Additives, such as antioxidants, are used to improve the oxidation stability of base stocks to meet the performance requirements of different lubricant products.

The hot temperature storage stability at 100°C of different base stocks containing the phenolic type antioxidants is shown in Table 12.13. The solvent-refined paraffinic base stock, containing the phenolic type antioxidant #1, was reported to have the hot temperature storage stability at 100°C for over 8 weeks but a poor stability of less than 4 weeks in the presence of another phenolic type antioxidant #2. The solvent-refined mineral base stocks do not meet the performance requirements of some lubricant products but have good solvency properties. The HC base stock was reported to have the hot temperature storage stability at 100°C for 4–8 weeks in the presence of the phenolic type antioxidant #1 and the phenolic type antioxidant #2. The hydroprocessed base stocks have lower solvency properties. The commercial base stock #1, containing the phenolic type antioxidant #1, was reported to have the hot temperature storage stability at 100°C for less than 4 weeks but improved stability of 4–8 weeks in the presence of the phenolic type antioxidant #2. Another commercial base stock #2, containing the phenolic type antioxidant

TABLE 12.13

Hot Temperature Stability at 100°C of Base Stocks Containing Antioxidants

Different Base Stocks	Solvent Refined (for Turbine Oils)	HC of HVGO HC Oil	Commercial Oil #1	Commercial Oil #2
KV at 100°F (SUS)	438	378	559	468
KV at 38°C (cSt)	95	82	121	101
VI	102	103	95	103
Pour point (°C)	−4	−15	−15	−15
Paraffins (wt%)	26.9	16.3	14.8	20.9
Naphthenes (wt%)	46.1	79.1	60.4	77.1
Aromatics (wt%)	27.0	4.6	24.8	2.0
Sulfur (ppm)	6800	8	170	12
Nitrogen (ppm)	42	3	5	11
Hot storage at 100°C				
Phenolic antioxidant #1				
After 4 weeks	Clear	Clear	Haze	Sediment
After 8 weeks	Clear	Sediment	Sediment	Sediment
Phenolic antioxidant #2				
After 4 weeks	Haze	Clear	Clear	Clear
After 8 weeks	Sediment	Sediment	Sediment	Sediment

Source: Galiano-Roth, A.S. and Page, N.M., *Lubr. Eng.*, 50(8), 659, 1993.

#1, was also reported to have the hot temperature storage stability at 100°C for less than 4 weeks but improved stability of 4–8 weeks in the presence of the phenolic type antioxidant #2. The variation in the composition of base stocks and the variation in the chemistry and the physical properties of antioxidants can lead to a variation in their hot temperature storage stability.

REFERENCES

Apar Industries, http://www.aparind.com (accessed 2003).

Carter, B.H., Marine lubricants, in *Chemistry and Technology of Lubricants*, 2nd ed., Mortier, R.M. and Orszulik, S.T., Eds., Blackie Academic & Professional, London, U.K., 1997.

Colyer, C.C. and Gergel, W.C., Detergents and dispersants, in *Chemistry and Technology of Lubricants*, 2nd ed., Mortier, R.M. and Orszulik, S.T., Eds., Blackie Academic & Professional, London, U.K., 1997.

Crawford, J., Psaila, A., and Orszulik, S.T., Miscellaneous additives and vegetable oils, in *Chemistry and Technology of Lubricants*, 2nd ed., Mortier, R.M. and Orszulik, S.T., Eds., Blackie Academic & Professional, London, U.K., 1997.

Dagom, D.N. et al., Process for improving lubricating base oil quality, U.S. Patent 5,554,307, 1996.

De Araujo Monte, V.L., *Boletim Tecnico da Petrobras*, 41(1/2), 3 (1998).

Denis, J., Briant, J., and Hipeaux, J.C., *Lubricant Properties Analysis and Testing*, Editions Technip, Paris, 2000.

Duncanson, M., *Journal of the Society of Tribologists and Lubrication Engineers*, May, 9, (2003).

Galiano-Roth, A.S. and Page, N.M., *Lubrication Engineering*, 50(8), 659 (1993).

Henderson, H.E., Chemically modified mineral oils, in *Synthetics, Mineral Oils, and Bio-Based Lubricants Chemistry and Technology*, Rudnick, L.R., Ed., CRC Press, Boca Raton, FL, 2005.

Inoue, K. and Watanabe, H., *Journal of the Japan Petroleum Institute*, 25, 106 (1982).

Kajdas, C., Industrial lubricants, in *Chemistry and Technology of Lubricants*, 2nd ed., Mortier, R.M. and Orszulik, S.T., Eds., Blackie Academic & Professional, London, U.K., 1997.

Lansdown, A.R., Aviation lubricants, in *Chemistry and Technology of Lubricants*, 2nd ed., Mortier, R.M. and Orszulik, S.T., Eds., Blackie Academic & Professional, London, U.K., 1997.

Ottewill, R.H. et al., *Colloid and Polymer Science*, 270(6), 602 (1992).

Pillon, L.Z., *Interfacial Properties of Petroleum Products*, CRC Press, Boca Raton, FL, 2007.

Pillon, L.Z., Asselin, A.E., and Vernon, P.D.F., Method for reducing foaming of lubricating oils, U.S. Patent 6,090,758, 2000.

Pirro, D.M. and Wessol, A.A., *Lubrication Fundamentals*, 2nd ed., Marcel Dekker, New York, 2001.

Raab, M.J. and Hamid, S., Additives for food-grade lubricant applications, in *Lubricant Additives Chemistry and Applications*, Rudnick, L.R., Ed., Marcel Dekker, New York, 2003.

Rhodes, R.B., *Society of Automotive Engineers* (Special Publ.), SP-996, 219 (1993).

Rudnick, L.R. and Shubkin, R.L., Poly(alfa-olefins), in *Synthetic Lubricants and High-Performance Functional Fluids*, 2nd ed., Marcel Dekker, New York, 1999.

Shell Lubricants, http://www.shell-lubricants.com (accessed 2001).

Tourret, R. and White, N., *Aircraft Engineering*, 24(279), 122 (1952).

Index

Printed and bound by CPI Group (UK) Ltd, Croydon, CR0 4YY

21/10/2024

01777089-0012